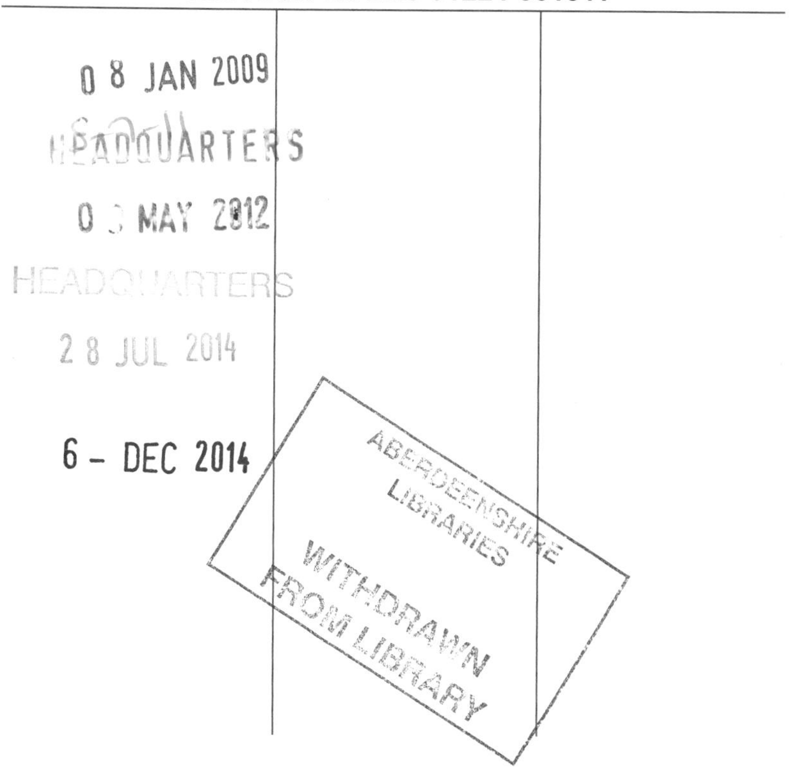
0 8 JAN 2009
HEADQUARTERS
0 3 MAY 2012
HEADQUARTERS
2 8 JUL 2014
6 - DEC 2014
ABERDEENSHIRE LIBRARIES
WITHDRAWN FROM LIBRARY

A L I S
1610510

Discovering Dorothea

Discovering Dorothea

The Life of the Pioneering Fossil-Hunter
Dorothea Bate

KAROLYN SHINDLER

HarperCollins*Publishers*

HarperCollins*Publishers*
77–85 Fulham Palace Road,
Hammersmith, London, W6 8JB

www.harpercollins.co.uk

Published by HarperCollins*Publishers* 2005
1

A catalogue record for this book
is available from the British Library

ISBN 0-00-257138-2

Set in Minion with Photina MT display by
Rowland Phototypesetting Ltd, Bury St Edmunds, Suffolk

Printed and bound in Great Britain by
Clays Ltd, St Ives plc

CONTENTS

ILLUSTRATIONS

ACKNOWLEDGEMENTS

This book could not have been written without the help and support of so many generous friends, scientists, librarians, archivists, and of course, Dorothea's relatives.

I am very grateful to the following institutions and individuals for their kind permission to quote from papers and documents in their care: the Trustees of the Natural History Museum, the Librarian of the Wellcome Library for the History and Understanding of Medicine, Archives of the University of Pennsylvania Museum of Archaeology and Anthropology, the Linnean Society of London, the Palestine Exploration Fund Collections, The National Archives (Public Record Office), Cambridge University Museum of Archaeology and Anthropology, Smith College Archives, University College London Library Services, Dyfed-Powys Police Museum and Giles Oakley.

My deep thanks to Dorothea's nephew Sir David Bate for his invaluable memories of his aunt and for permitting the use of family papers and photographs, to Leila Luddington's relatives, Malcolm McGougan and David Luddington, and to the late Maria Brown, her mother Anna and to Bridget Clarke.

I should also like to thank the staff of the British Library at St Pancras and at the newspaper reading rooms in Colindale, the Bodleian and Ashmolean Libraries, the libraries of the Institute of Archaeology, the Royal College of Surgeons and Royal Ontario Museum, the National Library of Wales, and the staff of all the many public libraries and record offices in England, Scotland, Wales and the Republic of Ireland who dealt so efficiently with my queries.

At the Natural History Museum, the curator of Quaternary mammals, Andy Currant, who describes himself as Dorothea's linear successor, has been amazing. He dealt with all my questions with unfailing humour and kindness, has read more drafts of the book than can be imagined and has been my guide and mentor through Dorothea's scientific world. Needless to say, all mistakes are my own. My deep thanks too to Chris Mills and to Ann Lum for providing me with a perfect research environment. Particular thanks to successive archivists at the Museum. I owe a great debt of gratitude to the late John Thackray for his encouragement, enthusiasm and introducing me to the rich resource of the NHM's archives, to Sam Collinette, and to Susan Snell and Polly Tucker for dealing with my ceaseless requests for yet more material and for brilliant work in checking archive references. My thanks to all the staff of the Earth Sciences Library, past and present.

For the chapters on Cyprus, I am very grateful to David S. Reese for his generosity in sending me so many of his papers on Cyprus fossil mammals and to Rhoda Gaze and her son Michael Milton-Gaze for permitting me to reproduce photographs of the Wodehouse family in their keeping.

For the chapters on Crete, I am indebted to Dr David Gill, Natalia Vogeikoff, Vasso Fotou and Marina Panagiotaki for sharing their research and knowledge so generously with me, and to the archivists Alessandro Pezzati of the University of Pennsylvania Museum of Archaeology and Anthropology and Nanci Young of Smith College, who found the invaluable Edith Hall Dohan, Harriet Boyd Hawes and Richard Seager references to Dorothea.

Josep Antoni Alcover has been most generous in guiding me through the vast quantity of research on *Myotragus* and advising me on the fossil record of the Balearics, while Jane Callander has been similarly generous with her research on Dorothy Garrod and archaeology in pre-World War II Palestine.

I have received so much help and encouragement from so many people that I do not have space to express my gratitude to them all individually here, but I wish particularly to thank the following people: Eddie Bell, Aimee Brown, Canon Brian Carne, Jane Clarke, the late Professor John Desmond Clark, Viviana Fain-Binda, Brigadier Alistair Fyfe, Edward Fitzwilliam, Richard Fitzwilliams, the late Robert Fitzwilliams, Tom Hassett, Daphne Hills, Susan Hora, Mary Johnston, Howard Jones, Richard MacCartney, Ellen McDonald, Marguerita Petrie, Peter Rollinson, Richard Sabin, Pamela Jane Smith and Sir Alan Muir Wood.

To Sophia Allana at Golders Green Crematorium who, just a week before the proofs of the book arrived, at last found the record of

Dorothea's cremation when I had almost given up hope, and to Eric Willis, a special thank you. I had looked so long for her last resting place, and extraordinarily, it was just a few miles from where I live.

At HarperCollins, my thanks to my editors Richard Johnson and Kate Hyde for their wisdom, encouragement and enthusiasm.

And finally to Henry and Alexander. Henry, who found himself living for several years with two women and the Natural History Museum, read, made countless encouraging suggestions, and re-read the manuscript with good humour and great patience. Both he and Alexander came with me to Wales, Ireland and the Mediterranean in pursuit of Dorothea. Alexander, who with great fortitude also explored the Wye Valley with me over a cold and sodden February half term – and survived, and who brilliantly and uncomplainingly accepted as normal that school holidays would always include Dorothea; to both of them I can only say, with love, a profound thank you.

PROLOGUE

A few months after I began work on this book, a friend said to me: 'I've wondered about this so often: I took my children to see the *Myths and Monsters* exhibition at the Natural History Museum and saw the reference to Dorothea Bate. I remember thinking how enterprising she sounded and what an extraordinary girl she must have been, and then I walked on. Thousands of other people must have done the same, but you didn't!' It is very curious, this process of how biographers choose their subjects, and it is still a mystery to me. Perhaps it is a sort of instinctive recognition, that something in this as yet scarcely known other person – their character, profession, fate or circumstance – triggers an almost visceral response. That, at least, is what happened to me.

It was in the spring of 1998 that I took my then nine-year-old son to that exhibition at the Natural History Museum in South Kensington. It explored how mythological creatures, from a unicorn to a yeti, might have had their origins in scientific reality. Among the fantastic monsters and the Museum's own wonderful specimens was a huge and bloodthirsty model of Cyclops, the one-eyed giant of Greek mythology whose favourite diet was humans. One of the specimens supporting the Cyclops myth was the skull of an extinct elephant and the origin of the myth at once seemed evident. If you were an ancient Greek who had no knowledge of elephants, what would you make of a skull you had just encountered on a Mediterranean island with apparently just one huge central eye socket? You certainly wouldn't know that this was the nasal opening where once a trunk had been; all perhaps you could do is clothe with monstrous flesh those inhuman bones and strange skulls and create a myth to interpret the unknown.

The display case also contained an open diary. It had belonged, I read, to Dorothea Minola Alice Bate. In 1901, at the age of twenty-two, she had travelled to Cyprus, and over the course of the next eighteen months, exploring alone, she became the first scientist to discover and excavate fossil pygmy elephants and pygmy hippos on the island. My ignorance quotient was growing; I had been to Cyprus and visited its archaeological sites, but I had never thought about its prehistory. I have to admit that I had no idea that either elephants or hippos had ever lived on Cyprus or anywhere else in the Mediterranean, nor that these great creatures had somehow evolved on islands to become dwarf species, no larger than pigs or Shetland ponies. As for Miss Bate herself, apparently already a recognized scientist in spite of her youth and sex, what was she doing exploring and excavating on her own in 1901? She seemed something so much more than another intrepid Victorian lady explorer. I was intrigued, by both Dorothea and her discoveries. The exhibition staff could not immediately suggest where I might find out more about her, but said they would let me know.

The following morning in my local library, I swiftly found that in spite of those pioneering discoveries, Dorothea Bate was in none of the obvious sources, not the *Encyclopaedia Britannica*, the *Dictionary of National Biography*,* nor any biographies of scientists or women. Her obituary, however, was published in *The Times*, on 23 January 1951, ten days after her death. It told me the astonishing story of how she had talked her way into working at the Natural History Museum, an association that lasted more than half a century, that not only was she a distinguished palaeontologist and ornithologist, but that she also had wide archaeological interests. It listed some of her major scientific achievements and described her as a woman who was determined, charming, loved gossip and 'pungent comment', as well as being a true and giving friend. On the subject of her family and background, the obituary was completely silent.

This tiny, preliminary lifting of the curtain over Dorothea Bate's life brought with it an extraordinary sensation of excitement and trepidation. This I was to experience hundreds of times over the next few years as new information came to light. Sometimes it might be in a single crucial letter, or, as on one unforgettable day, after I had completed the second draft of the manuscript, *ten* box files of Bate material were discovered on top of a cupboard in the Palaeontology Department at the Natural History Museum. These moments of discovery were to be counterbalanced by as many disappointments. A trunk containing family photographs and

* I have now rectified that, in the *Oxford Dictionary of National Biography*, published in 2004.

papers found in the attic of her great-nephew promised much but revealed just one brief reference to Dorothea. A photograph album belonging to the family with whom Dorothea stayed in Cyprus was wonderful from the point of view of the family, but contained not one photograph of Dorothea. Time and again, apparently promising discoveries about her life crumbled like poorly preserved fossil bones to nothing. A year into my research, when I was utterly obsessed by my elusive subject, came the worst disappointment of all. After Dorothea's death, her personal effects had gone to her sister Leila. In 1954, a devastating fire completely destroyed Leila's home, and with it everything that is so vital to the biographer – family photos, personal letters, journals. Nothing of Dorothea's intimate life appeared to have survived.

It was too late by then for me to abandon Dorothea. Her personal papers may have been extinguished, but I had been living for the past year with her adventures and her science, preserved in her work diaries, reports and letters in the Natural History Museum's wonderfully rich libraries and archives. A week after I had been to the *Myths and Monsters* exhibition, the Museum had rung to say they had all her papers and would I like to see them. For any researcher, that is complete happiness, but it is all the more so with a subject whose life is quite unknown. And so, just a couple of weeks after encountering Dorothea Bate, I first went behind the scenes at the Natural History Museum, one of my favourite places in the world. Laid out ready for me to examine on a large, leather-covered table in the Earth Sciences Library, were her intellectual remains – her letters and papers of more than fifty years of scientific work, preserved in a heart-stopping pile of large grey manuscript boxes and files.

Discovering Dorothea, her outstanding contribution to science and the magnificent institution where she worked, was about to take over my life.

CHAPTER 1

A Cathedral to Nature

The Natural History Museum in South Kensington stretches in all its Romanesque glory for nearly seven hundred feet along the Cromwell Road in west London. To the late Victorians, it was a palace for the people, a magnificent, terracotta-clad structure of soaring towers and arching windows. From early in the day, crowds would gather outside the great wrought-iron gates, waiting for the magical hour of ten o'clock when the gates would swing open and allow them in. They had time to wonder at the sculpted beasts, extinct and living, which enriched the façade; pterodactyls and sabre-toothed tigers, eagles and great lions, dragon gargoyles and griffin cats. In the morning sun, the creamy beige of the walls deepened to honey, offset by bands of pale grey-blue terracotta. It was seen to best effect across the bustling and noisy Cromwell Road, so busy with cacophonous horse-drawn vehicles and the occasional gleaming motor car that even then a traffic island was necessary for the safety of pedestrians.

On a morning in early autumn, 1898, a young woman waited with the crowd, then moved with them through the gates, across the forecourt and up the sweeping flight of steps to the entrance of the Museum. Ahead, deeply recessed, was the immense arched doorway, suggestive of the opening to some fabulous cave, framed by slim columns like sticks of barely sugar or slender stalactites. Above the doorway were two small, finely worked heads: a monkey, which represented evolution, and that symbol of patriotism, an English bulldog. The carved wood and glass doors swung inwards, heavier than they looked, held open by braided and capped attendants. Once inside, the crowd quickly dispersed in the vastness of the Museum, but the young woman paused by the attendants and announced herself. She was Miss Dorothea Bate and she would like to see Dr Richard Bowdler Sharpe, who was in charge of the Bird Room. In an age when only a tiny percentage of women worked in the professions and the ambition of most of her contemporaries was marriage, Dorothea Bate's was to work as a scientist in the Museum.

What happened next became the stuff of legend. From the great Central Hall of the Museum, Dorothea was escorted through the public galleries to a high panelled door. Behind it was the Bird Room, part of the Department of Zoology. The surprise of the assembled ornithologists at her arrival can only be imagined. This was absolutely a male preserve, a place of black frock coats, silk hats and weighty degrees from distinguished universities. Dorothea, still in her teens, was a self-taught natural historian and ornithologist. That she had aspirations to join them must have been one of the most astonishing ideas that any of the scientists had ever confronted. An early portrait of Dorothea by her sister shows a young woman of almost palpable authority, her gaze direct, her face resolutely tilted upwards, while her chin indicates utter determination. The pose is a conventional profile, her dress a soft afternoon gown, revealing white plump shoulders and a bosom of almost royal proportions, while her mass of dark hair is swept up, caught in a neat tartan ribbon. Round her throat lies a delicate necklace. That morning in the Bird Room, Dorothea's hair, of which she was very proud, was neatly stowed under her hat, her figure concealed beneath a long tailored coat and skirt.

Bowdler Sharpe himself, the curator of birds, was in his fifties, with the de rigueur neat pointed beard and substantial moustache of most of the Museum's scientists. He was a man whose personality was as formidable as his reputation. Initially he told Dorothea to go away; he had no time to be bothered with anyone. Her request was quite preposterous. Not only was she just nineteen but, more importantly, no women at all were employed in a scientific capacity by the Museum. Yet within a few minutes, Dorothea was standing at a table, sorting bird skins into species with assurance and skill. Perhaps as he watched her, Bowdler Sharpe thought of his own children. He was the father of ten daughters, one of whom, with great enterprise, was running her own entomological agency, buying, selling, arranging and naming collections of butterflies, from premises she rented in the Fulham Road. In Dorothea he must have seen an equally determined girl with a character as formidable as his own, but who dominated by charm and humour rather than irascibility. Even allowing for that, Dorothea's achievement in being accepted by Bowdler Sharpe was truly remarkable. Everything should have been against her: her extreme youth, her lack of formal education and, most of all, her sex. Instead she forged a relationship with the Museum that was to last fifty years. It became her university and her life.[1]

The Natural History Museum had been open to the public for just seventeen years when Dorothea arrived there. That it existed at all was a tribute to the wiles and perseverance of the brilliant, although intensely disliked, palaeontologist and superintendent of the natural history departments of the British Museum, Sir Richard Owen. He was responsible (among many other things) for immortally naming the order of great extinct reptiles *deino* (fearful, terrible, strange, wondrous)-*sauros* (lizard). The British Museum had originated in 1753, when the government purchased for £20,000, for the benefit of the nation, Sir Hans Sloane's spectacular collection of the 'nick-nackatory'[2] of nature and mankind. This was a fantastic bargain, as the collection had cost Sloane more than £100,000 (many millions in today's money), but his intention, as stated in his will, was for this great assemblage to be 'preserved and maintained, not only for the inspection and entertainment of the learned and the curious, but for the general use and benefit of the public to all posterity'.[3]

Over the decades, the collections expanded and, by the middle of the nineteenth century, the British Museum's site in Bloomsbury was woefully inadequate for the display, storage or scientific research of the natural history departments. The Zoology Department, for example, was housed in the cellars with windows knocked into the walls. Bowdler Sharpe, who worked there for nine years before the move to South Kensington, called it an 'underground dungeon', with two small studies for the Keeper and Assistant Keeper; everyone else – scientists, messengers and servants – were crowded together in a room so small it was irreverently known as the 'Insect Room'.[4] As for the collections, it was the custom until late into the nineteenth century, both at the British Museum and museums throughout Europe, for the most valuable specimens, no matter how rare or unique, to be mounted and displayed in the public galleries. For mammals and birds in particular, prolonged exposure to light, dust and insects is disastrous: fur and feathers bleach and discolour; the specimen deteriorates and becomes useless both for science and as an exhibit. Those specimens that were unmounted fared no better; they were piled 'a hundred or more' into boxes and, as Bowdler Sharpe wrote, had all to be turned out every time a particular specimen was required, 'and great injury to the skins resulted'.[5]

Owen took every opportunity to speak on the importance of creating a separate museum of natural history and in 1861 had the brilliant idea of taking the Prime Minister, William Gladstone, to see the shameful state of the national treasures. Gladstone saw, was duly impressed with Owen's arguments and in 1863 Parliament approved the purchase of the site in South Kensington. When asked to state his requirements for the new museum, Owen is said to have replied, 'I shall want space for seventy whales, to begin with.'[6]

The following year a competition was held for the design of the Museum and it was won by Captain Francis Fowke who had designed the Royal Scottish Museum in Edinburgh. His concept was a grand Renaissance-style building, clad in terracotta with domes and cupolas, somewhere between a *palazzo* and a great cathedral, but the Trustees of the Museum demanded changes to its internal structure. While Fowke was working on this, he died unexpectedly and the commission was given to Alfred Waterhouse, architect of the ornately Gothic Manchester Assize Courts. With the wisdom of a man who has previously designed buildings for committees, Waterhouse retained significant parts of Fowke's original design, including its cathedral-like aspects, but in order to incorporate the internal changes required, he 'abandoned', as he put it, the Renaissance plans, turning instead to the earlier Romanesque style of the tenth to twelfth centuries of 'Lombardy and the Rhineland'.[7] This allowed him to design great rounded arches, massive piers, and intricate and exquisite three-dimensional terracotta decorations, both inside and out, to echo its exhibits. Great towers replaced Fowke's domes and cupolas. By 1871, the design was finally agreed with the Trustees, although the whole process had been dogged by the question of cost. At one stage the Treasury had demanded that the two splendid towers at the front, deemed an extravagance at 192 feet high, should be considerably shortened. It would have been an architectural disaster, but this essential element of Waterhouse's design was saved by, of all people, the fire-safety officers. The water tanks essential for fighting fires had to be stored in the towers and the pressure would only be powerful enough if the towers were built to Waterhouse's original specifications. The towers survived but so did the cost problem. Waterhouse's solution seemed masterly: the museum would be built in two stages, first the magnificent front, then later the back and two wings along Exhibition Road at one end and Queen's Gate at the other. Work finally began in 1873 and the contract with the builders was for £352,000. Eight years later in 1881, the Museum opened to the public, just two years before Owen's retirement. It immediately became popularly known as the Natural History Museum, although its title officially was the British Museum (Natural History). Although it had its own director, it was still part of the British Museum, answerable to the same Board of Trustees.* With the front completed and the Museum functioning, the will to fund the second stage vanished and Waterhouse's planned wings were never built.†

Waterhouse's purpose had been to make the visitor's first impression of the Museum as exciting as possible and he succeeded brilliantly. When Dorothea walked through the great doors, she must have felt – as was intended – that she was entering a cathedral to nature. Even today, most first-time visitors (and many who know the Museum well) really do stop and stare. At one end of the magnificent Central Hall is a huge spectacular staircase, at the other a high arched bridge linking the upper galleries. Great arches lead into what in a church would be side chapels but here are recesses for display cases. Above, light floods down from the glass and iron roof with central panels of delicately painted shrubs and trees and from tall, narrow, stained-glass windows. It is a glorious space, its terracotta-clad walls an art gallery of exquisite carvings of a myriad plants, birds and beasts. In keeping with the arrangement of the exhibits, those decorating the east side of the building, which housed the Geology and

* Even after the Museum became autonomous, it remained the BM (NH) until 1988 when, after much debate, the name was formally changed to the Natural History Museum.

† In the 1970s, in the style of the time, an extension along the Exhibition Road was built to house the Palaeontology Department, whilst in the 1990s the old Geology Museum next door was transformed into the Earth Galleries. In autumn 2002 at the Queen's Gate end, the state-of-the-art Darwin Centre opened. It contains the Zoology collections and also offers visitors a unique opportunity to see behind the scenes in the Museum.

Palaeontology Galleries, were extinct; on the west side, the Zoology Galleries, the species of flora and fauna depicted were living. Behind the terracotta panels are the bones of the building, a vast skeleton of iron beams and columns. Not everyone was enamoured of the design. A year after the Museum opened, the authoritative journal *Nature* thought that the building did not work, that the architecture and exhibits clashed, while the position of the display cabinets ruined the architectural intentions. What *Nature* did approve of, however, was that instead of scientists having to work in the exhibition galleries on days when the Museum was closed to the public,* they now had galleries and rooms in every department where they could pursue their studies undisturbed.[8]

This idea that a museum has more than one function, that it is not simply an exhibition hall that displays specimens and artefacts but that it also must *advance* science, was new enough to need an explanation in the Museum's *General Guide* for visitors. The greater part of the collections, the general reader learned, was kept behind the scenes for the purposes of study:

> It is to this part of the collection that zoologists and botanists resort to compare and name the animals and plants collected in expeditions sent to explore unknown lands, to work out biological problems of the highest scientific importance, and generally to advance the knowledge of the science. In fact these reserve collections . . . constitute, from a scientific point of view, the most important part of the Museum, for by their means new knowledge is obtained.[9]

To be part of all that was Dorothea's ambition and her first step to achieving it was convincing the gentlemen of the Bird Room of her ability and dedication.

The Bird Room itself was a long room with central tables and wooden cupboards around the walls. Even in early autumn, the place could become uncomfortably warm from the unshaded sun scorching through the skylights. There was an almost overpowering smell of camphor, used

* For many years visitors were only allowed into the Museum on certain days and for a limited number of hours.

to protect the collections from insect infestations. Open cupboards revealed tiers of narrow drawers. Inside, carefully stored, were bird skins, each with a small label attached. These were glorious specimens, the intense, rich hues of the plumage preserved in their dark, protected environment. 'Nowhere else in the world,' Bowdler Sharpe was fond of telling visitors, 'will you find so many birds in so small a space as here.' The bird collection was estimated by Dr Bowdler Sharpe to contain some four hundred thousand specimens, including 100,000 eggs.* As the bird skins were moved, a faint cloud of white dust would rise, perhaps caught in a sunbeam before slowly dispersing, settling with deceptive gentleness over every surface. This was arsenic, used with camphor to preserve the specimens. Bowdler Sharpe and his colleagues were well aware of the danger but chose to ignore it. 'I have suffered on and off for the last dozen years now from arsenic poisoning,' Bowdler Sharpe casually told a visiting journalist, the symptoms being 'much the same as in lead poisoning. An intolerable itching in the corner of the eye, and before now I have suffered a continual pain in the stomach.'[10] Some collections sent in by the public, according to Bowdler Sharpe, appeared to have had an entire box of the poison poured over them.

There were at least ten visitors a day to the Bird Room, academic and amateur, to compare species they had found or were studying, with those in the collection. Some came in the hope that their specimen would be that Holy Grail for Victorian and Edwardian naturalists, a new species; occasionally, after exhaustive and lengthy research in both the collections and the library, it was found to be so. In many cases the finder's name would then be immortalized as part of the scientific name of the newly identified bird. Each specimen had a vital, detailed label, giving the date of acquisition, who had donated it or how much it had cost to purchase, where it had come from and when collected, its name and its sex.

As Dorothea left the Bird Room, she must have paused often as she walked through the great Bird Gallery, which ran for 300 feet along the west side of the Museum. Displayed there in splendid mahogany and glass cases were thousands of birds from all parts of the world, from the great condor of the Andes to the tiniest hummingbird. There was a

* That speaks volumes of his industry as curator. When he took charge of the collection in 1872, the total number of specimens was around thirty-five thousand.

sea-eagle from the Bering Straits, the great eagle owl of Europe and the dwarf falcon, not much larger than a sparrow. There were vivid birds of paradise, flamingos, a great auk, newly extinct, a kiwi and an Emperor penguin, birds frozen in all their glory, a tribute to the Victorian taxidermists' art. Of particular interest to Dorothea, and thousands of others who came to the Museum especially to see them, were the famous exhibits of British birds in their nests. 'Every particle of these groups is literally true to nature,' said Dr Bowdler Sharpe in an interview, 'for the nest is transplanted bodily to the Museum as it was *in situ*, and every leaf, every plant, is exactly reproduced, so that you have the identical birds which made the nest, the eggs they laid in it or the young birds which were hatched in it, and the very surroundings reproduced as they existed on the day the nest was taken.'[11] These displays were enormously popular, 'masterpieces of scientific art', the *Daily Telegraph* called them, noting that all day long the gallery was filled with admiring visitors: 'and indeed the vocabulary of admiration is never exhausted. "Wonderful!" "Marvellous!" "Exquisite!" are the exclamations every side.'[12] Now, a century or so later, in a neat reversal of the evolutionary process, dinosaurs occupy that gallery. Just a few of the original bird displays are still exhibited, just as Dorothea would have seen them. Now they are in a small side gallery together with two ornithological display cases almost exactly as they were when first exhibited in the Central Hall in the 1880s.

For many months after I first encountered Dorothea, it seemed as if, like the goddess Athena, she had sprung fully formed from the head of Zeus or, perhaps more appropriately in Dorothea's case, from a cupboard in the Bird Room but with no identifying label. Nothing whatsoever was known of her family except that she had a sister with the married name of Cuddington and that she had been born in 1879. Where was I to start? I had no convenient published synopsis of her life and background, not even the most basic, on which I could begin an attempt to recreate her life. I didn't even know where she had been born or where she had died. But as I worked my way through the great boxes of her scientific papers, clues emerged. Her sister had lived near Saffron Walden; she had a brother in the armed forces; and, from the reverse of scraps of paper Dorothea had saved and reused to make notes in the 1930s and 1940s, I learned that in 1900 she and her father had birdwatched near Lydbrook Junction in a remote part of Gloucestershire. The most vital clue, however, was that her childhood had apparently been spent in Carmarthen. Her birth certificate (which revealed that she had been born not in 1879 but in 1878) told me who her parents were, and with this crucial information I eventually discovered that Dorothea had a nephew living in a small town in western Canada. More months went past as my letters to him went unanswered. In desperation I rang the post office and then the police station in the town where he lived. He had moved, I was told, a few months before my first letter had arrived. They did not know where. Luck in these matters does sometimes play a part and, almost immediately, I learned that he had moved across the world to a remote farm in southern Africa (this, incidentally, was a man in his eighties). Contact was finally made.

As for Dorothea's sister, further research failed to reveal any marriage certificate for a Miss Bate and a Mr Cuddington. Expecting little, I rang directory enquiries. There was no record of any Cuddington. The very helpful operator then asked if I had the correct spelling, and we went through various permutations and then the alphabet (it must have been a quiet afternoon). There were no Buddingtons or Duddingtons, then suddenly, 'I have two Luddingtons, could it be that?' And thus through a Canadian Mountie and this saint of an operator, I located what remained of Dorothea's family and set off westwards, to Gloucestershire, Wales and Ireland, in search of her ancestry. It was already evident that Dorothea was a woman of great courage and resolve, and these qualities, I was to discover, seem to have been inherited.

CHAPTER 2

This Noble Family

The night of 31 January 1839 was fearful. Huge waves battered the Wexford coast of southern Ireland, driven by a northeast gale. It was all the worse for being unexpected. Murphy's Weather Almanack had forecast a 'fair and pleasant day' – a welcome break from the more usual fare of terrific and damaging gales that regularly lashed the Wexford coastline in the winter months.[1] The schooner *Thistle*, caught in the unsheltered sweep of South Bay, was pitched onto the infamous Cammeen sandbanks, which lay off the shore. According to the local newspaper, Mr Bate, chief officer of Coast Guards at Ballygearey, set sail immediately for the wreck with his crew and five gallant volunteers. After much difficulty, 'during which this brave officer displayed great intrepidity and perseverance', they succeeded in rescuing the crew of three men and two boys.[2] The brave officer, who was awarded the service's silver medal for his action, was Thomas Bate, Dorothea's grandfather, who, if nothing else, bequeathed to his granddaughter those qualities of intrepidity, perseverance, and considerable bravery.

Dorothea's father, Henry Reginald, was born in Wexford a few months after the great storm, the third of ten children of Thomas and his wife, Dorothea Maria Whitney. In 1855 at the age of sixteen he joined the Army as an ensign 'without purchase'* in the 77th Regiment of Foot,[3]

* In 1855 it was still the norm for commissions to be bought and sold. There was considerable prestige in being an officer, and just about anyone could become one, provided that he, or

transferring in 1862 to the 2nd Battalion, 13th Light Infantry. His career, with postings to Australia and India – and despite bouts of illness – progressed reasonably well and by 1874 he had been promoted to captain. The following year, the battalion was posted to Glasgow. Henry now was thirty-six, of medium height, with a sallow complexion, grey eyes, light brown hair and a scar on his left cheek. His figure was 'proportionate'.[4] In the inevitable social round followed by a presentable Army captain, he met Elizabeth Fraser Whitehill, the daughter of a prosperous shawl manufacturer, Matthew Whitehill, and his wife Jane Arthur. Elizabeth, whose family called her Lily, was in her early twenties; an accomplished musician, she was dark-haired, vivacious and pretty. In September 1876 they married and ten months later, in July 1877, their first child, Lilly Arthur, was born. Named for her mother and maternal grandmother, she was from birth brought up in Elizabeth's image and was destined to be her favourite child. At the end of the year Henry retired from the Army with the honorary rank of major. An officer he may have been, but as the son of an Irish coastguard he lacked the means to live as a comfortably retired gentleman at the age of thirty-eight. He needed a job and, in November 1877, was appointed Superintendent of Police in the Carmarthenshire force in South Wales and the family duly moved to Carmarthen.

On the day his appointment was announced, the Chief Constable of Carmarthenshire, A. B. O. Stokes, reported that 'the county is happily free from crime of a serious nature'.[5] Carmarthen itself had long had a reputation for drunkenness and disorderly behaviour, but by the last half of the nineteenth century, there was relatively little violent crime. The cases that routinely came before the quarter sessions reveal considerable poverty and a harsh penal code. For stealing a pair of stockings, for example, a woman was sentenced to six weeks' hard labour; the crimes of stealing twenty-one pieces of old iron, a rabbit trap or, pathetically, two flannel vests and some coal, attracted sentences of six months' hard labour.[6] That drunkenness should have been a problem is scarcely surprising as in Spilman Street alone, one of the main thoroughfares, there were

more generally his father, could afford it. Education, aptitude, and ability had very little to do with it. Exceptions to purchasing commissions were made for those who had passed the examination for a commission, or whose father was or had been an officer himself. Henry qualified on both counts. It was a system that aroused great passions and, despite many attempts, it was not until 1871 that commission and promotion by purchase was abolished.

ten pubs and alehouses, as well as an old coaching inn, the Ivy Bush. Next door to the Ivy Bush and set back a little from the road, is an early nineteenth-century, three-storey, brick-built house. I saw it first on a wet July day, a slightly shabby building converted now to office space, but for me it was of enormous significance. It was here in Napier House, on 8 November 1878, that Henry and Elizabeth's second daughter was born. Four weeks later at St Peter's Church down the road, she was baptized and christened Dorothea Minola Alice Bate. As her sister was named for her mother, so Dorothea was named after her paternal grandmother and one of her father's sisters, Alice (or Alicia). Minola, a popular name with Victorians, has no immediate family link.*

When Dorothea was two, the Bates moved again, about a mile out of the centre of town to an old farmhouse called Green Hall. Dorothea and Lilly Arthur, whose name soon and permanently was contracted to Leila, were cared for by a nurse, agreeably named Temperance Rees. The following year Elizabeth gave birth to her only son, Thomas Reginald Fraser, this time the name combining those of both families. Six years later Henry retired from the force and the family almost immediately moved about twenty-five miles away to rent a yet larger house, Gellidywyll† at Cenarth, near the town of Newcastle Emlyn. Concealed from the road, it took me hours to find Gellidywyll, or, at least, what is left of it. The main house was demolished as derelict in 1973; all that remains is one ancient wing, now a farmhouse, and the memory of what it once had been. Preserved under a great square of grass in front of the farmhouse is the outline of rooms long gone but still easily traceable. I walked where once there had been a porticoed entrance, a large hall, dining, drawing, and sitting rooms. Beneath them were extensive cellars and storerooms. Upstairs there had been eleven bedrooms and accommodation for housekeeper, butler, and other servants. In all there must have been about thirty-six rooms. In 1888 a long drive had wound through grounds, which contained outhouses, a garden, pleasure ground, two fish ponds, a coach house, stables for eight horses, and dog kennels. Altogether, it seemed to me,

* It is an Italian name, attached most famously, perhaps, to Katharina Minola, the eponymous heroine of Shakespeare's *The Taming of the Shrew*. Apart from a certain obstinacy and determination, Dorothea's character could not have been more different!

† This is the most common spelling, but it also appears variously as Gelly Dowl, Gellydowl, Gellydywyll and so on.

very curious accommodation for a retired police superintendent with just his Army and police pensions and little inheritance.

For Dorothea, who was now ten, and Leila and Thomas, who were eleven and seven, Gellidywyll provided an idyllic place to play. It stands high in the hills over the valley of the river Teifi, above the famous Cenarth falls. From the house, the children could run through the lush green fields and woods down to the river. When the water was low, the rocks left exposed above the falls formed little magical islands for childhood imaginings. When it ran fast, salmon and sewin leaped the falls; in full spate, the torrents of water crashing down over the rocks reverberated round the valley and up to the mansion high above. Farther along the valley stood the ruined castle of Newcastle Emlyn, its tumbled walls perfectly inviting.

Almost from the moment they moved to Gellidywyll, the family became prominent members of the Tivy-side community as the area along the river Teifi or Tivy was known. The term applied to the great houses and hamlets that bordered a twenty-mile stretch of the river, their local towns being Cardigan and Newcastle Emlyn. Within weeks of moving to Gellidywyll, Henry and Elizabeth attended the high point of the season, the hunt ball. All Tivy-side society was there, and the Bates subsequently appeared on the guest lists for all the best parties, weddings and balls – although Elizabeth attended rather more of these than her husband. How did I discover this with no family papers to consult? To my amazed delight, my trawl through the local newspapers in the British Library's newspaper reading rooms struck gold. Social events were chronicled in one particular newspaper, *The Cardigan and Tivy-side Advertiser*, and week after week (I could hardly believe it as I followed their progress) the activities of the Bate family, newly arrived as they were, were reported with extraordinary and satisfying frequency.

It was through the newspaper that I was able to identify two families who, I already knew, were of particular importance to Dorothea although I had no idea how they had met. In the end it was so simple. They were neighbours. The first of these was the Wodehouses. Clarence Wodehouse had been an officer in the 77th Regiment of Foot in which Henry had first served. His wife Francie, whose family owned a mansion and great estate in the neighbourhood, was, like Elizabeth, an accomplished musician and a stalwart of the charity concerts that so occupied local society. They also had three children. Jack, the eldest, was three years younger than Dorothea; and there were two little children, a daughter, Nessie, and a son, Arthur Cecil, known as Billy. Clarence, who had private means, had rented in 1886 a large estate in the area, Llwynbedw. There, he farmed and bred pedigree cattle, sheep and pigs, and kept donkeys and short-haired rabbits as pets.

Even closer to Gellidywyll than the Wodehouses' estate was a great red-brick Victorian mansion called Cilgwyn, owned and built by the Fitzwilliams family. There were ten children, eight boys and two girls, and Dorothea, Leila and Thomas were often invited over to play. Cilgwyn was so much more magnificent than Gellidywyll, and Thomas, perhaps through a touch of jealousy and the grandeur of it all, would get his own back by pulling the plaits of the youngest and prettiest Fitzwilliams daughter, Mary Ulrica Alicia – but that did not stop the invitations. It was an association that would lead to Thomas's marriage, to Dorothea losing her inheritance, and to a rift in the Bate family which was never healed.[7]

In these early years, however, the Fitzwilliamses were simply splendid and hospitable neighbours. On New Year's Eve 1889, they held a Grand Ball. High on the guest list were the Bates. As *The Cardigan and Tivy-side Advertiser* reported, it was a fabulous affair. In the great oak-panelled hall and the drawing room, there was dancing 'to the strains of Hulley's excellent and well-known band'. A table extended the length of the Fives Court, spread with countless dishes of meat and fish, whole sirloins of beef, succulent hams, galantines of turkey, pies, quenelles and sumptuous desserts, while tea and light refreshments were served in the dining room. The rooms were swathed in brilliantly coloured draperies with trails of ivy and Christmas roses. At midnight, then as now, everyone sang 'Auld Lang Syne'.[8]

But it was a community that believed in its social obligations as well as its rights and pleasures. Hardly a week went by without a charity concert, a National School outing or a fund-raising tea for a deserving cause. Elizabeth Bate had clearly found her *métier* and Dorothea, Leila and Thomas were expected to play their full part. According to the indefatigable *Cardigan and Tivy-side Advertiser*, 'Mrs Bate is always ready and willing to help on any and every good or festive cause in the district and a perfect host of thanks is due to her for the thousand-and-one services she has rendered so cheerfully in the district.'[9] Many hours of

Dorothea's childhood were spent with her mother and sister, assisting the vicar's wife (known to the local press as 'Mrs D. H. Davies the Vicarage') in decorating with flowers and ivy the church at Christmas and Easter or the ballrooms for the Tivy-side Hunt or Lawn Tennis Club Ball, long before Dorothea and Leila were old enough to attend the ball itself.

Elizabeth also taught her children responsibilities as well as decorative skills. For a number of years, just after Christmas, the whole Bate family entertained the children of the National School at a party either in the schoolroom or more often at Gellidywyll itself, fully reported, of course, in the local paper. At two o'clock the schoolchildren assembled in the schoolroom and were then marched up the hill 'in soldier-like fashion' to the mansion, led by the vicar, his wife and the schoolmaster. At precisely three o'clock, 'at a word of command from Major Bate', the children were comfortably seated and served a splendid tea, attended to by the whole family. The great Christmas tree 'was ingeniously made to turn on a pivot, attached to which was a musical box from which proceeded a volume of harmonious sounds'. The schoolchildren sang, one of them read out their thanks for the feast, and then they were given gifts from the tree, 'dolls, knives, drums, workboxes, Noah's ark and so on'. The children gave three hearty cheers for the Bate family and departed laden with cakes, sweets, and their presents. 'Great credit is due to this noble family,' enthused the paper, 'for the great interest they take in everything that tends to elevate the morals and refine the taste of the people.'[10]

Leila and Thomas may have enjoyed these accolades, but for Dorothea, who liked nothing better than to be left alone to explore the glorious Welsh countryside, many of her family's activities must have been agony and none more so than the charity concerts. These were regularly performed by the local amateur musicians, of varying degrees of talent, to packed audiences, the proceeds going to a deserving local cause. For years the same names constantly appear, including Francie Wodehouse as chief piano accompanist or soprano and Mrs Bate organizing, or playing, or both. Elizabeth played the harmonium, piano, violin, and double bass and taught her children, who from an early age were expected to perform in public. Her beautiful, lively, eldest child Leila, was, like her mother, a gifted violinist. Little Thomas also showed promise on the violin, but Dorothea, awkwardly the middle child and neither as beautiful nor as vivacious as her sister, seems not to have been musical either.

Dorothea's first ball was much less agonizing as she loved to dance. Especially arranged for the children of the Tivy-side gentry, it took place on an April afternoon in 1894 in the ballroom of a local hotel. They danced polkas, lancers, gallops, and waltzes, ending with the exuberant 'Sir Roger de Coverley'. For the younger children it was wonderful fun, but Leila and her friends, who were nearly seventeen, felt themselves far too grown up for such an event and decided to organize their own. Within a few weeks, according to the local paper, 'The spinsters of Tivy-side, finding things as the commercial phrase goes, dull', had founded a committee and planned the Tivy-side Spinsters' Ball for December that year. 'Of course,' the paper archly reported, 'when the project first got wind, there were not wanting those who prophesied that the thing could never be: that a committee of petticoats would quarrel among themselves' and would be forced to ask 'some of the male sex to assist them in organising their entertainment'.[11] That of course did not happen and the ball was a brilliant success, not ending until 4.30 a.m. Once again, it was a triumph for Leila.

In these early years, Dorothea – always Dorothy to her family and close friends – seems deep in shadow. Her mother was so involved in the life of the neighbourhood, with nurturing Leila's talents and with making sure young Thomas went to the right schools, that Dorothea hardly seems to have been thought about. Leila's ability won her a place at the Academy of Music in London. Thomas, at the age of eleven, was sent away to Bilton Grange prep school in Warwickshire, housed in a splendid mansion designed by Pugin who drew on some of his designs there for his work on the Houses of Parliament. It could not have been easy for Henry to afford the fees, although the school's links with military families meant it offered sizeable bursaries, which helpfully reduced the £50 (about £3,000 in today's money) required each term. At thirteen, Thomas went on to Wellington College in Berkshire, which two of his cousins and one of the Fitzwilliams boys had also attended. Again there was a reduction in fees for the sons of officers and in addition he was awarded a scholarship of £16 a term. Alone of the three children, Dorothea remained at home. The only advantages bestowed on her were of her own creation.

She used to laugh that her education was only briefly interrupted by school, but she was surrounded by opportunities to acquire knowledge and, with or without governesses or teachers, she did. Natural history was the pastime that obsessed Victorians of all classes and ages, not just observing, but collecting. Insects, stones, fossils, ferns, flowers, birds' eggs, shells and starfish: all were desirable. Birds and mammals were also collected, shot for their skins and skeletons to be examined, or the finest examples stuffed and displayed. Few were squeamish about the process of killing, skinning, or dissecting creatures; it differed little, after all, to the everyday business that went on in many kitchens. The advance in microscope technology in the nineteenth century created great excitement, as the tiniest specimens could now be viewed in unimagined detail, while the magnifying glass could be found in every collector's pocket.

Dorothea would spend her days in the woods and fields and down by the river, fascinated by the local wildlife. That much I knew from her colleagues' memories, but what I didn't know, without her letters or private journals, was what had drawn her to science. For many natural scientists, it is a sudden certainty and at quite a young age. They can be looking at a flower, a beetle, a stone, a fossil shell or bone and just know that that is what they want to spend their lives investigating. It becomes a passion and an obsession, often around the age of nine or ten. The collecting instinct comes first, then the need for establishing order among the specimens, then, with maturity, the need to explore and uncover the unknown, and to experience the almost indescribable joy of discovery of something new and unique. Her father appears to have encouraged her, pointing out particular birds and plants, and teaching her country pursuits. She fished in all weathers and became an excellent shot. She learned how to skin birds and small mammals, and pored over the different shapes of beak and claw, observing how surprisingly different were shrews, voles and mice when examined in detail. She must have had a special place at home for her finds, a few shelves or a cupboard filled with neatly arranged specimens: feathers and bird skins, tiny shrews and delicate mammal bones a fraction of an inch in length, shiny beetles or perfect eggshells, and perhaps her own paintings or drawings of this wildlife she was learning so much about.

In June 1888, just after the Bates moved to Gellidywyll, Wombwell's Royal National Mammoth Menagerie, a great travelling circus, swept into nearby Cardigan with one hundred mischievous monkeys, a hippopotamus, an elephant and her baby, chimpanzees, lions, tigers, jaguars, leopards, pumas, bears, wolves, hyenas – in all, more than 700 animals.

Logistics aside, it was an amazing operation, offering not only entertainment but lectures about the animals. Also on view were the freak shows – the largest lady who weighed more than forty stone, the tallest man and the heaviest baby, allegedly ten months old and weighing more than six stone. With no modesty whatsoever, the menagerie claimed to be 'as near perfection as a travelling zoological collection can possibly be'.[12] But for a child already interested in natural history, it must have been quite wonderful.

There was also plenty for Dorothea to read. Elizabeth, whose energies did not stop with her music, had established a reading room at Cenarth, stocking it with books, newspapers and periodicals on all manner of topics, both educational and entertaining. There was *Science for All*, a magazine that cost just 7*d.* per month, *The World of Wonders, Chum* for boys, *Natural History* and journals on geography, history, and needlework, and collections of short stories. Board games and cards were also available. As Elizabeth had established the reading room, she would have made sure her family set an example by supporting it, and for Dorothea at least that was no hardship. In 1891 when Dorothea was twelve, *The Cardigan and Tivy-side Advertiser* started a column entitled 'Our Young Folk's Letter'. By reading that column every week, she would have acquired a wide-ranging and highly eclectic store of knowledge. It carried articles about Darwin, the death of a giraffe and the ingenuity of a thrush. In other weeks she might have learned about the white stork, a surgical use for ants, the mysteries of growth, why bees make honey, the solar system, food preparation of Australian aborigines and how limpets hold on. She could have read about the shoemaker naturalist in Germany who had an immense collection of beetles, discovered that there was a wonderful aviary at the Zoological Gardens in London, and that highly satisfactory progress was being made with collections at 'the splendid Natural History Museum in Cromwell Road which is a part of the British Museum'.[13] She may well have persuaded Henry to take her to London by the very efficient railway to see for herself.

There was one article in particular that seemed to me to draw a curious parallel with the path Dorothea's own life was to take in the next few years. As I read it, I had the strangest sense that this may have been of determining importance to Dorothea, yet I have no evidence that she even saw it. It was a piece titled 'A Famous Boy Traveller', about an

American explorer, Ernest Morris, who had become well known for his remarkable travels as a youth in South America. Like Dorothea, a keen natural historian, Morris persuaded collectors to fund his explorations into the Amazon basin to collect orchids and other botanical specimens. He was highly successful and made a number of hazardous expeditions characterized by personal danger, appalling food and dreadful living conditions, but with the reward of splendid specimens. 'He had the happy faculty,' according to the article, 'so essential to the best explorers, of being able to make the best of circumstances . . . Morris knew very well the taste of monkey stew, and like the natives among whom he wandered, he more than once appeased his hunger with a meal of snake meat, which he declared to be not at all disagreeable.'[14] Dorothea's subsequent experiences in Mediterranean islands echoed those of Master Morris. She too had the 'happy faculty' of making the best of circumstances, enduring considerable hardship and some danger, and, while not quite reduced to eating monkey stew and snake meat, she tolerated some pretty unpalatable stuff, at times facing real hunger. That article about Ernest Morris may well have sown the seeds of her own ambition, although it must have then seemed to her simply a marvellous adventure story.

Two years later, the departure from Tivy-side of the Wodehouse family was to lead to Dorothea's own opportunity for adventure and scientific exploration. In 1890, the Wodehouses' five-year-old son Billy died. Perhaps after this they found life in Tivy-side unbearable for, in June 1893, they auctioned everything, furniture, china, paintings and all the animals. After just seven years, Wodehouse abandoned the life of a country gentleman farmer and joined the Colonial Service. His grandfather, the Earl of Kimberley, had been Secretary of State for Foreign Affairs, but Clarence accepted a less exalted position: as a local commandant of police in Cyprus. By the beginning of 1894 they had gone. Elizabeth particularly had lost a dear friend. Although the families remained in touch, as Jack was at Harrow School and Francie would return for holidays with her children at her parents' house, the close camaraderie of the charity concerts was no more. For Dorothea, although she did not then know it, the Wodehouses' move to Cyprus was to present her with the means by which she established her career and her reputation.

A year or so after the Wodehouses' departure, a curious thing happened. Elizabeth and her family almost vanished from the local society columns. From being prominent at almost every social event, Elizabeth in particular appeared to have attended scarcely any and there is virtually no mention of the Bates for about two years. Then, in August 1898, *The Cardigan and Tivy-side Advertiser* carried a preliminary notice of a three-day sale to be held in late September of 'furniture etc' from Gellidywyll. In the same week, Dorothea and Thomas took part in the tennis championship at the Tivy-side Lawn Tennis Club and both reached the finals. 'We must congratulate the recruits generally,' commented the local paper, 'and Miss D. Bate in particular, on the good fights made against their more experienced opponents.'[15] For both of them, their performances may have been sheer bravado. When the full advertisement for the sale at Gellidywyll was published, the 'furniture etc' was in fact 'The whole of the Valuable Modern and Antique Household Furniture, Pictures, Glass, China, Books etc etc the property of Major Bates [*sic*]'.[16] After ten years, the Bates were leaving Tivy-side.

I have found no records to show why they sold everything and moved away from Tivy-side, and no family papers survived that devastating fire at Leila's house in the 1950s. The cause could be as simple as the ending of the lease on Gellidywyll, but that would not explain why they sold all their possessions. A financial crisis is a more likely explanation (running a fully staffed, thirty-six-roomed house on Henry's fixed income must have been a challenge), but there is something else. In the folklore of the Bate family there is a suggestion that Elizabeth's behaviour may have been the cause. She was wonderfully energetic and clearly talented, and could contribute greatly to the society around her. But she could also, apparently, go too far. Her tireless attendance at so many balls and parties with or without her husband may have begun to give rise to gossip. Her sudden disappearance from the centre of Tivy-side society is certainly curious. According to her family, she came to be regarded as 'a bit of a handful' and 'a restless and rather irresponsible person',[17] although whether this led to a scandal that made the move inevitable is unrecorded. Whatever happened, from Tivy-side they moved more than one hundred miles away, to a large house near the Forest of Dean in Gloucestershire.

This was in any event a difficult time for Elizabeth as her children began to establish their own lives. Thomas was sent to a crammer's in London to prepare for the entrance examination to the Royal Military Academy in Woolwich and a subsequent career in the Royal Artillery.

Leila was often away, staying with friends and studying music in London; despite the attractions of a musical life, she would doubtless make a suitable marriage. That left Dorothea as Elizabeth's only companion, and while her brother and sister were both making the most of life in London, the role she could see being created for her was one that was all too common at the time: the dutiful daughter at home, left to look after her ageing parents. That simply would not do. Not only did Dorothea want a life for herself, she knew exactly what that life was to be. By now she had acquired considerable knowledge of the local flora and fauna, particularly ornithology, and however odd it might seem to her family and her friends, she wanted to learn more. Wasting no more time, in the early autumn of 1898 Dorothea made that historic visit to the Bird Room at the Natural History Museum in South Kensington.

On her return home, she wrote to Bowdler Sharpe. This is her earliest known letter, preserved in the Museum's archives. It was written just after the family moved from Gellidywyll to Gloucestershire in a flowing, even hand, with a sublime disregard for punctuation. Why use a full stop when a dash expresses so much more?

> 15.10.1898 Bicknor Court, Coleford, Gloucestershire
>
> Dear Dr Sharpe
>
> I was very pleased to hear from you the other day – so glad you did not discover much bad work of mine or I should have been afraid to go back to the Museum – as I hope to do some day – We have all been very busy moving in here and unpacking our various goods – We have only been here for about a fortnight I think we shall like it. The country is very pretty, but I think the famous Wye is not a patch on our river at home [the Teifi]. It is not a good bicycling district – we have a long hill up to the house, I have only been out once to Ross – and then had to leave my bike behind – fortunately got a train part of the way home. I am afraid you will think me a great bother – this time I am going to ask you if you could spare me a couple of zoo tickets for my brother who is still in London – Is the new Director a particular pal of yours?
>
> Yours very sincerely
>
> Dorothea M. A. Bate[18]

The woman she became is already evident: there is great charm and enthusiasm, directness in asking for favours, thoughtfulness for others and recognition of the importance of knowing the right people. There is another notable point about her letter to Bowdler Sharpe. He had obviously written to her in terms quite flattering about her work, and she, equally clearly, felt easy in their relationship. Yet Dr Bowdler Sharpe had a reputation for being irascible and brusque, and the Bird Room seems not to have been a place of total harmony, as is evident in his outburst to the Museum's Director against a member of his staff:

> The question arises whether Mr Oates is a fit and proper person to be allowed in the Bird-room. He has shown such persistent spite and malice during the past few years that it is allowable to believe that he is off his head, but a man who has done his best to ruin me in the eyes of my superior officer and the Trustees may be capable of anything such as mutilating the specimens or breaking eggs in order to get us into trouble.
>
> I therefore think that an account is due from him (if of any worth) and an apology given to yourself for his conduct. This is the least that we can expect. I shall prefer his not being allowed to use the Bird-room. He has been a terrible incubus on me for a long time.[19]

The archives of the Natural History Museum reverberate with such passions; the degree of dislike evident in some members of staff for their colleagues or superiors is astonishing – but what original and innovative scholarship ever evolved from tranquillity?

Dorothea's letter is the only record of her first appearance at the Museum and from what she wrote it seems that it was something of a trial run. For the moment she remained at home, exploring this new and very different countryside. Bicknor Court, the house that the Bates had rented on the edge of the Forest of Dean, was again surprisingly large. As I drove up to it on a dark February day, it looked grey and austere, half hidden by skeletal winter trees, but behind its formal eighteenth-century front was a rambling, comfortable house dating back to the Middle Ages. It stands isolated on top of a hill so steep it tested my car's gears and brakes; how Dorothea cycled at all in the neighbourhood with exhausting hills and valleys in every direction is a wonder. Yet the Forest of Dean, an area of great beauty although rich in iron ore and coal, soon became as familiar and fascinating to her as Carmarthenshire. Dorothea kept notes of the various birds in the neighbourhood, recording what she saw and where.* 'Wheatear. Father saw one [of] these birds close to the line near Lydbrook junction – May 1900' and 'Lesser whitethroat. Found its nest with five eggs on May 25th 1900 – close to Bicknor Court – in a wild rose bush.' 'Garden Warbler/Sylvia. Found a nest and identified bird in the High meadow woods – June 1st 1900.'[20]

From the attic windows in Bicknor Court, I could just see the wooded course of the river Wye. Dorothea must have stood where I was, gazing out over the hills and treetops, following the course of the river as it meandered in great loops through its valley, and it became a magnet to her. Its limestone rocks were rich with caves, and it was here that she discovered the branch of natural history that would dominate her life. Beneath her feet, in some of those caves she visited, and locked in the surrounding walls were fossils, not shells or great Jurassic creatures millions of years old, but the fossilized remains of more recent extinct animals, small mammals dating back perhaps 10,000 years. These creatures, such as mice or voles, were seemingly the same as the modern species she knew so well, but with just enough difference to set her wondering. Iron-ore miners had excavated many of the caves, and she may well have found her first fossils simply lying loose on the disturbed cave floor. Within a few months of her first visit to the Museum, Dorothea must have returned, although there is no record, taking a few fossil specimens with her. There she met the Keeper and scientists from the Department of Geology, and they encouraged her to undertake her first systematic palaeontological exploration in the caves of the Wye valley.

The Keeper of Geology was Dr Henry Woodward, a bustling, balding, confident man who had overseen his department's move from Bloomsbury to South Kensington, and was responsible for vastly increasing the collections in his charge. As editor of the *Geological Magazine*, which he had co-founded in 1864, he had a reputation for encouraging young

* These few notes – literally torn scraps of paper – survive in Dorothea's papers in the Museum because she used the backs of them to scribble page references and so on for her later work.

geologists to publish their earliest work, guiding them through the essentials of a scientific paper and editing drastically but constructively where necessary.[21] To some of his staff in the Museum he could be much less congenial. His successor as Keeper, Sir Arthur Smith Woodward, recollected a 'hostile atmosphere and attitude of his chief' when he was appointed to the Geology Department in 1882, as Henry Woodward had preferred another candidate 'who was very good at writing labels'. The clash of names also irritated Henry Woodward and with what now seems outrageous arrogance, as Smith Woodward (then plain Arthur Woodward), recalled: 'He told me he had been considering the possibility of my changing my name; he thought however that if I called myself Smith Woodward, this solution of the difficulty would suffice.'[22]

As Keeper of Geology, Henry Woodward had charge of the palaeontology collections. Although at the end of the nineteenth century palaeontology was recognized as a vital and independent science, it was considered by some the 'hand-maiden' of the encompassing science of geology. Simply what to call the department was a matter of debate. As a 'very learned official' of the Museum told a journalist in 1897, Palaeontology 'would be a more correct term, but Geology is the term in general use in such museums as ours, so we have retained it'.[23]* The *Geological Magazine*, in an article in 1902 to mark Woodward's retirement as Keeper, even referred to 'its geological and zoological readers (no one at the present day, we presume, desires to be called a palaeontologist)';[24] and as Woodward was still the editor, those, presumably, were his views. It was not that he denigrated the science itself, far from it; it was simply that he saw palaeontology as part of a wider picture. Geology dealt with the great concepts of the earth, its history, composition, formation and chronology. Palaeontology, literally the study of ancient beings, is the science of ancient life-forms, animal or plant, whose remains have become fossilized in rock. It is the essential key for the geologist in identifying the structure and chronology of rocks, not simply to augment knowledge of the earth's history, but for essential practical purposes, such as construction of canals, roads, bridges and great buildings, or the discovery of fossil fuels.

That the nature of particular strata could be determined from the type

* It was not until 1956 that the Geology Department was redesignated as Palaeontology.

of fossils embedded in them is thanks to the pioneering geologist and canal engineer, William Smith, who made the link in the early part of the nineteenth century. He was the first to observe that rock strata lie chronologically, with the youngest on top. As his work took him all over England, he discovered that there were similarities in fossils in particular strata throughout the country. His most famous work, the first ever geological map of England and Wales, was published in 1815.

But fossils were not just a means to a geological end. The study of these strange remains of organic matter purely for their innate interest was advancing at an equal rate. Geologists found themselves acquiring botanical, zoological and anatomical skills to determine the nature of these extinct creatures and plants, while anatomists, zoologists and botanists acquired the skills of the geologists to understand the rocks in which the fossils were embedded. The *Manual of Palaeontology*, which Henry Woodward greatly admired and would have recommended to Dorothea to read, described palaeontology succinctly as 'the ancient life-history of the earth', which could 'be defined as the Zoology and Botany of the past'.[25] Woodward himself wrote that palaeontology 'strives, by comparison with living forms, to restore the successive faunas and floras which have passed away . . . and thus to show the evolution of life on the earth from the earliest times to our own'.[26]

The idea that from a fragment of a fossilized skeleton it is possible to discover what the beast looked like, how it moved and functioned in its ancient environment and how it might link to living creatures, did more than simply captivate Dorothea's imagination. It also marked the beginnings of a significant change in her intellectual development. From a broad desire to know what things were, there were now the first stirrings of how that knowledge might be applied. In the early nineteenth century, with their new-found understanding of the stratigraphy of rocks, geologists had divided geological time into fifteen divisions, from the earliest, which was Cambrian, to the Pleistocene, the period immediately preceding our own and characterized as the Ice Age. The age of these divisions was relative. Then there was no way of knowing the absolute age of the earth, only which rock strata were the earliest and which the most recent. The fossil fauna from the limestone caves of the Wye valley dated from the Pleistocene. With her knowledge of recent birds and mammals, Dorothea found them of enormous interest. As they come from the period that immediately preceded our own, Pleistocene fossil mammals give the best chance to study why animals become extinct and how others survive and evolve into the creatures we know today.

In addition to the *Manual of Palaeontology*, there was a wealth of literature to help her both in this and in the very practical and essential area of how to identify a site where fossils might be found. The Museum published one such pamphlet, *Directions for Collecting Specimens of Geology and Mineralogy*. It was written 'for the use of such persons as are supposed to be entirely unpractised in Geology and Mineralogy,' and described which rocks would be of interest and which 'had much better be left where they are than made the source of embarrassment'.[27] Those that might provide objects of real scientific importance to the intelligent but 'unscientific' observer included boulders, rolled pieces, rubble stones, gravel, sand and silt. Determining their origins was important, such as whether they had been washed down by rain or rivers from higher levels, and it was in this material that the collector might well find the 'well-preserved' fossilized remains of teeth and bones of 'elephants, hippopotamus, rhinoceros, petrified wood etc'.[28] Caves in limestone rock were 'uncommonly interesting', the pamphlet stated, as bone deposits frequently occur there.[29]

The pamphlet also temptingly suggested that the accidental discovery of a small portion of bone rising through the rock may lead to that of entire skeletons 'if sufficient care be employed'.[30] This is an optimistic scenario. While it may happen, it is in reality quite rare. For the bones and teeth of an animal to survive long enough to become fossilized, the creature has to be covered swiftly in some form of sediment, such as a lake bed or volcanic ash, or the tar pools of California, otherwise it will either rot or be eaten by local scavengers. Most fossil-bone remains from the Pleistocene in Europe are found in caves or cave sites where the original cavern has collapsed over time. The animal might have been swept there by a river or flood, or killed and carried there by a carnivore or by early man, or the cave might have long been used as a shelter, in which case animal remains would have accumulated there over thousands of years. Under these circumstances, the chances of a skeleton becoming fossilized in its entirety are pretty low. Similarly, the selection of extinct animals that survive as fossils is quite random. Out of the millions of creatures that have inhabited the earth over hundreds of millions of years and have become extinct, the fossil remains of just a few hundred thousand have been discovered.

From Bicknor Court, Dorothea would bicycle or walk over to Symonds Yat, just a couple of miles away. Occasionally she would take the train from Lydford Junction, a bicycle ride away from Bicknor Court, to the station on the banks of the Wye at Symonds Yat East, a popular destination for Sunday afternoon outings. Line and station have now gone, but the valley there is still glorious, a deep majestic curve downstream of red and gold in autumn, a myriad shades of green in spring. I followed Dorothea's path along the river on a morning of dripping trees with mud thick and sticky underfoot, the aftermath of a night of torrential rain. As Dorothea had done, I took the ancient hand-ferry across to the far bank at the point that had been used as a crossing since Roman times. The ferry is still owned and operated by successive landlords of the Saracen's Head where the ferry slipway runs into the river. She came to know this stretch of the valley intimately: the calmness of the river at the crossing point until it rounds a bend in a flurry of rapids; Lord's Wood and Lady Park Wood, outlying parts of the Forest of Dean that swept down each of the precipitous sides of the valley; the green meadow that bordered the right bank and then intriguingly, two miles or so from the hand-ferry, high up on the hillside to her right, three or four outcrops of white limestone cliffs thrusting through the trees. Past the open meadow and once more into the thickness of the trees, in one or two places limestone rocks and boulders of all sizes had tumbled down, a river of white lying in the steepness of the forest. From the far side of the river, though invisible from the path immediately below, she could see the rocks had fallen from a great cliff, in which were numerous cracks, holes and small caves.

Dorothea had been told by iron-ore miners of one particular cave that had never been excavated for fossil remains. To reach it meant climbing more than a hundred and fifty feet up an unforgivingly steep slope, avoiding the treacherousness of the scree and clinging to the trees for support. At the top of the slope the limestone crags rise sheer, and hidden in a south-facing corner, invisible even from the base of the cliff and fifteen feet up it, is the cave. So sheer and dangerous is the slope that Dorothea would have been risking her life to climb – or, particularly, to descend it – without the help of a rope; to reach the cave itself, a ladder was essential. She would also have been encumbered with long, heavy skirts, as well as her collecting bag and tools. Even with a guide, probably a local miner without whom she could not have attempted this, it was an extraordinarily hazardous undertaking and one that fills modern palaeontologists with admiration. It is considered one of the most inaccessible and dangerous sites to excavate, even with climbing equipment, though I did not discover this until later. With the foolhardiness of ignorance, I climbed the slope without ropes or any other equipment, an experience I can only describe as character-building. Every step has to be tested; secure-looking branches turn out to be treacherously fragile and one small slip on the way down can (and did) turn into a terrifying and uncontrolled slide.

About halfway up the limestone cliff which rose from the top of the slope was a dense growth of trees and shrubs. Behind this and completely concealed from below was the mouth of the cave, known today as Merlin's Cave. So difficult had it been to reach that Dorothea noted in her report that it was 'evidently inaccessible to foxes and badgers, as there are no holes used by them here, although they are to be seen in almost every other cave I have visited in the district'.[31] The cave consisted of two chambers, the first and largest going back about thirty yards into the cliff. In the uncertain light of her candle lantern, she saw, embedded in the walls and floor, numerous fossilized bones of small mammals and birds. They dated back to the very end of the Pleistocene, about ten thousand years ago, which saw the end of the last period of glaciation. We are now in a warm period, an interglacial. From the early twentieth century until fairly recently, it was thought that the great glaciers and ice sheets of the Ice Age swept over and then receded from northern and central Europe four times. Now it is calculated that, since the beginning of the Ice Age about 1.7 million years ago, this happened at least fifteen times, and in spite of fears of man-induced global warming, there is nothing to say that it may not happen again. The tiny fossilized creatures Dorothea found were mainly ones that thrived in a cold climate.

The inner chamber of the cave was utterly dark, with no natural light at all and a ceiling so low she had to crawl on hands and knees. In here, embedded in every shadowy ledge and crevice, were small bones, most of them belonging to one or other of the smaller species of voles and mice. It was cold and, although water dripped at the back, it was mostly dry, otherwise the bones would not have been so well preserved. Much of the floor had been dug up some time ago by iron-ore miners, but the walls were untouched. She needed two geologist's hammers to extract the fossils: one that weighed between two and four and a half pounds for breaking the rock, and a smaller one for trimming the extracted specimens. When she made her first uncertain attempts at prising the tiny fragile bones from the hard limestone, they fractured and broke. However, she quickly discovered that if she chipped away pieces of rock, 'similar bones were certain to be found loose in any soft or crumbly places'.[32]

Once she had extracted the bones, intact or otherwise, she packed the larger ones in paper, and used wool and cotton for the more delicate fossils, gluing or pasting a small label to each. She and her guide then somehow managed that treacherous descent, carrying both tools and fossils. The palaeontologists who excavated there nearly a hundred years later procured a plastic cider barrel from a local pub and ran a rope through the middle to haul their equipment and finds up and down.[33]

Extracting these tiny bones and teeth from the cave was one thing. Even with Dorothea's growing knowledge of anatomy, identifying this fossil fauna was quite another. From a single fragment of bone six inches long, which had come from New Zealand, Richard Owen, the prime mover behind the creation of the Natural History Museum, had in 1839 identified and described a large flightless bird that was 'a heavier and more sluggish species than the ostrich', yet there was no evidence that such a bird had ever existed.[34] He had compared the bone fragment with fourteen other species, from man to a giant tortoise, taking in an ostrich and kangaroo on the way, before he made his confident assertion that he had discovered a new species of bird. Four years later he was sent from New Zealand complete bones that confirmed his identification. The Maoris called the great extinct creature a moa.[35] The scientists called it *Dinornis*. Owen was the greatest palaeontologist of his day. Dorothea was just twenty-one when she excavated the bone cave. She lacked experience, but she could, and did, refer to the Geology Department at the Museum for help in identifying her specimens and writing her report. She applied to the Director of the Museum, Edwin Ray Lankester, for permission, as a student, to study her fossil specimens there, using its collections for comparisons. With Woodward's support, this was granted, and on 27 September 1900 she arrived with her carefully packed boxes of fossil fauna. The doorkeeper signed her in as 'Miss D. M. A. Bates (student)', a mistake never to be repeated. The intruding 's' was immediately deleted.[36]

With guidance and advice from two vertebrate palaeontologists in the Geology Department, Dr Charles Andrews and Dr Charles Forsyth Major, Dorothea began work on the specimens. They showed her how to mend the tiny bones where necessary, using a little glycerine or glue, what distinguishing features to look for, and how to compare them with specimens in the collections, with reports in scientific journals and in the great volumes of palaeontological studies. All the fossil bones she had found were from known species and the two experts instructed her in the essential business of naming them correctly.

In the eighteenth century, the Swedish naturalist, Carl Linnaeus, had devised a system of classifying all plants and animals, living and extinct. His binomial system still remains the basis for modern taxonomy and systematics, the naming and classification of organisms. Linnaeus gave every organism two Latin names, the genus or generic name which is always written with a capital letter, and the trivial or specific name which never has one. Dorothea would then have used the term 'trivial', although now it has all but fallen out of use in favour of 'specific'. The Latin form is still used, its universality attempting to prevent mistranslations and misunderstandings between scientists around the world. So, for example, among the fauna she identified were the fossilized bones of a mole that still thrives in Britain. Its binomial is *Talpa europaea*, literally and simply, the European mole. She identified a bat, *Rhinolophus hipposideros*, the genus of bat with a loph-shaped growth on its nose (from the Greek *lophos*, meaning crest), associated with the high-frequency sounds it emits. Together with the specific name (literally, horse-iron), its species is identified as that of a horseshoe bat. Almost all the other small mammals she identified – voles, lemmings and lagomorphs (member of the rabbit family) – are now extinct in Britain, though they survive in central and northern Europe, Russia and Siberia.* It was, as she was learning, a typical

* These included five species of vole (the sixth, *Microtus agrestis*, still survives in Britain); the Norwegian and Arctic lemming (*Lemmus lemmus* and *Dicrostonyx torquatus*); two types of lagomorph – the mountain hare, *Lepus timidus*, and the tailless hare or pika, *Ochotona pusilla*. Among later remains were those of a dog, sheep and small deer as well as a few artefacts, a bone needle or hairpin, and a portion of a copper ring from human habitation of the cave in the Bronze Age.

assemblage of cave fossil fauna from the end of the Pleistocene, creatures that had thrived in the Ice Age in Britain, dying out as the climate warmed. The bird remains were identified for her by Dr Charles Andrews, the Museum's foremost expert on fossil birds.

With her identifications complete, Andrews and Forsyth Major encouraged her to write her report for publication. Her style is clear and fluent, her descriptions of the cave and its contents detailed and fresh. Henry Woodward found it of such interest that in keeping with his policy of encouraging young geologists, male *and* female, he published it in the *Geological Magazine.* To her enormous pleasure and pride, it appeared as 'A short account of a bone cave in the Carboniferous limestone of the Wye valley', with the byline 'Dorothy M. A. Bate'. It is only in her earliest papers that she allows the familiar form of her name to be used for a publication.

Andrews and Forsyth Major did more than guide Dorothea's research. As she had won the respect of the ornithologists for her work in the Bird Room, so she won that of the geologists for her work on the Wye valley and they inspired in her a passion for each of their areas of expertise. From Dr Andrews she acquired a lifelong interest in fossil birds; she was, of course, already a considerable ornithologist. Dr Forsyth Major's great interest was the fossil and recent fauna of the Mediterranean islands, where he spent as much time as he could afford, particularly in the western islands of Corsica and Sardinia. Some palaeontological excavations had taken place in Malta and Sicily, but little was known of the fauna, fossil or recent, of the more eastern islands of Cyprus and Crete.

With splendid timing, at the end of 1900, the Wodehouses returned for a few months from Cyprus to stay with Francie's parents at their estate on Tivy-side. Francie Wodehouse was pregnant and the baby was due in early spring. The Wodehouses invited Dorothea to return to Cyprus with them, creating the perfect opportunity for her. In the spring of 1901, with the encouragement and support of the Natural History Museum in the person of Henry Woodward, Dorothea, aged twenty-two, embarked on the first of her pioneering expeditions to the Mediterranean islands.

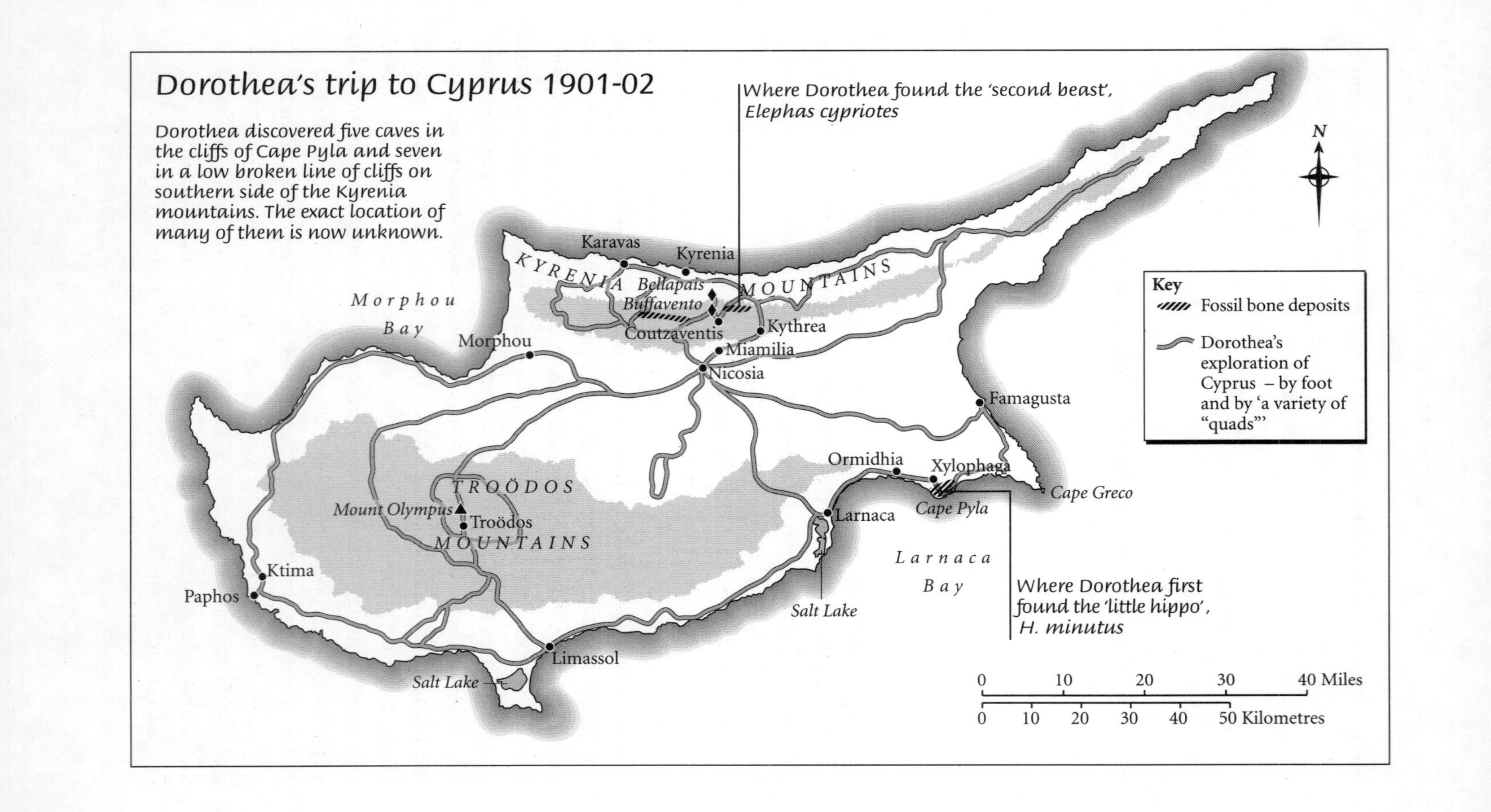
Dorothea's trip to Cyprus 1901-02
Dorothea discovered five caves in the cliffs of Cape Pyla and seven in a low broken line of cliffs on southern side of the Kyrenia mountains. The exact location of many of them is now unknown.
Where Dorothea found the 'second beast', Elephas cypriotes
Where Dorothea first found the 'little hippo', H. minutus
Key
Fossil bone deposits
Dorothea's exploration of Cyprus – by foot and by 'a variety of "quads"'
N
Karavas
Kyrenia
KYRENIA MOUNTAINS
Bellapais
Buffavento
Coutzaventis
Kythrea
Miamilia
Nicosia
Morphou Bay
Morphou
Famagusta
Ormidhia
Xylophaga
Cape Greco
Cape Pyla
TROÖDOS MOUNTAINS
Mount Olympus
Troödos
Larnaca
Larnaca Bay
Salt Lake
Ktima
Paphos
Limassol
Salt Lake
0 10 20 30 40 Miles
0 10 20 30 40 50 Kilometres

CHAPTER 3

Leave to Collect

On the morning of Wednesday, 17 April 1901, a very large party left Charing Cross Station on the boat train to Paris. With Dorothea were Clarence and Francie Wodehouse; their children Jack, Kitty, and baby Francie Petronel (who was just five weeks old and whose birth had brought the family back to Britain); the baby's nurse and an Italian manservant, Vigano. For Dorothea it was an extraordinary undertaking. She was now a published palaeontologist with the respect of eminent geologists and zoologists at the Natural History Museum, yet her only experience of palaeontology so far had been excavating tiny mammals in the Wye valley. Now she was about to become an explorer, the first scientist ever to attempt a systematic search for fossil mammals in the limestone caves of Cyprus (cave-hunting, as it was popularly known). As Charles Forsyth Major told her, 'nothing was known in this respect about the island, but much was to be expected'.[1] Before she could begin to excavate the fossil fauna, she had first to find it, and in this large, wild and mountainous island, there were literally thousands of caves. She would in every sense be cave-hunting.

But there was another side to her life in Cyprus. Her cabin trunks contained boots and practical clothes for climbing, hunting, collecting, and 'fossilizing' as she termed it, but, as the guest of a senior official of the British administration, she also had a role in society. Even with her limited resources, she would need dinner, ball and afternoon tea gowns as well as all the accessories for tennis, croquet and riding parties at Government House and with the ex-patriot community of Cyprus. In another trunk was her equipment for collecting both fossil and recent fauna, as she also proposed to observe and collect for the Museum's Zoology Department. She had a collecting bag and tools for excavating; a net for catching insects and the pins and cases for mounting them; a camera, labels for her finds – both fossil and live species; preserving and packaging materials; and her gun and ammunition. The collecting materials came from the Museum; before she left, Dorothea had agreed to donate on her return £2 worth of specimens to the Museum – the value of the materials she had received. She brought with her a 'bible' in the heavy shape of the *Palaeontological Memoirs and Notes of H. Falconer.* Published in 1868, this two-volume work by the botanist and palaeontologist Hugh Falconer describes and illustrates a vast number of fossil mammals – essential for the new and the experienced collector. The same volumes travelled with her throughout her life, over the years the string that tied them together wearing deep grooves in the covers. She also took with her a map of the island, which detailed its physical features and distinguished Christian villages with a cross and Muslim villages with a crescent.

Most important of all were her essential notebooks, substantial and leather-bound, to record her expedition. During her eighteen months in Cyprus, she filled three of these. She wrote on just one side of the page, leaving the other available for sketches and additional notes. Her exuberant handwriting records with wonderful directness the places she visited, the wildlife of Cyprus, details of her finds, and the appalling difficulties she experienced in discovering them. There are tantalizing glimpses of the life she led and people she met, but almost nothing of her private thoughts or emotions. These are work diaries and she could record nothing too personal; others might need to refer to them. The result is both fascinating to read and utterly frustrating, as time and again she seems to come close to revealing something of herself, and then pulls back. Nonetheless, there is such verve and energy in her writing that her voice and speech patterns seem captured in the page, so alive that I would find myself almost in silent conversation with her as I worked my way through her diaries.

The party arrived in Paris in the late afternoon of 17 April and caught the overnight train to Marseilles. 'No questions raised in customs,' Dorothea noted, 'about my gun.'[2] She recorded the passing scenery as they sped through France: the fruit trees in full blossom, great numbers of magpies, snow-capped hills and, as the train ran parallel to the Rhône, the countryside that had flooded for miles where the river had burst its banks. At Marseilles they boarded the SS *Saghalien*, an elderly 4,000-ton ship of the French Messageries Maritimes Company, bound for Beirut, then part of Syria. As the ship did not stop at Cyprus on the outward voyage, they would have to take another vessel from Piraeus to the island. At dawn on 20 April, they arrived at Naples. Even before they anchored, the boat was boarded by crowds of Italians, 'selling every imaginable article', Dorothea wrote; 'most of them spoke a little English'.[3] It was a glorious morning as they went ashore; flowers and flowering shrubs seemed to be everywhere, while oranges and lemons hung heavily on their trees. 'The town looked awfully gay and jolly,' she wrote, 'built in a semi-circle all up the side of a hill so the streets are very steep and generally narrow – Goats seem to wander as they like.' As they left the Bay of Naples she noted the beauty of the surrounding hills and that Vesuvius was smoking away – it erupted five years later. Everything for her is new and fascinating; she observes and records with at times an almost childlike excitement, her voice in these diaries very clear.

The weather changed as they sailed south. It grew cold and the wind rose. Dorothea lay miserably in her bunk for a whole day as the boat pitched and rolled, her eyes closed as she fought the seasickness that plagued her all her life. The springless carts on the many almost impassable tracks of her later excursions were to have much the same wretched effect. On 22 April they reached Piraeus and went ashore at once. The town was very dusty but bright, scented with 'mimosa trees in blossom and any amount of roses'. She noted, with the detail that characterizes all her observations, faunal and human, that there was a 'great variety of nationalities to be seen about – most picturesque were the Moreans dressed something like Albanians, their legs and feet covered with a closely fitting cloth that looked like white flannel, caps with long black tassels, ditto on the points of their shoes – a rough coat, shirt and a blue cloth kilt pleated at the back just like a highlander's'.[4]

They climbed a small hill and gazed over the plain of Marathon, while in the distance they could see Athens and the Acropolis. At that moment, far in the landscape below them, Dr Arthur Smith Woodward, the Assistant Keeper in the Geology Department, was preparing to excavate at the famous fossil-bone deposit at Pikermi, halfway between Athens and Marathon where he had arrived just a few days previously. The site, exposed by a small stream in the ravine of Pikermi, had been discovered in the 1830s. Over the years a magnificent collection of creatures from the Miocene or early Pliocene world of around six million years ago had been uncovered. The most important finds and 'brilliant generalizations' of the site, as Smith Woodward termed them, had been made by the great French geologist, Jean Albert Gaudry. He had discovered extinct forms of gazelle, hyaena, monkey, rhinoceros, large and small mastodons (primitive forms of elephant), a number of big cats, and relatives of the giraffe. All these, as Gaudry noted, were related to animals that now only live in India and Africa. He also found remains of *Hipparion*, an early horse with three toes on each foot.[5] Gaudry's discoveries, Smith Woodward wrote, had 'made known nearly all the essential facts' concerning this extinct fauna,[6] and although Smith Woodward's own work at Pikermi would reveal further important specimens, he found little that was new to science. In the limestone caves of Cyprus, Dorothea would be the first in everything she did, in the discovery of fossil-bone deposits and in establishing for the first time what extinct fauna had once inhabited the island.

Returning from their walk to collect their luggage from the *Saghalien* in preparation for boarding the ship to Cyprus, Dorothea and the Wodehouses were told that the incoming vessel had been delayed by medical authorities and they were, as Dorothea vexedly wrote, 'stuck with the Messageries boat'.[7] In addition to mosquito-borne malaria, contagious sickness, particularly plague and cholera, was endemic in the region and ships had to clear quarantine. It meant a much-extended voyage. The *Saghalien* was bound for Beirut by way of Turkey and frustratingly passed within ten miles of the Cyprus coast. However, it would not put in to the island to set down its unwilling passengers until the return journey. The situation was so unsatisfactory that, in October 1901, the Cyprus High Commissioner, Sir William Haynes Smith, in a despatch to the Colonial Office in London, asked whether the Messageries Maritimes Company 'could be induced' to call at Cyprus to drop mail and passengers *before* proceeding to 'Beyrout' where their ships remained for two days and only then sailed for Cyprus.[8] His request came to nothing. The consequences for Dorothea of staying on the SS *Saghalien* were quite severe, 'as our roundabout journey cost me very nearly £10 extra [more than £500 in today's money] which has made a rather serious hole in my supplies'.[9] Her parents were in no position to support Dorothea with anything like sufficient funds for her expedition; as far as the Museum was concerned, she would be paid for any specimens it wanted, but she had no official position there and it was unthinkable that it should contribute to her venture.

They reached the port of Smyrna the following afternoon, 23 April. 'It is a fine bay and the town lies on the shore and is surrounded by hills,' Dorothea wrote, and there is an almost audible intake of breath as she adds, 'infested with brigands'. Their plans to go ashore were delayed as they had no passports with them. That was not uncommon before the First World War, as passports were usually only necessary if the traveller was visiting somewhere potentially dangerous, or with military installations, or where relations with a country were not quite what they might have been. Had the party known that their journey would take them to Turkey with its strategically crucial position between Europe and Asia, and every European power jockeying for position, they would undoubtedly have brought them. The British Consul came to their aid, sending 'one of his men to take us round'. It was a cool, rainy afternoon as they visited the bazaar, walking through streets so narrow that passing carts forced them 'to take refuge in a doorway'. At a café where they stopped to rest, Dorothea had her first 'real Turkish coffee – four of us for a halfpenny! They hand round tumblers of water with it.'[10]

By the following morning they had reached the Dardanelles. As they sailed through the strait, Dorothea revelled in the beauty of the countryside on either side, exclaimed at the shoals of porpoises playing alongside the ship, and wondered at the great number of windmills: 'little stone towers with a wheel but could see no sails on them'. She noted flocks of 'cigar-shaped' birds flying up and down just above the surface of the water at such a speed they were a blur of brown above and white beneath. As for Constantinople, which seemed to her to stretch for an enormous distance along the waterfront and up the hill behind, it was 'quite the finest place I have ever seen . . . As we got close in the sun was fairly low and passing over the town in a broad band of light – illuminating in turn each mosque with its minarets.' Once they neared the port, Turkey's strategic importance became very clear to Dorothea as she observed that there were 'warships of almost every nation – the French and ourselves having two each besides a despatch boat'. From Turkey their unscheduled cruise to Beirut continued, during which, she notes tersely, 'Passed within about 10 miles of Cyprus during the night.'[11]

They finally reached Larnaca on the southern coast of Cyprus on the hot, sunny morning of 1 May, the heat made just bearable by a gentle breeze. What was less bearable were the 'exceedingly inconvenient' quarantine arrangements at Larnaca, which Sir William Haynes Smith raised robustly with the Colonial Office: 'There are no means of landing at the [quarantine] station,' he wrote, 'so that passengers and goods have to be brought ashore on the backs of porters. Great irritation is occasioned and female passengers regard having to be so handled as an outrage.'[12] The problems were not confined to the quarantine station. Even after twenty-three years of British administration, Cyprus possessed not one harbour of any size or depth. Few steamer companies were prepared to lose time while their ships transferred passengers to smaller craft to land them on the island, a process that in any event was impossible in rough weather. Hapless visitors to Cyprus would frequently find themselves taken on to Egypt.* Trade and communications suffered as a result, while tourism, in spite of the island's key position in the eastern Mediterranean and its wealth of antiquities, was virtually unknown.

Cyprus had been conquered by the Ottoman Empire in the sixteenth century and had become, according to a French traveller in 1807, 'a place of infection and death. We left it as quickly as possible.'[13] During the nineteenth century, there were attempts at reforming the administration, but little was achieved. Britain, however, was anxious to establish a secure base in the eastern Mediterranean and, at the Congress of Berlin in 1878 following the Russo-Turkish war, Britain agreed a defensive alliance with Turkey. In return for protecting Turkey against a future threat from

* One such visitor in 1896 was Colonel A. O. Green, who wrote of his experiences in *Cyprus: A Short Account of its History and Present State*. Attempting to sail from Alexandria to Limassol, he booked a passage on a ship advertised as sailing on a Wednesday. He was advised by the shipping agents that it would not in fact leave until the Saturday, but it actually sailed on the Thursday. He finally boarded the vessel the following week, but departure was delayed a further thirty-six hours while it waited for a consignment of cargo. Even by 1914 little had changed.

Russia, Britain would occupy and take over the administration of Cyprus. It was, nevertheless, a most tortuous arrangement, with the island still under the sovereignty of Turkey and its inhabitants still subjects of the Sultan. Cypriots, both Greek and Turkish, were included in the administration, although senior positions were, of course, held by the British. Government was similar to that of a Crown Colony, with a High Commissioner, a Legislative Council and an Executive Council. For administrative purposes, Cyprus was divided into six districts, controlled by a commissioner who for most purposes was the head of all executive departments in each district.

Clarence Wodehouse was now District Commissioner for Papho, as it was then called; in Larnaca, where the party had to stay for two days until a boat arrived to take them westwards to Paphos, the District Commissioner was Claude Delaval Cobham, a highly cultivated and hospitable man whose arthritic knee and resulting limp had earned him the soubriquet from some 'wag' in the British community of 'Dot and carry one'.[14] This was according to the journalist and traveller, Edward Vizetelly, whose view of the British in Cyprus in the late 1890s was not wholly flattering. Of the sizeable British community in Larnaca, Vizetelly noted they were dressed 'all in white – bodies in white suits, feet in white canvas shoes, heads capped with white helmets or white pulp hats of many different shapes and dimensions'.[15] Rather than try the Cypriot diet of goat, wine, anchovies, olives, tomatoes, 'stringy cheese', cucumber and figs, the British dined on 'bully beef', bacon, ham, whisky and beer.

> When he [the Englishman] dashed off to Cyprus to make his fortune [wrote Vizetelly], he was accompanied by tons upon tons of tinned provisions . . . by thousands upon thousands of cases of jam, marmalade, condensed milk, tea, coffee, chicory, spirits, wine and beer; by hundreds of barrels of soda water and other non-toxicants; by parcels without number of British cutlery, glass and crockery, hardware, paraffin lamps, tins of lobster, iron bedsteads, pipes, tobacco, patent medicines, agricultural implements and most of the other articles that he is accustomed to.[16]

The Wodehouses and Dorothea stayed with Cobham in his palatial house, built a century previously by the English Consul. It had a great central salon, a splendid colonnade, and a garden filled with flowering shrubs, palms and fruit trees. Cobham had been commissioner for more than twenty years, during which time he had collected a fine library of books and manuscripts, both antiquarian and modern, in a multitude of languages, all on Cyprus and the region. It was, according to the writer H. Rider Haggard who stayed with Cobham in 1900, a 'unique collection'.[17]

From the moment she arrived, Dorothea noted every living creature she saw, writing on her first day in Cyprus:

> There are lizards everywhere of several kinds, and I caught a chameleon on a branch in a hedge, put it close to some green glass dishes and it became the same colour in about 5 or 10 minutes. Caught a metallic green beetle in a field . . . Got a few chrysalises on grasses etc. They had just turned – saw some of the caterpillars – they had three tufts of yellow hair on their backs. Saw only some common butterflies about – painted ladies, clouded yellows blues and coppers.[18]

At the salt lake, two and a half miles from Larnaca, she noted snail shells, sandpipers, and yellow and black flies, which settled on the reeds that grew thickly on its east side. An aqueduct wall farther round the lake was home to numerous birds including jackdaws and 'a good many lesser kestrels – that seemed to keep the other birds in order a bit'.[19]

In just two days Dorothea noted nearly thirty different species of birds, small animals and insects, but she had expected many more. She found this part of the island 'distinctly discouraging'; much of it, she thought, seemed to be flat with some low hills 'parched up and almost destitute of vegetation for the greater part of the year'. Cyprus had once been famous for its forests, and although there was a policy of reforestation, young trees were rarely allowed to mature, either being cut down prematurely for firewood or succumbing to the insatiable appetites of innumerable goats. She had hoped to add to the *List of the Birds of Cyprus*, which Lord Lilford, the noted ornithologist, had published in the ornithological magazine, *Ibis*, in 1889. Lord Lilford's list, which identified 231 birds, was reprinted at the back of Dorothea's well-thumbed copy of the 1901 *Handbook of Cyprus*, the co-author of which was Claude Delaval Cobham. As Dorothea later wrote, it might have been possible, 'During a good winter, and with ample time for observation',[20] to add to the list, but she had arrived at the beginning of the hot, dry, Cyprus summer, and although birds were plentiful, she discovered in all her time on the island just one new species, a type of wren.

On 3 May, the party left Larnaca. Their transport was the one vessel that plied the coastal route, an old English-built yacht called the *Thyra*, of some forty-two tons. Nine hours later they arrived in squally weather at Paphos. Their journey from London had taken almost three weeks. The Wodehouses lived two miles from the port, at Akrotiri House near Ktima. It was, according to a visitor a few years earlier, 'a bright and beautiful English house', standing on a high limestone cliff overlooking the sea. The drawing room, where a coal fire burned in the winter months, opened onto a 'fine arcaded balcony-terrace', with a flight of stone steps down to a well-watered garden, shaded with orange and almond trees. On the sands below, the English officials had a 'golf ground', where the holes had to be renewed after every game.[21]

When Clarence Wodehouse was promoted to District Commissioner for Papho in 1895, he reported to the Colonial Office that 'the District was consistently the scene of murders and other acts of violence of the most barbarous nature. Though I do not venture to take to myself the entire credit for the suppression of crime in my District, I think it but just to myself to mention that more has been done during my tenure of office . . . than during the preceding seventeen years.'[22] Violent crime was a problem throughout the island and Dorothea was proposing to journey – alone – on her cave-hunting to some of the most remote and dangerous areas.

For these first few months, with hot weather approaching, she devoted herself to collecting live species. At every possible moment she was outside, often before breakfast with the sun scarcely up. In a field near the house she watched men apparently catching locusts, which they put into large sacks, weighed and then buried under earth mixed with lime in a three-foot-deep pit, also lined with lime. The men were paid according to the weight of the locusts. For peasant farmers this was a useful source of income, although how effective it was as a means of controlling locusts is questionable, as Dorothea observed: 'All I saw seemed to be just the grasshoppers that are all over the fields – several kinds.'[23] But locusts were a real problem, regularly stripping bare large areas of land. A locust tax had been imposed in 1881, revenue from which was supposed to be used wholly for their destruction, but it was a source of grievance among the Cypriot population that this money tended to be appropriated for any emergency. It had even been used by the British administration in the purchase of the *Thyra*, which in addition to its role as coastal transport was also meant to guard the shores of Cyprus and act as a quarantine boat.[24]

On 13 May Dorothea wrote to William Ogilvie-Grant, the ornithologist in the Zoology Department at the Museum for whom she was observing and collecting birds:

> I like this place immensely but have not done much collecting yet as my tin has not arrived and my gun is still in the customs – They have a very strict close season here but I have applied for leave to collect. We have had lovely weather and I have been riding about on a variety of 'quads' – a donkey, mule and pony . . . I have been trying to get up a little Greek but when tried on the natives it is not very successful! Fortunately Mr Wodehouse is a great hand at it – knows some Turkish as well.[25]

Mr Wodehouse's uses went further than his language skills; until her own gun arrived, he lent her one of his. To collect faunal specimens for the Natural History Museum, Dorothea needed to shoot and then skin them; a description of the creature was not enough. The needs of science demanded the evidence of the creature itself and in the best possible condition, together with photographs or sketches and descriptions of its habitat.

By now, Dorothea had become a notable figure as she explored the countryside, a striking young woman carrying an elegant sunshade, a collecting bag, and the essential gun. The residents of Paphos had never seen anything quite like her and, towards the end of May, the word went round that the young English lady at the Commissioner's house was interested in the local wildlife – and she would pay for specimens. Not only was information sent to her, but examples of the local flora and fauna began to arrive, both alive and dead; in the latter case, often pungently so. Two boys brought her a young fox, 'but I was not able to keep it'. She thought it was a month old, 'quite tame but a bit nervous – it lapped up milk and they said it ate cheese. Its eyes were bluish like a spaniel pups before they turn brown. ¾ of its ears and its four paws were black . . . its whole colour sandier than our foxes.'[26] She was brought a hedgehog for which she had to pay *6d.*, three small bats, which had 'a heavy, musty and disagreeable smell', a couple of bee-eater's eggs, and a griffin vulture, 'which had been gorging on a dead donkey . . . it was a sort of sunburnt brown with black tail and wing feathers – its head and neck covered with short whitish feathers, eyes grey blue – Had no time to skin it so had it buried for its skeleton.'[27] A shepherd told her of a francolin's nest ('a most excellent bird on the table',[28] according to the *Handbook of Cyprus*), which she photographed. The hen, Dorothea noted, 'sat awfully tight – allowed me to fix up my camera just in front of it and only left when I moved aside some of the corn right in front of the nest . . . there were eleven eggs about the size and colour of a pheasant but with some whitish blotches dotted irregularly over them – the bird just ran off the nest and kept quite close all the time I was there, making a shrill peeping noise.'[29] Even Mr Wodehouse brought her goldfinches.

She began to develop a method to her collecting, numbering each specimen listed in her diary. Each evening she set traps for rats, mice and voles, while her insect collection was also expanding. One that particularly fascinated her she had first found crawling on the tennis court at Akrotiri House. It was a sphalanga, which looked like a 'huge hairy ant' with no wings, its head and abdomen black with yellow markings, and its thorax a dull red.[30] It was said, she noted, 'to sting like a wasp and be extremely poisonous'. An unnamed acquaintance of hers, from his use of English probably Greek Cypriot, sent Dorothea a shocking description of the effect on a farmer who had apparently been stung as he slept on his 'threshing floor'. She might not have caught the insect quite so casually had she read this truly horrendous account first:

> Poor fellow! he was a monster of ugliness . . . his face was so dreadfully swollen that his eyes were pussed up in a grayish blistered mass of flesh, and the unfortunate fellow could not see at all . . . The sick [as the writer calls the victim] was trembling from head to feet . . . There was a continuous tumultuous snuffing, sniffing, panting, snoring and snorting with him. He could not even speak . . . his lips and tongue were parched, his mouth always wide open and his tongue lolling out.

The doctor came, the writer continued, but nothing could be done. 'The sick had already put one of his feet in his narrow home, and death was hovering over him . . . Some time after, he expired in the most frantic agony of pains, throes and sufferings.'[31]

Dorothea would not accept that the sphalanga sting alone could cause this agonizing death. She discovered that the district medical officer of Larnaca, Dr George Williamson, had studied cases of supposed sphalanga stings and had concluded that anthrax was in fact the cause.[32] This devastating disease commonly affects sheep, goats, horses, and mules, which are infected by grazing on contaminated pasture and can then in turn infect whoever is in contact with them. The link with the insect, according to Williamson, was that sphalanga bred in anthrax-infected carrion beetle nests. Consequently when the sphalanga left the nests early in the year, they were very poisonous, 'but gradually become less so', Dorothea wrote, 'and possibly eventually quite harmless to people in a healthy state'. In fact, as she was relieved to hear from Alfred Lascelles, the King's Advocate on Cyprus, the wife of one of his clerks had been stung 'without any ill effects following'.[33]

Dorothea collected and described many more benign and spectacular-looking insects. She found 'a cockroach sort of beast in the Wodehouse's bedroom';[34] dragonflies that were lavender, crimson, dull yellow, or had a bright red streak down the back;[35] a huge caterpillar, three and a half inches long, yellowish-green in colour, with two white stripes with white spots along its sides, a faint purple tinge along its back and a bright yellow tail. She observed it becoming 'flabby and shrivelled-looking' before turning into a chrysalis and she charted its transformation into a moth.[36]

Some of the time she explored on her own, but very often Jack Wodehouse went with her. Dorothea was twenty-two, Jack was now nineteen, a tall, charming product of Harrow School, The second question I am always asked about Dorothea concerns her relationships and I don't think I know the answer. The diaries, the main source for this part of her life, are totally discreet and neither family has any information. I am inclined to think there was an attraction on both sides, and that Jack's apparent enthusiasm for collecting may have had more to do with Dorothea than natural history, but that (alas) can only be speculation. Dorothea is silent in her work diaries about everything emotional – even, as I discovered, the most distressing events. For the last week of June 1901 there are no entries at all in her diary, and for a long time I could not understand why. It was only when checking a fact about Clarence Wodehouse that I discovered that the baby of the Wodehouse family, three-month-old Francie Petronel, died during that week. It was eleven years, almost to the day, since the death of their five-year-old son Billy. No Wodehouse letters survive and if Dorothea did record the devastation of the family or her own emotions, as surely she must have done, it would have been in a letter to her family and now lost.

By July the weather was unbearably hot, and, like Leo, the boy in L. P. Hartley's *The Go-Between*, Dorothea is mesmerized by the ever-rising thermometer. '3 pm – 91° Fahr in the shade', '92° in the shade in the middle of the day – greatest heat we have had yet.' Then, 'Thermometer just over 94° in the shade early this afternoon.'[37] In Cyprus, as in India, the British escaped to the hills in the intense heat of high summer. In mid-July the entire colonial administration transferred to the Troödos mountains where there were government offices and a summer camp for the troops quartered on the island. Society centred on Government Cottage, as the High Commissioner's summer residence was known. It was a substantial house, half hidden in pine trees and with the obligatory tennis courts. Nearly everyone else lived either under canvas or in simple huts.

Just three weeks after the baby's death, at 4 a.m. on 18 July, the Wodehouses and Dorothea began their trek to the mountains. 'With servants etc,' she wrote, 'we were a party of ten.' Their considerable baggage was sent on ahead. The journey took the whole day, plodding on mules through vineyards and lemon groves, along a river valley and then the slow climb through the hills. At noon they stopped in a village for lunch and to feed the animals. They rested in a house provided by the *mukhtar* of the village, the headman, whose task it was, on prepayment, to provide any accommodation or food for travellers on government business and their animals. There they stayed until 'Jack found a bug biting him and then I discovered one on my boot'.[38] The last part of the journey was up 'a very steep and tough zigzag path to this place Asprokremno which is right in the Forest'. This was to be their home for the next three months; their quarters were neither tents nor a hut, but a good-sized house complete with drawing room.

Dorothea was swept at once into the hectic social life in the hills – and she loved it. There were full-dress dinners, dances and concerts, lunch and tea parties and croquet. Dorothea was a good tennis player – she had done well in tournaments on Tivy-side – and with Jack became a regular visitor to the tennis club, which, she notes disapprovingly, has 'asphalte [*sic*] courts'. Although it was cooler at nights, the days were still very hot, and the cloud of red dust that was swirled up by the lightest wind or passing feet was, Dorothea noted, 'simply awful. The air here is so dry that one can hardly perspire and ones lips get dry and cracked.'[39]

She explored as much as possible, riding with Jack and the Lascelles to the summit of Mount Olympus, the highest point in the island, and almost every day, often alone, climbing fearlessly over the Troödos mountains in pursuit of mammals, birds and insects. It was dangerous terrain and she soon acquired a bruise '7 inches by 3!' on her arm, which meant it was too sore to use a gun; all she could do was collect insects. Not long afterwards, on an early-morning expedition, she slipped on a steep and rocky path and sprained her ankle badly. For three weeks she was laid up and for some time after that could only hobble about. When she recovered she drove herself hard to make up for the missing weeks.

Almost every day she would get up 'about 5am or a little after and have to use a candle to dress by', and use these early mornings for shooting. It was on one of these forays near a stream below Government House that she found the new species of wren, which she called on her return home *Anothura cypriotes* Bate.[40] These activities made her as much a curiosity to her compatriots as she was to the Cypriots. There could not have been too many young English women at the tennis club with her extraordinary sense of dedication and enquiry, who could happily handle the contents of the traps she set every evening after dinner, catch in a net a sizeable quantity of bats, and devise a very efficient method of catching moths: 'I used some native stuff called "Epsima" which is concentrated grape juice – with a very little treacle in it.'[41] She called this 'sugaring' and would paint it on trees and shrubs or leave little dishes strategically placed. She was delighted with its success. She went to a *mandra*, a shelter for shepherds and goats, to photograph the men at work.

> They are marvellous at whistling – doing without fingers in their mouths or anything. You can hear it simply shrieking and echoing

> up the valleys. Sounds well at a distance what with the goats bells, their shouting and dogs barking. They have great powers of shouting and making themselves understood from one hill to another. They make a primitive sort of flute which they play seemingly just for their own amusement. It is merely part of a large reed with a mouth piece filled in and several holes pierced by burning – a few crude patterns cut out with a knife.[42]

By the end of September the weather was considerably cooler and it rained nearly every day. 'Everybody,' she wrote on the 29th, 'is thinking of leaving Troödos now.' For Dorothea this meant the considerable task of packing her many specimens – bats, rats, mice, shrews and other small mammals, the birds and, of course, insects and spiders. She had prepared them as she went along, measuring, labelling, skinning and preserving them. For the mammals and birds to be of any scientific value to the Museum, she had to prepare them almost immediately because of the heat, and with great care. Dorothea was an instinctive scientist; her knowledge of anatomy and the skills she had demonstrated in the Bird Room of the Natural History Museum were vital here. The Museum's instructions for the preservation of birds counsel that 'it should be remembered that no more time is wanted to make a *good* skin than a *bad one*,'[43] but Dorothea was already an expert. She used a small, sharp knife and scissors to skin her specimens. She then brushed the inside of the skins with arsenical soap to preserve them, taking care not to get any on the fur or feathers. If she had cotton wool or bleached cotton, then she might stuff the specimen into shape, but in most cases she probably simply packed the skins flat.* To keep moths away, she used ordinary mothballs, made of camphor. As for packing the insects, Dorothea had so many specimens that she had to ask William Ogilvie-Grant at the Museum to send out more insect cases and pins.

In these last few days in the mountains, Dorothea went out collecting

* The Mammal Section of the NHM's Zoology Department was kind enough to show me how to skin a small mammal. I was given the victim of a road accident, a small rabbit, slightly squashed, and spent the next three hours with my admiration for Dorothea increasing exponentially. Not that I was squeamish; it was the extraordinary skill that is necessary to produce an undamaged skin, and she dealt successfully with hundreds, not in a laboratory, but in the field.

as much as possible, between farewell lunches and teas and games of badminton. On their last night at Asprokremno, since everything had been packed away, including all the bed linen, 'we had a big fire going in the drawing room all night,' Dorothea wrote in her diary, 'and slept on the sofas.' The mules arrived in the early hours and the party left at eight o'clock. The journey down to Paphos was uneventful, although it took nearly twelve hours. At Akrotiri House, Dorothea unpacked and then 'began rearranging my goods and chattels'.[44] She had spent five months in the west of Cyprus collecting wildlife; now, with the cooler weather, she planned to widen her search for the live fauna of the island and begin her hunt for 'extinct beasts'.

CHAPTER 4

In Search of Extinct Beasts

Since no one had ever looked for the extinct fauna of Cyprus and absolutely nothing was known about it, Dorothea faced a bewilderingly difficult task. With nothing to guide her, she was attempting to discover extinct mammal-bone deposits in what was the third-largest Mediterranean island, covering an area of 3,500 square miles, with two mountain systems, precipitous sea cliffs, and innumerable caves. It may well have been with the help of Claude Cobham and his splendid library that she discovered two possible leads.

The first was in the only book then published on the geology of Cyprus, *Géologie de l'Ile de Chypre*, by Jean Albert Gaudry, who surveyed the island in the 1850s shortly before his great discoveries at Pikermi. Gaudry refers to the Dutch traveller, Corneille le Bruyn, who visited Cyprus and the Levant in 1698 and wrote of finding a bed of bones near the monastery of Chrysostomo to the north of Nicosia, in the foothills of the Kyrenian mountains. Le Bruyn wrote that these were believed to be the bones of saints and were worshipped by the Greek inhabitants of Cyprus.

The second lead was a reference to numerous petrified human bones found by the archaeologist Luigi Palma di Cesnola in 1877 in two caves at Cape Pyla, about fifteen miles from Larnaca in the south of the island.*

* A self-styled general, explorer, and excavator of Cypriot archaeological sites on a grand scale, Cesnola was also the United States Consul. No attempt by the Turkish government of the time was made to prevent him exporting antiquities. Under an Ottoman law of 1874, one-third of

Dorothea wrote to Charles Forsyth Major for his advice. He told her, as she must have suspected, that the bones were highly unlikely to be human and, he wrote, 'would probably prove to be the remains of more humble mammals'.[1]

And that was all Dorothea had to go on when she left Akrotiri House. Cape Pyla was her first destination, a remote and rocky promontory on the eastern side of Larnaca Bay. Jack accompanied her to the port and saw her onto the *Thyra*. Dorothea was now, for the first time, travelling alone.

The *Thyra* arrived late in the afternoon of 11 October at Larnaca, where Dorothea 'Got the custom's man on board to look after me and my baggage.' Almost immediately, however, her inexperience as a traveller caused her difficulties over how much to tip the porters. 'Went to the Royal Hotel where I nearly had a row with the men who brought up my luggage – but shut my door in their faces and told them to go to the custom's officer if they wanted more – after a time they ceased shouting and departed.'[2] This was not the most auspicious start, but she had more fundamental problems to confront. She knew that there were bone caves near Cape Pyla, but she had no idea of their exact location. The countryside was wild and remote; she had little knowledge of geology, less of the language; and had no idea what reception she might get from local villagers.

In the morning she sought advice from Claude Cobham, the Commissioner for Larnaca. He recommended a Mr Perks, but when they met that afternoon at her hotel, he was not quite what she had expected. 'He is a broken down old Englishman,' Dorothea wrote unhappily and surprisingly unsympathetically, 'who has been here off and on since previous to the English occupation in '79 [*sic*]. I believe he has seen better days but now is very badly off and lives among the Turks – though poor

---

finds belonged to the government, one-third to the landowner, and one-third to the finder, although Cesnola appears to have appropriated considerably more than that. The Metropolitan Museum of Art in New York, of which he was to become Director, acquired the greater part of his magnificent collection for $138,866. The 1901 *Handbook of Cyprus* tartly remarks that 'the uncertain provenance and haphazard grouping of many of the objects detract not a little from its scientific value.' The Met opened new Cypriot Galleries in 2001, exhibiting 600 of the 6,000 or so splendid artefacts discovered by Cesnola. Even the wisest curators, however, cannot rectify 'uncertain provenance', and many of the objects had to be labelled with the caveat, 'said to be from' a particular site.

he is proud – have arranged to go out with him tomorrow. I think he knows something about geology but is a frightfully rambling old beggar.'[3] Whether she liked him or not, he had an ability that Dorothea needed even more than his knowledge of geology: he spoke Greek and Turkish, and she did not.

Dorothea then embarked on the sort of journey into the unknown that many people would think twice about today. Clarence Wodehouse may have been proud of the decrease in crime in the district of Paphos, but elsewhere in the island murder and barbarous violence were all too common. The very terrain was inhospitable, much of it an empty, stark and rocky landscape, with villages few and scattered, and countless places where the unwary might easily be ambushed. At daybreak Dorothea was impatient to set off, anxious to use every minute of daylight, but had to wait for both the mules and Mr Perks. It was nearly eight o'clock before they left for the village of Ormidhia, thirteen miles east of Larnaca across open, empty country on the other side of the bay. The October sun was still fierce, although the autumn rains had come and the countryside was turning green. At Ormidhia she found a guide who claimed to know of the caves with bones at Cape Pyla and would take them there. Two or three miles farther on, they came to the small village of Xylophaga, where 'we got a shepherd also to accompany us. It is frightfully barren country here,' she wrote, 'a certain amount of cultivated land and the rest dotted with bushy junipers.' As they rode towards the cliffs the going became very rough, with the mules stumbling on rocks and loose stones. 'The whole promontory,' she noted, 'seems to be of a different formation to the surrounding country and is composed of a hard limestone. Mr Perks tells me it is a brecciated limestone.'[4] Breccia, as Dorothea was learning, is composed of angular fragments of rock. Bone breccia is where fragments of bone and rock have mingled together.

The first cave was halfway down a cliff, with two entrances facing the sea, only one of which was accessible from land. They scrambled down the cliff, occasional spray from the waves breaking beneath them stinging her skin. At this western entrance to the cave, bones were instantly visible, 'embedded in the rock', she wrote, 'which is excessively hard being exposed to the air and doubtless to the spray from the sea in winter'. Here at her feet were remains of Cyprus's ancient fauna. It was the first indication of what she had travelled more than a thousand miles to find.

With no known explanation for these petrified bones, the cave had become known as Aghios Sarounda, the Cave of the Forty Saints.* Pilgrimages had been made there for centuries and, as she noted, 'candles burned in honour of the supposed sacred remains of saints'.[5] At the back of the cave the floor was covered in earth, but her preliminary dig revealed only recent goat bones. The fossilized bones at the cave mouth interested her far more, although if these were the bones of saints, she must have thought as she examined them, they would have walked on all fours. They were, as Forsyth Major had supposed, the bones of 'humble mammals'. What Dorothea needed to find were teeth, particularly molars or cheek teeth, which are immensely important in identifying fossil mammals and determining their age, diet, and whether, in the case of herbivores, they were browsers or grazers. The grinding surface of a tooth is unique to each species, as distinctive, as Henry Woodward expressed it, 'as finger prints of individuals'.[6] She examined the rock carefully, but saw no sign of teeth (excavating is rarely that easy, although Dorothea on occasion had an extraordinary amount of luck). With few daylight hours left, she pressed on to the second cave mentioned by Cesnola. This was the Spelio Macaria, about three-quarters of a mile to the west. It was a difficult walk, followed by another climb down cliffs to the mouth of this cave, which was just above the sea, but there was no indication whatever of any remains. Furthermore, it did not tally with the description given by Cesnola, nor did any other cave in the area. She returned exhausted to Larnaca in the dark, now all too aware of the true enormity of her task. The mountains and cliffs of Cyprus were riddled with thousands of caves; the locality of those said to contain bones often vague and misleading. It suddenly seemed to her remarkable that, on her very first day, she should have found a fossil-bone deposit at all.

Eager to make an early start, she ordered the mules for daybreak. She was acutely concerned at the length of her journey. It had taken about four hours each way, leaving her little more than two hours to work; the hazards of travelling at night in that wild terrain were too great to risk. Time for Dorothea really was money: she had to pay for her hotel in Larnaca, for guides, for a man – or sometimes men – engaged to do the

* Dorothea's spelling of names in all her journeys is phonetic and inconsistent. For clarity I have given accepted modern phonetic spellings as far as possible.

initial hard digging, for the mules or donkeys to get them to the site and, of course, for Mr Perks. Finding somewhere to stay near the caves would save her travelling time and increase the hours available to work, but accommodation was scarce. There were just six hotels in the whole of the island, although rooms were available in most villages. Travellers, however, had generally to bring with them their own food and cooking appliance, 'and [a] cook', according to Dorothea's *Handbook of Cyprus*. It then helpfully added that the traveller was 'lucky if he finds a room free from bugs and fleas; and sanitary arrangements, where they exist at all, are generally primitive and filthy'.[7] The nearest village to the Cape Pyla caves was Xylophaga, where the houses were built of mud brick. 'It boasts a church and a windmill,' she wrote, 'but too wretched a place for me to be able to stay at.'[8] It would take her a little longer to realize the demands and privations of cave-hunting in remote regions, and to grasp that using her time for digging rather than travelling had to be the overriding concern, not the wretchedness of her accommodation.

In spite of her order for the mules at daybreak, they did not turn up at all and she had to go and fetch them. It was lunchtime when Dorothea, Mr Perks and a man she had engaged to dig arrived at Xylophaga, where they stopped at the café. 'Being Sunday there were a lot of people,' she noted with amusement, 'and I was soon the centre of a crowd of men all jabbering about me and my affairs – but they were all perfectly civil.'[9] What an extraordinary character this intelligent and resourceful girl was; few circumstances intimidated her and she seems to have found nothing odd about marching into an unfamiliar place and enquiring, with a charming smile and a very determined air, whether anyone knew of a cave with bones in it, nor about offering money to strange men in return for information or specimens of live animals and birds. When a man claimed to know of another cave with bones, she promptly hired him as a guide to take her there. He in turn recruited 'a couple of native sportsmen to come with us – each with his dog, muzzle loader and goatskin bag on his back'.

Leaving her mule at Xylophaga, Dorothea set off from the café with the guide, the two native sportsmen and their dogs, the man she had already engaged to dig and, of course, Mr Perks in her wake. This is my iconic image of Dorothea, striding out ahead of her attendant men, or climbing down the most dangerous cliff, focused completely on her quest for extinct beasts. The passion that most people direct to relationships was for Dorothea turned to her science.

The cave was at a place called Kochinos Kremnos, or Red Cliffs, and, as she noted, dangerous to reach. 'To get to the cave, which is invisible I should think from anywhere except the sea, we had to scramble down, under and over large boulders of rock.' A slip could result in a broken limb or worse; at the very least it might lead to an unscheduled swim. It was a small cave, its mouth almost on a level with the sea. This time, the rewards were immediate. To the right of the entrance was a small patch of the red earth characteristic of bone caves, although it was now as hard as rock. It was filled with fossilized bones and this time there were teeth as well. Reading Dorothea's straightforward account of this discovery, it is astonishing to remember that her only previous experience of excavating was of tiny Ice Age voles and rabbits in the Wye valley, small bones that had been fairly easy to extract. These bones were large enough to be mistaken for human and were embedded in weather-hardened rock. It is difficult to believe that the men she hired knew anything about excavating fossil bones other than what she told them to do, and Mr Perks, she later wrote, certainly did not. Yet with no apparent doubts or uncertainties, she just got on with it, excavating as many of the teeth and bones as she could in the few hours available. On her second day of cave-hunting, she had found something, she hoped, 'that would admit of identification'. On the way back to Larnaca, one of the native sportsmen shot a quail for her and gave her a fox he had shot that morning.[10]

In the dim light of her room she examined the teeth, flicking through the pages of Falconer's *Palaeontological Memoirs* to try to find a description or illustration that might identify them, but they were fragments, too worn and broken for her inexperience. How could she even know where to start? She had no education, no training, and the only fossil fauna she really knew about were rodents and small mammals. This was something very different. Only Charles Forsyth Major at the Natural History Museum, she thought, could make sense of them. At first light she packed the teeth fragments and the few bits of bone she had excavated, and arranged for their dispatch to London. She followed meticulously the Museum's own *Instructions to Collectors*. The boxes were to be small and numerous rather than 'large and few' and should be tightly packed so the contents could not move around. 'The opening of cases by Customs officers in docks and on the frontier of foreign states,' the *Instructions* caution, 'is often more fatal than a very long journey to the contents of boxes. Bribery in such cases seems permissible, to ensure lenient treatment of collections.'[11] In later editions, the reference to bribery has been replaced with: '*Every effort should be made* [my italics] to ensure lenient treatment of collections.'[12] All boxes had to be addressed to the Director of the Museum.

Now she had found two bone caves at Cape Pyla, she was keen to travel north, to try to find the fossil-bone deposits discovered by Corneille le Bruyn on the southern slopes of the Kyrenia mountains. Leaving Mr Perks in Larnaca, she travelled by carriage to Nicosia, some twenty-six miles away across the vast, dusty plain of Mesaoria, a brownish-gold semi-desert in high summer, a sea of green with crops in the cooler months. She had arranged to stay with various friends of the Wodehouses, beginning with the Frere family. Bartle Henry Temple Frere was President of the District Court of Nicosia. Staying with his family involved her in the social life of the administration. As on Troödos, this was an endless round of tennis and croquet parties, hockey and cricket matches, impromptu tea parties, concerts, theatricals and dinners, both formal and informal. Much of this took place at Government House, the residence of the High Commissioner, Sir William Haynes Smith. The residence was, by common consent, ugly, uncomfortable and inappropriate. The writer W. H. Mallock, who spent a winter in Cyprus in the late 1880s, described it as an enormous wooden shed with a red-tiled roof.[13] Mrs Lewis, an English traveller who stayed there on a visit to Cyprus in 1893, called it a 'magnified bungalow [which] scarcely meets one's ideas of a suitable installation for our High Commissioner'.[14] The standard-issue dwelling for the hotter parts of the British Empire, it had been shipped out from England in kit form, under the mistaken belief that the climate of Cyprus was tropical. It was built round three sides of a quadrangle, with open verandahs that were the only access to the bedrooms, rain or shine. Considering the number of fires that had to burn during the cooler months, Mrs Lewis remarked that it was a wonder that the wooden house had not burned down. What she found most inexplicable was that 'perfectly good' building stone was available just a few hundred yards away. The quadrangle contained a tennis court surrounded by flower beds, and Dorothea came to know the residence well.

Frere, an enthusiastic naturalist and collector, on her first morning took her southwards from Nicosia to where there was 'a bed of oyster and other sea shells', she wrote, 'just lying about loose on the ground'.[15] These fossilized shells, millions of years old, had been noted by both le Bruyn and Gaudry. They were remains from the bed of the ancient Tethys ocean, which 300 million years ago stretched from Europe to Asia. As the continents moved and collided over millions of years, the Tethys was narrowed and squeezed. The Troödos mountains in the west of Cyprus were first thrown up from the ocean floor; many millions of years later at around the same time as the Alps were formed, the Kyrenia mountains in the north of the island emerged. As the Tethys was obliterated, a great basin was left at its western end, which became the Mediterranean. Land bridges linked the continents and animals could wander freely, allowing the migration north from Africa and Asia of species such as those that Gaudry and Smith Woodward had excavated at Pikermi.

Over time, the collision of the continents and fluctuating global sea levels sealed off the Mediterranean, first at its east end, then the west. Once it was landlocked, it took probably no more than a thousand years for the sun's heat to evaporate the water. About six million years ago, the Mediterranean sea dried up, leaving a vast salt deposit, which was discovered under the sea floor in the 1970s. For a million years, the Mediterranean is thought to have been a 'deep, dry, hot hell hole', as Kenneth Hsü, the discoverer of this desiccation, described it, descending some two to three thousand metres below sea level.[16] As temperatures on earth fluctuated, the Mediterranean seems to have experienced cycles of refilling and evaporation.

About five million years ago, the land barrier to the west of the Mediterranean fell, allowing the waters of the Atlantic to surge through what we now know as the Strait of Gibraltar, and the basin refilled for the final time. During the Pleistocene, however, global sea levels fluctuated dramatically with the climate as the great ice caps grew and then receded, releasing unimaginable quantities of water to flood coastal and low-lying regions. As the sea levels rose during the last interglacial (the last time that it was as warm as it is now), about 125,000 years ago, some creatures found themselves stranded on islands as the land bridges to the mainland were submerged. In the early 1900s, it was generally thought that Cyprus had been connected to Syria in the early Pleistocene, but it is now believed that no such bridge existed; to reach Cyprus, animals would have had to swim or drift there, carried by currents.

On 20 October, Dorothea, accompanied by the Freres and a shepherd she had hired to do the heavy digging, set off to look for le Bruyn's fossil-bone deposits in the southern foothills of the Kyrenia mountains below the eleventh-century monastery of Chrysostomo. Their journey took them through a curious formation known locally as 'the hummocks', which ran parallel with the south side of the mountains and looked, she wrote, 'as if it has been squeezed by some immense force into the contorted and fantastic shapes it is now in . . . these look as if they had been crumpled up like brown paper.'[17] It was a strange and desolate place, uncomfortable to ride through as much for its loneliness as its hazardous terrain.

Like Aghios Sarounda at Cape Pyla, the bone deposit was a well-known site. When the English traveller Mrs Lewis visited Cyprus in 1893, she was told that le Bruyn had discovered 'masses of bones of men and animals petrified and imbedded in the rock, and hard to extract unbroken. It is said that even the marrow is to be found in the bones; and there are teeth of enormous size.'[18] Without the work of le Bruyn and Gaudry, Dorothea notes, 'One would not be likely to find them now as they are just embedded in some rock and some boulders on the hillside.' There must once have been a cave there, she thought, 'though now it is hardly possible to imagine it'.[19] As at Cape Pyla, this was a deposit of fossil animals; of man, saint or otherwise, there was no sign.

In a remarkably short time Dorothea had found three fossil-bone sites and had not the faintest idea, as yet, what the creature or creatures could be. At this deposit at Chrysostomo, the bones and teeth were exposed to the weather and very fragile, while the rock was excessively hard, splitting under the force of the shepherd's hammer, often right through 'the desired spec[imen]!'[20] But as she patiently scraped away at the matrix, a variety of bones emerged, including one or two vertebrae, a jaw with some teeth still attached, then portions of two massive jawbones. She also found something else. There were some pieces 'which,' she wrote, 'I feel sure must be the tusks of a pig of sorts'.[21] Her reading of Falconer became more specific, but what she needed most was to hear from Charles Forsyth Major.

Torn between the competing demands of zoology and palaeontology, she spent some of her time at the bone deposit, the rest skinning the birds and animals she had collected, as well as looking for more.* At Chrysostomo the work was slow. After days of finding only fragments of bone and teeth, she noted acerbically: 'had a shepherd who worked most energetically – for a few minutes now and again'.[22]

Then at last, in mid-November, more than a month after she had sent the teeth, she received a letter from Dr Woodward and Forsyth Major. She read it not knowing whether to feel despair or hope. The specimens, particularly the teeth she had sent, were too 'worn and broken' for a positive identification. But Charles Forsyth Major, an expert on the fossil fauna of islands, did have an idea of what they might be, in spite of their condition. He thought the teeth 'recalled distantly the Hippo, but were remarkably small and lacked the characteristic trefoil-shaped pattern'.[23] This was an intriguing prospect; Dorothea knew that a species of pygmy hippo still lived in West Africa, while fossil dwarf hippos had been found in Malta and Sicily. Why should they not also have once lived on an eastern Mediterranean island?

Selecting a few of the choicest specimens from Chrysostomo, Dorothea sent them off to Charles Forsyth Major and, with renewed intensity, went back to work. For three days she excavated, finding splendid material including 'perfect specimens of teeth'. By the end of the third day, she was working in an area more than two feet below the cave floor in an earth bed and the finds were becoming fewer. 'The men,' she noted, 'do not work nearly so well when [they do] not find good things continually.'[24] She needed a rest and it was provided by the arrival of her cousin Claude. He and his brother Ronald had gone to Wellington at the same time as her brother Thomas, and they had often spent their holidays at Gellidywyll. Claude was now a lieutenant in the Royal Navy. '17th Sunday,' she wrote. 'Went to Church! – All day was expecting Claude to arrive as part of Med[iterranean] Squadron has come into Larnaka [*sic*] – he did not

* Her observations on live species are very engaging; under her pen, some of the creatures acquire very definite personalities in addition to their physical description. In Nicosia, for example, she was given a young hedgehog with a pale chestnut face and forehead. It was 'about half-grown', she wrote in her diary on 21 October; 'never saw anything so fearless – We put it on a table and gave it a saucer of milk and it walked about in the most unconcerned manner. It drank so much milk that it was sick eventually – put its whole snout into it and gulped it down – walked right into the saucer and sat in it. One immediately remarked its very long ears sticking right above his bristles which seemed to have no hairs among them.'

arrive till about 6pm.'[25] The next day she took Claude to the bazaar 'for supplies'; then, with Claude and another officer, she went for a drive into the countryside. It was a glorious morning – not simply in terms of the weather but because she could for a few hours utterly enjoy herself with the first member of her family she had seen for seven months. How exhilarated she must have felt, sitting up on the box of the carriage between the two officers, as they drove towards the distant hills in the warm November sun. What she had not expected was that Claude would have to return to his ship at lunchtime. After he had gone, she went up to the tennis club and 'In the evening went to at home at Gov[ernment] House – would have been very dull but most of the sailors and some of us went out and danced on the asphalte [*sic*] tennis ground.'[26] Claude's early departure also left her with a spare day: 'Shopped and packed bones in morning,' she wrote, adding wistfully, 'Expecting C[laude] to be here had told men not to come to Chrysostomo.'[27]

Two days later she went with a party of friends to the Crusader castle of St Hilarion, some two thousand feet up in the Kyrenia hills. She had given up another day at Chrysostomo for the excursion, and by doing so may have avoided at best witnessing a most horrific scene and, at worst, becoming part of it. 'When I got back this evening,' she wrote in her diary, 'heard that the old abbot of Chrysostomo had been murdered at 9am this morning – in the hummocks between Miamilia and the monastery – where I pass nearly every day.'[28] The details of the murder were sent to the Colonial Office by the High Commissioner, Sir William Haynes Smith. 'The Abbot of Agios Chrysostomo,' Haynes Smith reported, 'while on his way to Nicosia with an Acolyte was shot at between 8 and 9 o'clock in the morning by two men [and] his head was then almost completely severed from his body and the property on him taken with the exception of his watch.'[29] That was exactly the time that Dorothea would have been there. Her diary entry for 22 November is a laconic: 'Did not like to go to Chrysostomo today.'[30] Although the two men, both Turkish Cypriots, were caught within a few days, it would be some time before Dorothea dared to travel again to Chrysostomo through the hummocks, leaving her with a longer and more circuitous journey.

By 25 November she had excavated to a level three feet below the original floor at Chrysostomo and could hardly find any remains at all. Not wanting to waste time or money digging further here, she decided to return to the caves in the south of the island at Cape Pyla. Without delay she organized a man to load the mules with all the specimens and bring them to Nicosia. What her latest host and hostess, Inspector (of police) and Mrs Spencer with whom she was now staying, thought of this invasion of fossil fauna in their home can only be imagined. Packing the specimens was an immense and tiring task; by her last day in Nicosia, she still 'Had a frightful lot of packing to do – got heartily sick of wrapping up bones!'[31] It was past midnight before she finished. Just a few hours later, on a bitterly cold and rainy December morning, she took a carriage to Larnaca. Pausing only to pick up a few provisions, some equipment, and Mr Perks, she was driven along a rocky, potholed and very muddy road to Ormidhia where she had taken lodgings. She now understood the importance of being near her excavations, even if she still could not bring herself to stay in a mud-brick house in Xylophaga. 'I have got quite a nice unused room,' she wrote, although 'rather cold as it has a stone floor 2 doors and five windows or window places as there is no glass, only shutters.'[32] Her landlord (and landlady) lived there, but there was no place for Mr Perks. He had a room a little way off in the village. In spite of the cold, she was so exhausted that she slept for nearly eleven hours.

After organizing a man and a boy she had previously employed, Dorothea began the day intending to inspect the caves she had already seen and 'arrange plan of campaign'. Her scheme fell apart at the first cave she visited, Kochinos Kremnos, the Red Cliffs cave. 'Was disgusted,' she wrote, 'to find the surf breaking into it and the greater part of the floor submerged.' The waves were washing over the bone deposit she had previously found at the front of the cave, and although when she waded farther in she saw more fossil bones, she could do no more than look. The sea was now surging in with such force that they had to leave. Dripping wet, cold and buffeted by the wind, they made their way over rough and rocky ground along the extremity of the cape to Aghios Sarounda, the cave of the Forty Saints. Here too the sea was surging in, but it was nothing like as bad. Almost immediately she found in the rock a number of specimens, including part of a skull with three perfect teeth. The rock, however, was so hard that without special tools she could get nothing out.[33]

The following morning Dorothea sent a man into Larnaca to get the tools and, with Perks and a villager from Xylophaga as a guide, she exhaustingly searched every likely cave along the coast for a bone deposit. They found none and, during the course of that frustrating and strenuous day, communication between Dorothea and Mr Perks evidently broke down. Whether there was an argument or Mr Perks just got bored, perhaps resenting being told what to do by a twenty-three-year-old woman, Dorothea does not say, but in her diary she wrote, 'Mr Perks departed,' and then, not very convincingly, 'tried to persuade him to stay on'. If she was as impatient with him in person as she was about him in her diary, then his departure is not to be wondered at. He had all along, she wrote, 'excused himself from doing any commissions for me',[34] such as taking messages or buying provisions. Now that she understood the rudiments of Cypriot geology, his only function seems to have been that of interpreter, but after seven months in Cyprus, she had more than a smattering of Greek and enough confidence to use it, and the loss of this 'ancient Englishman' did not seem to bother her. She was only thirteen miles from Larnaca, which, as she wryly noted, '= civilisation!' while 'the people of the house are very decent and do everything for me, or as much as I can ask them to!'

In fact, with the departure of Mr Perks, it was as if she had shed a burden. Work at Aghios Sarounda progressed well and on a return journey to Ormidhia she seemed not even to mind when, in a downpour, 'my mule took the opportunity of coming down flat and depositing me in the mud!' The cave itself was very wet. Waves now regularly washed in, while water ran down the walls and dripped from the roof. If these conditions bothered her, she did not reveal them in her spirited letter of 9 December to Ogilvie-Grant at the Museum. It is a letter, however, that seems designed to pacify her conscience as much as anything else. 'I am dreadfully ashamed,' she wrote, 'of not having written months ago to thank you for the insect cases and pins you sent me – I hope you will forgive my horrid rudeness.' She then gives Ogilvie-Grant a rather curious account of her activities: 'I hope you won't be very disgusted with me! I have had to give up all hope of even getting a decent lot of birds as since the beginning of October I have devoted myself entirely to extinct beasts as there is a much better chance of getting something worth having in that line.' From her own diaries, it is patently clear that she had been actively continuing her search for live fauna, but perhaps it seemed to her a way of trying to persuade Ogilvie-Grant to help finance her search:

'If I only had time and was able to go to the right places,' she wrote with all the art she could decently muster, 'I am sure it would not be at all difficult to add to Lord Lilford's list of birds – I will try and get a few more before I leave.'[35]

These were tempting possibilities that she offered, but Ogilvie-Grant did not respond. As for the 'extinct beasts', she was still waiting for the post to bring her news from Forsyth Major. A week later it arrived and was everything she hoped for. From the second batch of specimens she had sent, Forsyth Major had now identified the creature. The teeth, less worn than the first batch, were, he wrote, those of 'a Hippo which in its teeth has not yet acquired the characteristic feature of the genus', that is, the trefoil-shaped pattern. Furthermore, as he told Henry Woodward, 'For me it is now quite doubtless – and this is one of the important points of Miss Bate's discovery – that the Hippopotamus is derived from some early member of the Suidae.'[36] Dorothea's initial description of a 'pig of sorts' was exactly right. Before her discovery, it was thought (as she would have read in Falconer) that hippos were related to an extinct aquatic creature called a *Merycopotamus*. The animal she had found showed that in fact they were part of the family of swine.

For a hippo, the creature was tiny, smaller than other known fossil dwarf hippos, and there was every possibility that Dorothea had discovered a new species. It was only about two and a half feet tall and less than five feet long, about the size of a pig. Normal-sized hippos can be up to fourteen feet long and over five feet tall. They live by rivers and lakes in warm, wet climates and their diet is mainly varieties of grass. This small animal lived on a mountainous island that now has a notable scarcity of both freshwater lakes and rivers, although during the Pleistocene the climate may have been wetter. Normal-sized hippos were native to most of the Mediterranean region during the Pleistocene and first appear to have arrived on Cyprus more than 100,000 years ago. In the colder phases, as the ice spread as far as Germany and southern England, sea levels would have dropped dramatically, but animals would still have had to swim thirty miles or so from the nearest mainland to Cyprus; no land bridge emerged from the retreating sea. What disaster, if that is what it was, caused them to migrate to Cyprus is not known. Nor is it known why they should have become smaller once they were island dwellers, although there is no shortage of theories. Fauna on islands, as Forsyth

Major would have told Dorothea, appear to have their own laws, quite separate from similar creatures – both fossil and recent – on continents, and demonstrate just how ably animals can adapt to new and very different environments.

On mainlands, it is possible that great size for hippopotami is necessary as a defence against predators. For some reason these predators did not also make the journey to Cyprus and other Mediterranean islands. As the islands were so mountainous, and food probably less plentiful, the hippos' shape began to adapt to this new environment. Successive generations became smaller and stockier, and their preferred method of movement became not swimming but walking, apparently on the tips of their toes, rather than on footpads, presumably to cope with the mountainous and rocky terrain. Dwarf hippos died out on Cyprus at the end of the Pleistocene, about 10,000 years ago, possibly because the climate became hotter and drier, although there are theories that link the demise of the pygmy hippo to the arrival of man on the island. However, there are no certainties. In the same way that great hippos became smaller, there was a phenomenon of small mammals on islands, such as dormice and shrews, growing larger. Dorothea had read of fantastic creatures on Malta whose fossilized remains had been discovered: an extinct dormouse the size of a squirrel, a gigantic swan possibly too large to fly, and giant tortoises. Although this phenomenon of dwarfism and gigantism on islands is imperfectly understood, it is probable that an absence of predators meant that small creatures had less need to be tiny; with few enemies, mammals and birds had less need to escape into burrows or holes or, literally, take flight.

Forsyth Major was delighted with her discovery, as he wrote to Henry Woodward:

> It is evident, that if the diminutive Malta island has yielded up to now at least eight or nine Pleistocene mammals (including two species of Hippo) a great deal more may be expected from the giant Cyprus, which is larger than Corsica . . . She has [he concluded] discovered a fossil which quite certainly will go a long way in clearing up the ancestry of the Hippopotamus, which up to now has been quite an isolated type and it is quite certain that by intelligently conducted researches other quite as interesting forms may be brought to light.

This now seems extraordinarily dogmatic. Malta, like other islands of the western Mediterranean, had once been connected to the mainland and that accounted for its 'eight or nine' extinct species. That Forsyth Major could be 'quite certain' that more could be expected from an unexplored island in the easternmost corner of the Mediterranean reflects the belief of the time that Cyprus had once been joined to Syria. 'It is only to be hoped,' Forsyth Major added in his letter to Woodward, 'that she is doing the excavations properly [and] on as large a scale as the means at her disposal allow.'[37]

The pressure that Forsyth Major's expectations put on Dorothea was enormous. She continued her cave-hunting with extraordinary energy. In between excavating at Aghios Sarounda, she pursued every possible lead to find other bone caves that might contain a more varied fauna or more perfect specimens of the little hippo. Through cold, rain, and mud, over wet and treacherous rocks, she trekked along the cliffs, the grim terrain making journeys of three or four miles take three hours. Sometimes the effort was for nothing. After one such expedition, with amazing good humour she wrote: 'drew blank. People seem to think I shall be able to find bones at any rocks called after a saint!'[38] Sometimes the effort was rewarded. Between Aghios Sarounda and the Red Cliffs cave, after a walk of half an hour and a precarious climb down the cliffs, about forty feet above the sea, at a point where the cliffs plunged 'perpendicularly', she found a bone cave with more hippo remains. Once again a saint was involved, this time 'Llanee' or 'Ay Yanee', as Dorothea phonetically attempts to spell it; in other words, Aghios Ioannis, St John.[39]

For the last two weeks in December, she worked without a break, excavating alternately at Aghios Ioannis and Aghios Sarounda. Refusing all invitations, she even worked at Aghios Ioannis on Christmas Day, quite alone in the cave. By now she had amassed a great quantity of fossil bones and teeth, much still embedded in blocks of rock. Storage was a major problem; she did not like leaving the material in the caves and her room was scarcely suitable. Claude Cobham came to her rescue, offering her the use of a government grain store in Larnaca until she had time to send her fossils to the Museum. 'Got up extra early,' she wrote on 28 December, 'packed all my stones from Aghios Sarounda on to five donkeys and started off to Larnaka [*sic*] with my two men – took four hours to get there!'[40] There was no suggestion that any of her finds should remain in Cyprus. Rider Haggard was told that export of all antiquities was forbidden, except to the British Museum, which of course was the destination of Dorothea's fossils, but there was, nonetheless, a thriving trade in antiquities. Forsyth Major warned her not to talk of her discoveries as there was 'a Natural History dealer somewhere out there',[41] and the export of artefacts from archaeological sites infuriated Cypriots. A member of the legislative assembly, George Chacalli, was outraged that antiquities were removed and traded every day 'as if they were dealings in barley, wheat and manufactured goods'. He wanted a complete ban on their export in the same way as he knew the Cretan government had prohibited all export of antiquities from that island.[42]

Once the fossils were safely stored, Dorothea returned to the cliffs of Cape Pyla and resumed her search for bone caves. She found two in one morning, to the west of Aghios Ioannis, one large and the other not much more than a rock shelter. Both contained hippo bones. She managed to excavate one tooth, but could do no more as her candles gave out. In keeping with the absence of any local name, she solemnly called them The Great Anonymous Cave and The Small Anonymous Cave.[43]

Dorothea had been working for nearly six weeks without a break throughout Christmas and into the new year. She was exhausted by the constant travelling and climbing, followed by the physically demanding and painstaking work of excavating fossil bones. Conditions were often cold and wet, and she returned to an unheated room. One night there was a thunderstorm and it poured with rain. 'Luckily had not turned in as the rain came through the roof being liquid mud by the time it got to the floor,' she wrote. 'Had to move bed and all belongings – more than half the roof was leaking.'[44] The following morning she complains of 'beastly winds lately and resulting toothache'. Two days later she felt very cold, 'toothache worse than ever, spent a miserable day at Ay Yanee'.[45] All she noted the next day was 'ditto only more so', although she managed to add that a boy brought her a lesser and a large horseshoe bat. She struggled to Aghios Ioannis the next morning, but felt so bad she could hardly climb to and from the cave. She was far too ill to work and needed medical help. With quite astonishing determination, she ordered her men to bring all the fossils and tools from the two caves and managed – somehow – to pack them up. 'Got all the things ready,' she wrote, 'and started off with four donkey loads soon after 8.30am – took just over 4 hours to get to Larnaka – paid off men.'[46]

Just before she set off, Hambi, one of her men, brought her not fruit or flowers but a 'basket of 8 fruit bats that he told me were from a cave at Pyla where he said there were thousands'. She booked herself into the Royal Hotel in Larnaca, feeling very ill and 'bitterly cold . . . Lay down till dinner and then was all I could do to get to bed – never felt so bad before.'[47] After suffering for another day, she called in Dr Williamson, the Medical Officer for Larnaca and expert on sphalangi. He diagnosed some sort of malarial fever and confined her to bed. It was another four days before she began to recover, but even when the fever went she was left very weak.

There was great concern for her among the English community. The Lascelles, who had been staying with Mr Cobham for the assizes, insisted that she return with them to Nicosia and stay until she was quite recovered. She needed no persuading. She was still weak and had to take things quietly – observing birds rather than shooting them – and making social calls, although that in itself could be demanding. By the end of January she had recovered enough for some not too gentle exercise: 'In aft[ernoon] played hockey for Limassol v. Nikosia – we won 5 to 4 goals – though Capt Smith put his knee out.'[48]

It was not only illness that had put a stop to Dorothea's work. Her six weeks of excavating at Cape Pyla had made deep inroads into her meagre funds and after paying off her men she had barely enough for travelling expenses. The invitation from the Lascelles had been most timely. At the beginning of the year she had written to Dr Henry Woodward, the former Keeper of Geology, to ask whether there was any possibility of a grant. Without it she would have no alternative but to return home. Aware that the more evidence she had of important fossil-bone deposits the stronger her case for a grant, she decided to explore the Kyrenia hills further, even though it was barely a week since her recovery. She had been told of caves at Dikomo to the north of Nicosia and on the morning of 30 January, accompanied by Miss Lascelles, she rode up to have a look round.

In cliffs just above Pano Dikomo she found a cave that had recently been used as a *mandra*, a shelter for shepherds and their goats. For Dorothea in her weakened state, its most important feature was that 'It opened on to the hillside no climbing required to get to it.' At its far end she found a few bones and a skull with a few teeth embedded in a crevice. 'Could not see it well,' she wrote with an air almost of ennui, 'but believe it to be the usual hippo.' She had by now found literally scores of fossil dwarf hippo bones and teeth, but it was still essential to excavate the remains; she wanted to find as much of the skeleton as possible and there was always the chance of discovering other species lying in the same deposit. She extracted a couple of pieces of bone, but could see little else; the cave floor was filthy from the goats and the rocks were encrusted with dust and spiders' webs. 'It has been so much used as a mandra,' she wrote resignedly, 'that it would require to be cleaned out a bit before much could be done there.' She looked at some of the other cliffs near by, 'but did not feel up to doing much yet'.[49]

Walking back to her mule, she heard coming from a cave in the cliffs a great squealing and screeching noise. On investigation, and perhaps this differentiates the naturalist from lesser mortals, she discovered the cave was filled with thousands and thousands of fruit bats, the species Hambi had brought her in a basket when she was ill. 'When I came in,' she wrote, 'lots of them started flying about – the roof was a good height – they hung head downwards and were packed together as close as possible, [looking like] a black mass squeaking and moving, every now and again one or two detached themselves flying to another crevice'.[50] Interesting though they may have been as a species, they were enormously destructive to fruit crops, scooping out the flesh from oranges while still on the tree and destroying date palms. They were, Dorothea noted, 'excessively common' in Cyprus.[51]

With neither money nor strength to excavate, Dorothea returned to Larnaca to pack up her great haul of fossils in the government grain store and dispatch them to London. Pursuing her from Paphos to Nicosia and now to Larnaca was a letter from Henry Woodward. Impressed by her endeavours and her discovery of the pygmy hippopotamus, he was, as she had requested, attempting to get her a grant from the Royal Society. She replied at once:

> It is very good of you taking so much trouble about trying to get me a grant. I do hope you will be successful, I shall be dreadfully disappointed if you are not.
>
> If this beast (or beasts, I believe there are two species) is really of

> great interest it is a pity not to get more of its remains. I could easily do so if I get the grant as up to this I have already found seven places* containing these remains, in the three best I have done a little work. It is quite possible I might find other things as I have only explored a very small part of the island.
>
> I am very sorry to hear that the result of the applications is not known for some time. If mine should be successful it is of the utmost importance that I should know as soon as possible – After the end of April one cannot depend on the weather being cool enough to do much work. So if I have luck and you hear soon that you have been successful in your efforts on my behalf it would be worthwhile to send me a cable – 'Bate, Cyprus Continue' would be quite sufficient and I should know that you had got me at any rate £20 and I would be able to hear all particulars from you by post before that was finished.[52]

Woodward could hardly fail to have been moved by her passion and determination. How she must have longed for that telegram, 'Bate, Cyprus, continue'.

The packing was a huge task. It took her three days to organize the first four crates of fossils and she sent them at once to the Museum. Apart from this, she could do little without money except wait to hear whether a grant had been approved. On 9 February she left Larnaca to stay with friends of the Wodehouses, Major and Mrs Girvin, at the military camp at Polimedia near Limassol. She filled her time with riding out to collect and observe, and joined in the various social activities. 'Met Mr Kitchener – a brother of Lord K he is a mining expert come to see the copper mines – after dinner played ping pong at the mess.'[53]† As well as the very new and popular table tennis, she played hockey, tennis, and cricket, 'Ladies v. men with shovel handles'[54] and sent more of her fossils back to London, writing to Henry Woodward's successor as Keeper of Geology, Dr Arthur Smith Woodward: 'Will you please send a postcard as soon as they reach you to say how the heavier of the two has travelled. I will not pack up or

* The five caves at Cape Pyla and two in the Kyrenia range.

† Ping Pong was just one of the trade names for this new game. Other manufacturers called it Gossima or Whiff-Waff, but these quickly vanished. (See Gerald Gurney, *Table Tennis: The Early Years*, International Table Tennis Federation, 1989.)

send off any more till I hear from you as I don't wish to run needless risk of having them smashed on the way. I have others heavier than those already sent.'[55]

On 13 February, accompanied by the Girvins, Dorothea went to the lighthouse on the Akrotiri peninsula. From there they walked along the sandstone cliffs where in places, she noted, 'the rock seems almost entirely composed of shells – never saw such numbers – chiefly oysters and other bivalves'.[56] Perhaps if Dorothea had been alone, she might have explored the cliffs further. Had she climbed down at a particularly steep point called Aetokremnos, known as Eagle or Vulture's Cliff, she would have found the site of a collapsed rock shelter above a vertical drop to the sea of about one hundred and fifty feet. It contained fossilized bones, mainly of pygmy hippos, and in truly vast quantities. The site was not discovered until 1961 (by the young son of a British serviceman), and it was not fully excavated until the 1980s. A quarter of a million pygmy hippo bones have been discovered there, the remains of well over two hundred animals.[57] As it was, Dorothea and her friends walked unknowingly along the cliff top above, finding, apart from the bivalves, nothing but the fossilized remains of a few sea urchins. Palaeontology maybe a science, but the discovery of fossil fauna is so often a matter of luck.

And then, at last, on 17 March, she heard from Henry Woodward, not by telegram but by letter, which had prolonged her wait by at least three weeks. But the news was wonderful. Her application to the Royal Society had been successful. That most prestigious of scientific organizations had awarded her a grant of £30 (about £1,700 in today's values), £10 more than she had hoped for.[58] It was the first public recognition of her very considerable achievement.

With her funding in place, Dorothea returned to the Wodehouses at Ktima while she organized this next stage in her exploration, but constant squalls of wind and rain delayed her departure. On 5 April, as the weather began to clear, Clarence Wodehouse left on one of his regular tours of his district, and Jack, who had now joined the police force, was based for much of the time at Morphou, in the northwest of the island. Dorothea and Francie Wodehouse were alone at Akrotiri House when a telegram arrived at 3 p.m. for Francie with the news that her mother, Frances Saunders-Davies, was dangerously ill. Dorothea immediately sent out a message to Mr Wodehouse but it did not reach him until late that night.

At 8 p.m. a second telegram followed, which devastatingly announced that Mrs Saunders-Davies 'had died very suddenly'.[59] It was an appalling shock for Francie Wodehouse. Not only did she adore her mother, but Frances Saunders-Davies was a great and dominant matriarchal figure, the absolute centre of her family, which she had ruled from their splendid estate of Pentre on Tivy-side. According to an almost Wildean newspaper report, Mrs Saunders-Davies had been visiting one of her sons in Sussex. Arriving at Victoria Station, she went into a waiting room, where she was 'observed to stagger and fall upon her knees, afterwards falling backwards. Finding the return half of her ticket and other documents in her pocket, the station master at Worthing was contacted, but her identity was made known by a card found in her pocket.'[60] Francie was distraught. In the space of ten months she had lost both her mother and her baby daughter.

Dorothea, who had planned to leave in a day or two, had no choice but to delay her departure. Jack was still in Morphou, and Clarence Wodehouse, who returned the following day, had organized a lunch at Ktima for over one hundred and thirty Turkish *mukhtar* – the headmen of the district. Clarence was, notes Dorothea, 'the only commissioner who does this'.[61] The lunch had of course to go ahead and Francie needed Dorothea with her. For a few more days Dorothea stayed, but the weather was growing warmer. 'Have started to perspire and expect to continue for next six months,'[62] she wrote. Even the bereft Francie realized that she could not keep Dorothea from her cave-hunting any longer.

CHAPTER 5

The Most Unpromising Place

On 12 April, Dorothea left Akrotiri House to begin her exploration of northern Cyprus and the Kyrenia hills. It was to be a severe test of her physical and mental stamina. Riding a mule, and with two others for the muleteer and baggage, she headed northeast. Her journey took her through Morphou where Jack was based, and she made a detour to an outlying village, expecting to see him there, but he had just left. In Nicosia she stayed with the Lascelles for a couple of nights while she organized her trip, then booked herself into the Armenian Hotel, the additional cost outweighed by her freedom to come and go as she pleased.

For a hundred miles the Kyrenia hills stretch along the north of the island, with great precipitous cliffs rising abruptly from the lower slopes. The highest points reach 3,000 feet, crowned by famous medieval ruins: the Crusader castles of St Hilarion, Kantara, and Buffavento, perched dizzyingly on craggy hilltops, and the stark Gothic beauty of the ruined abbey of Bellapais. Dorothea rode north over the mountains and down past Bellapais, with the distant sea sparkling and dancing in the spring sunshine while countless flowers swayed in the sea-wind: irises, anemones, cyclamens and a myriad shy wild orchids. Every day she climbed and crawled into dozens of caves, peering by the light of her candle into crevices and fissures, or scraped in hot sunshine in crumbled rock and earth at the foot of cliffs where the limestone looked deceptively promising, but she could only record, 'Found nothing – so only negative result of tramp.'[1]*

At a small village in the hills to the west of Nicosia, Dorothea offered a reward for information about bone deposits, showing the villagers specimens of bone and tooth, but no one knew of anything similar. She examined a group of (empty) caves, then rode over a rough and rugged pass to the tiny seaside village of Karavas, west of Kyrenia, where she found a room. 'Spent a beastly night – one of the poles of my bed broke and its place had to be taken by a couple of chairs also,' she added, 'I was devoured by fleas.'[2] When her path took her near the hummocks and the scene of the abbot's murder, she found herself with an escort of *zaptieh* (policemen). On her way back across the mountains to Karavas, she saw a cave high in a cliff, but, after a lengthy climb up to it, 'found it a flop like all I have been to so far'. After more than a week of mountain climbing, Dorothea had heard a cuckoo, caught a magpie, seen rollers, golden orioles, bee-eaters, crag and house martins, and white-bellied swifts; and had been presented with yet another fruit bat, but she had found not one bone cave. 'There are,' she concluded wearily, 'plenty of holes and cracks about but very few real caves like Dikomo.'[3] She revisited Dikomo Mandra, the cave north of Nicosia that she had discovered in January, and found it had not improved. 'The whole place is horribly dirty,' she wrote, 'and the "goaty" smell ubiquitous.'[4] However, it did contain 'the usual hippo' and she could not afford to be squeamish.

She returned to Nicosia, organized her workmen for Dikomo Mandra and began work there at once. The smell and the filth were worth it. A crevice at the back of the cave revealed a great quantity of hippo remains, including perfect specimens of vertebrae and limb bones, and a number of small bones, including parts of two fossil shrews' skulls, which appeared larger than the living species.[5] Furthermore, her offer of a reward had resulted in information about two more sites. At the first, a place called Imboähry, the bone deposit (pygmy hippos again) lay in an indentation in the ground at the foot of a low broken line of cliffs; the cave itself had long since vanished. The second site was at Coutzaventis, between Chrysostomo and the village of Kythraea. She had climbed steep cliffs to a cave that, once more, revealed nothing, but as she emerged into the

* Dorothea's third Cyprus diary, which begins on 12 April 1902, has, like the others, a leather binding, but it is actually a cashbook.

sunlight, she saw, just to the left, 'a place which I suppose to have been a cave but which now has no roof – Here all round at the bottom of the walls found bones and teeth of my hippo.' She dug out a few teeth and part of a lower jaw of 'my hippo', but wrote in her diary that night: 'seems no chance of finding anything else'.[6]

Yet she had told Henry Woodward that she believed there were two species and somehow she had to discover this second beast. Organizing workmen to continue digging at Dikomo Mandra, she embarked on another five days of cave-hunting, accompanied by Petrie, one of her men. As I read her diaries for these days, it seemed almost impossible that, as one disappointment followed another, she could still persevere in her quest with such energy and resolution. In the rising heat of early summer in unknown territory, with little to guide her except her developing instinct for which sites might be bone-bearing, she must have covered hundreds of miles in her determined search. She criss-crossed the Kyrenia hills, climbing 2,000 and 3,000 feet up to barely negotiable mountain passes, or rode eastwards along ridges as far as the 'pan-handle' of Cyprus through terrain of bare rock or dense bushes, which revealed a splendid variety of birds, but 'could not see or hear of any caves'.[7]

A man hoping for the reward had given her a basket of hippo bones and teeth, which he said had been unearthed near the village of Trapeza on the slopes of Mount Pentadactylos. In the emptiness of the hills and cliffs she became lost as she looked for the village and only found it 'with some difficulty – many paths in the hills and very few people about to direct me – Petrie,' she remarks with little sympathy, 'did not enjoy this.' When she did finally arrive there, no one knew anything about caves or bones. She could only conclude that the man with the basket had 'probably lied – disappointed,' she added, 'as I have not found anything this side of the hills'.[8] Too weary to negotiate the mountains again and return to Nicosia that night, she headed instead to the town of Kyrenia, about ten miles away on the coast.

In the morning as she walked round the harbour, she met Charles Cade, a local commandant of police from Nicosia. He accompanied her to Kyrenia's famous castle with its magnificent views over the town and harbour and its dramatic backdrop of the Kyrenian hills. From the crumbling battlements she described to him her travels of the last few days, pointing up to the peaks and passes where she had climbed and searched in vain for bone caves. Cade had heard a story of bones found in the rocks on the coast between Kyrenia and Karavas – the village with the fleas and the collapsing bed. Once again pious villagers believed they were saintly bones, in this case, those of St Phanouris. According to legend, St Phanouris was called by Christ to sail to Cyprus. On landing, he tried to climb the cliffs on horseback. The beast fell and both man and horse were killed. The saint's bones supposedly became embedded in the rock. For centuries, villagers had exhumed bones from the site, ground them to powder, mixed them with water and then drunk the concoction, believing it would cure all ills. Not knowing exactly where the bones were, Dorothea 'went a good bit along the shore but no good'.[9] Nor could she find anyone to ask and she had no time or money to spare to look further. Luck was still deserting her. In fact, the bones lie quite exposed on an open terrace just above the sea. Now there is a village, Aghios Giorgios, close by. In 1902 there was just empty coastline.

On 6 May she gave up and returned to Dikomo Mandra, only to discover that while she had been on her futile search, the two men she had left working there in her absence had found another cave, again in the same general locality as the others. It was in cliffs 'just above Pascali's chiftlik [farm]', she wrote. This cave, which is now well known, is small and narrow with a very low roof; it, too, had become another place of saintly worship. 'They called it St Elias and evidently consider it a sacred spot – a sort of entrance has been built round the mouth of the cave and in this there is an [altar] where they burn a votive lamp – a woman came to light it while I was there – My man laughed at the idea of the hippo's being saints bones and none of them minded my digging out some specimens . . . Got a few teeth and pieces of jaw bones.'[10] Dorothea's man may have laughed, but there are two astonishing realities here. The first is the very sincere belief, which lasted until recent times, that so many saints had chosen to bless Cyprus with their mortal remains. The other, almost more remarkable, is the sheer quantity of fossilized tusks, bones, and teeth of the little hippo that have somehow survived through so many thousands of years, entombed in limestone beds.

That night on her return to Nicosia, Dorothea received news so disappointing that it must have seemed as if St Elias was seeking revenge for her excavations that day. It was a letter from Charles Forsyth Major at the Museum. Painfully she records in her diary 'that after all the hippo is not new but was described by Cuvier [Baron Georges Cuvier, the great French zoologist and palaeontologist] more than a hundred years ago'.[11] She had been so sure that the hippo was a new species; it was smaller than any hippo found on other Mediterranean islands and appeared to show unique characteristics. Forsyth Major had examined the remains she had sent from Cyprus, comparing them with the Museum's collections of dwarf hippo bones and teeth found in other Mediterranean islands, then with the larger fossil hippos found in Tuscany and Egypt. In each case he had found material differences with the Cypriot animal. However, he had then looked in Cuvier's work, *Recherches sur les Ossemens Fossiles de Quadrupèdes*. There, over ten pages, he found a description of a 'petit Hippopotame fossile', which was, Forsyth Major wrote, in 'perfect agreement in shape as well as in size with the Cyprus creature'.[12] Cuvier, it appeared, had come across the bones in the basement of the Paris Museum; a few similar remains had lain for over a hundred years in other French museums. It had been believed that they came from 'some locality in the south of France', but, in truth, no one knew. When Cuvier found the little hippo, there was no identifying label nor any information to record its origins. All it had was a name, *Hippopotamus minutus*.*

In his paper on the Cyprus hippo, which he wrote for the Zoological Society of London, Dr Forsyth Major heaped both considerable and genuine praise on Dorothea. 'It gives me great pleasure,' he wrote, 'to announce that this exceedingly interesting first indication of a Pleistocene Mammalian fauna on the island is entirely due to the untiring energy of a young English lady, Miss Dorothy M. A. Bate, who started last year for Cyprus with the express purpose of discovering and exploring ossiferous caves.' The little hippo may have been a known species, but Dorothea's discovery was still an outstanding achievement, not least because the provenance of the 'petit Hippopotame fossile' in Paris would have remained unknown for many more years but for her discoveries.[13] How pleased the Museum was with her work was conveyed to her by Smith Woodward. The Trustees of the Museum, he wrote, had agreed to purchase from her the first selection of her pygmy hippopotamus remains for the sum of £50 (about £3,000 in today's values). It was more than she could have hoped for. She wrote in reply:

* In 1972 it was renamed *Phanourios minutus* after the saint.

> I am very pleased indeed with what you say I am going to get for the fossils – it is very good of you to look after my interests so well – I hope to be able to put some of it towards the next journey! . . . I am still digging away as luckily it is a late summer and it is not yet too hot . . . Do you want still more hippo bones? Just now I am devoting my time to trying to find something else – but can get more for you if you think it advisable. If so let me know by return as I shall have to stop in about three weeks or a month.[14]

She resumed her hunt for 'something else' at Dikomo Mandra, employing now three men to dig, trying to get below the hippo remains in the hope of finding a new mammal. After four more days and three and a half feet below the floor of the cave, she was still finding hippo bones, and she called a halt. Taking all the fossil bones with her, once more she headed south. In the borrowed government grain store in Larnaca, she packed the bones into four boxes, three containing Royal Society fossils. The fourth, as she wrote to Smith Woodward, 'contains odds and ends I have found in different parts of the island',[15] and she dispatched all four to London. Continuing her search for a second beast, she organized a boat to take her across the great sweep of Larnaca Bay to Cape Greco, the sharply pointed peninsula east of Cape Pyla. Her crew consisted of a man and a boy. 'They landed me in a cockle shell just held boy, self and baggage,' she wrote. But there were no bone caves in the limestone of Cape Greco; all she found were beds of fossil coral. With no means of transport back until the following day, she discovered a place in the cliffs 'where only few mosquitoes so bathed and loafed through day here. In evening beastly wind and dust storm.' She spent a restless night in the lighthouse, 'the mosquitoes simply awful.'[16]

With no more information of bone caves in the region of Larnaca, Dorothea returned to Nicosia, anxious now to complete her excavations at Dikomo Mandra before summer temperatures made digging unbearable. Instructing the men to resume work the following morning, Dorothea rode up to have another look at the not very promising site at Imboähry that she had seen three weeks previously. Fifty yards or so from the place she had first looked at was a small cave, its mouth about three feet off the ground. It ran back into the cliff and was very low, in places barely three feet high. This cave also contained fossil hippo remains, but nothing that seemed to demand her immediate attention. She rode on up into the hills, but found nothing more.

The rock level she had now reached in Dikomo Mandra proved impossible to dig out, 'so have started blasting with gunpowder', she wrote, 'on a very small scale'.[17] It now sounds extraordinary, but using gunpowder in excavations was not that uncommon. Dorothea had fairly easy access to it through her friends in the administration, and one of her men did the hard work. In a suitable area of rock, he would drill a few little holes using a narrow chisel and a hammer, pour a small amount of gunpowder into each one, pack it tightly and possibly surround it with clay, light the cordite and then run around the corner, or at least to a suitable distance, and wait. Even with blasting and employing now a boy as well as two men, she was still getting on very slowly 'as it is difficult work'.[18] By five feet down, the hippo bones had ceased entirely and at the end of two weeks, when she had excavated to a depth of eight feet, she reached the bedrock, and 'reluctantly' stopped work.[19] She had excavated here a considerable quantity of bones and teeth, many in excellent condition, but most were specimens of *H. minutus*. There was still no sign of a second beast.

Her men had found another cave, about a mile to the northeast of Dikomo Mandra and on about the same level. It was called Anoyero, 'which means balcony', she wrote, 'this from it being a sort of double cave one chamber above the other'.[20] Here too she found only hippo bones. At St Elias's cave near Pascali's *chiftlik* and at Coutzaventis, she found the story was the same: plentiful hippo bones, a few small mammals, but nothing else, while Imboähry seemed no more promising than before. Villagers told her of bones and enormous teeth like stone that they had seen in a field, but after investigating them she noted tersely, 'No good – horses.'[21] With her anxiety rising with the increasing temperatures of early June, she arranged once more to go cave-hunting farther afield.

On 9 June, 'a man brought me some hippo bones,' she wrote, 'which he said he had found in a cave close to Buffavento Castle so went off with him.' The castle stands on a vertiginous crag 3,000 feet up in the Kyrenia mountains, with the cave just below it; to reach it involved a strenuous climb over ground frequented by poisonous vipers. For five hours under the burning afternoon sun she searched the top of the mountains, but for all her efforts she found 'No bones at all which disgusted me'.[22] It was dark by the time she returned to Nicosia, in a state that she laconically describes as 'done'. For four days Dorothea again rode along the north side of the Kyrenia mountains, the sun painfully blistering her exposed skin. When she came upon caves, they were empty. She slept in villages and risked her health by drinking water that was not only dirty-looking and full of sediment, but smelled. Since her exploits were known even in these remote parts, villagers brought her antiques of dubious provenance, among them Egyptian figures and scarabs. These were not presents; a recognized – if questionable – source of income for Cypriot peasants was selling antiquities. As she wrote to Arthur Smith Woodward, 'It is quite impossible for an English person to try and do anything here without it being known.'[23]

Dorothea returned to Nicosia on 16 June, hot, exhausted, and despondent. The harder she looked, the more the 'something else' seemed to elude her and the Royal Society grant was dwindling rapidly. She found Jack Wodehouse waiting for her at her hotel and they dined together. The following day she went back to Imboähry, the unpromising site where she had found a few bones at the end of May. 'Started the men to dig in the piece of field just in front of what must have been the back of the cave,' she wrote; 'found a good many hippo remains.' And then all the wretchedness of her desperate search in the Kyrenia hills, the misery of the mosquito-infested lighthouse and the appalling climb up to Buffavento, evaporated. 'The boy found near the surface a tooth, or rather piece of one, as unluckily he broke it,' she wrote, 'which I feel must belong to a proboscidean.'[24] There was no ambiguity with this tooth fragment as there had been with the little hippo. Very clearly, the grinding surface of the tooth bore a characteristic ridge pattern and she recognized it at once. She had at last found 'something else'. It was the fossilized remains of an elephant, and it was tiny. As she was to write later in her report for the Royal Society, 'I was at length successful, although it was not until a certain amount of tentative digging had been carried out in four out of five newly discovered deposits that work was started on what appeared at first to be the most unpromising looking place which had been found, and,' she admitted, 'was consequently the last to receive attention.'[25]

At first light she was back at Imboähry. 'Set the men to look over every bit of stuff they had dug up,' she wrote, still hardly able to believe she had at last discovered a second extinct beast, 'to try and find the other half of the tooth but failed so I fear it must have got broken to atoms – The odd thing is I cannot come across another piece – but feel there can't be any doubt about it so must just go on until I do find something or the funds give out.'[26] The next day she found another elephant tooth, about half the size of the first one. 'Take some time,' she noted drily, 'if I only get one every two days.'[27]

All these weeks, engrossed as she was with work, she still found time to play her part in society, dining at Government House or with friends. On 22 June, social obligations – and pleasures – took over completely. It was the beginning of a week of rejoicing for Edward VII's coronation on 26 June and great celebrations were planned throughout the British Empire. Her cousin Claude was also due to arrive for a few days. 'Was expecting Claude all day,' she wrote on the 23rd, 'but he did not turn up.'[28] Nor did he arrive the next day. Then on 25 June came news that the King was ill and all coronation festivities were postponed indefinitely. The British community was stunned. So much had been planned for this first coronation for over sixty years and now everything was on hold. Claude finally arrived on the 27th but he could stay for just twenty-four hours, and all Dorothea could note was 'good to have seen something of him'.[29] This had not turned out to be the week's holiday that she had so looked forward to, and furthermore she was beginning to feel unwell. On 30 June she returned to Imboähry, but 'felt rather bad all day and when got back in evening had to go to bed – fever'. It was a recurrence of the malarial fever that she had suffered from in January. For three days she remained in bed, 'in varying stages of discomfort', and when the fever went, she felt for the next few days 'like the proverbial "rag"'.[30]

Throughout her life, Dorothea was impatient with sickness, always struggling back to work before she was fully recovered, and this was no exception, but she dared not delay longer. It took three more days of digging at Imboähry before she found any more of the elephant and that was a small, perfect tooth. 'Just seem to get one now and again to keep up a faint hope of coming on real bed of remains which I feel must be somewhere but,' she added, 'it is very tantalizing and disheartening.'[31] In two weeks of digging at Imboähry she found only ten or eleven cheek teeth of the elephant, and many of those were broken. She may have found her second beast, but she needed more and better specimens, and she had reached the end of her grant. 'Today is the last day I can manage with the Royal Society money,' she wrote in her diary on 15 July; 'was all day at Imboähry but was very disgusted did not get a single E[lephant] tooth – Am going on for another week on my own account.'[32]

By the end of that week she had found a few more teeth, a piece of jaw with one whole tooth and part of another, and a curved tusk about nine inches long, which was broken in several pieces. She measured the site and paid off her men. The heat was now too great and her financial state too depleted to continue. She left Imboähry, not knowing how she could find the money to resume work there in the autumn. So concerned was she that word of her great find might reach the ears of a natural history collector that she told no one at all of her discovery. After she had paid her hotel bill the next morning, she had very little money left, but had still to make the journey up to Troödos, to which the British community had already retreated. All she could afford was to spend the night 'in a good house but not very comfortable as, having no bed, I had to lie on the floor'.[33] Like her compatriots, Dorothea had longed to go up into the mountains at the beginning of July, as she had told Smith Woodward, but the only way that she could continue excavating the elephant remains was, as she wrote in her diary, 'by screwing up on Troödos'.[34] She did not use the vernacular without good reason. By continuing to work, she would have little money to spend over the summer; she was working in dreadful heat instead of enjoying the more bearable temperatures of the mountains; but, most importantly, she was desperately needed by Francie Wodehouse.

This was the first anniversary of baby Francie Petronel's death; it was barely twelve weeks since Francie's mother had died; and it was only a few months since Jack had become a police officer and left home. In the back of Dorothea's third Cyprus notebook is a letter from Francie, which Dorothea must have slipped in at the time and there it has remained. Written on black-bordered notepaper, this brief and scarcely legible letter shows Francie to have been in a highly emotional state. 'Dearest Dorothy,' Francie began, 'Come [to Asprokremno] whenever you like it will suit me any time . . . I am rather down on my luck and think all sorts of horrors [and] look upon everything as a bad omen – I wonder how long it will go on for.' Signing herself with 'Best love', she adds in a postscript: 'I shall be so glad to see you again. You are a firm and sensible person the best sort to drive these horrid fancies out of my head.'[35]

It is difficult to shout 'Eureka', or indeed anything, in a library, but this was the first glimpse I had had into how Dorothea was regarded by friends, away from her courageous explorations as a pioneering palaeontologist, and the letter made me pause. I knew so much about what Dorothea did and how she did it, but was I doing a disservice to Dorothea to attempt to portray her without direct access to her personal life? Yet when I read the only other personal letter to her that has survived (slipped into the diary at much the same time), I had no choice but to continue, partly because this second letter throws so much light on her family, but mainly because of the woman who emerges from those pages. I already knew she was witty, acerbic, clever and courageous, unafraid to demonstrate a pioneering spirit far ahead of her time. Now there emerged someone who was strong, loving, and as concerned for her friends and family as she was dedicated to her science. I liked her too much and found her life far too interesting simply to let her go.

The second letter is from her father, and she received it shortly after Francie's. This is the only letter at all from – or to – any member of her family that has survived. It is a long, remarkable outpouring and shows a close and loving relationship between father and daughter. However, it also reveals that life for the Bates was full of indecision and uncertainty. Its emotional effect on Dorothea was immense. Earlier in the year the family had moved again to another rented country house, Horsley Court at Nailsworth near Stroud in Gloucestershire. For Henry's sixty-third birthday on 10 July, Dorothea had sent him a present of a £5 note (nearly £300 in today's values). She also sent him a cheque for £10, money that he had given her for her trip, but which Dorothea, in her desire for independence, chose to regard as a loan. The finances of her family must have become even more precarious for her to feel the need to repay the money at the very point when her Royal Society grant had almost all been spent.

Henry addresses her as 'My Dear Dorothy', and begins by thanking her for her 'very great kindness and self-sacrifice' in sending him the money: 'as you know,' he continues,

> I am not a good hand at saying much, but you may feel quite sure that tho' I don't say much in this instance I quite fully understand how very good it was of you to think of sending it – at the same

> time I hope and think that you will not think it nasty of me to say that I could not think for a moment of depriving you of your hard-earned money: it is this way, if I took it and bought a bicycle or anything else with it I never could use it or see it even without thinking myself a Brute (with a large B) for taking your money: if you came in for a Colossal Fortune or something of the sort I might do so tho' even then I wouldn't promise, so you see it would not give me pleasure to take it and then I'm sure you will be quite satisfied that I should not.

He suggests instead that she should knit him a necktie similar to the one she gave him on his previous birthday. 'Your last one I'm rather saving, but if you give me another I may use it with greater freedom.' Having dealt with why he should return the £5, he broaches the matter of the cheque for £10. He makes it quite clear that even if Dorothea regarded it as a loan, he did not and declares he cannot receive it back; it was altogether hers to use as she pleased.

> If you had been at home I daresay your expenses one way or another would have come to much about the same amount as all the money you have had from me, so it isn't like as if I had been paying so much extra and I shouldn't imagine I am a halfpenny out of pocket by your trip, so you may take it all with quite an easy conscience . . . so to conclude the matter I send you a cheque for ten pounds instead of the one you sent me: it wouldn't do to return yours as it was made payable to me, so I have sent it to my Bankers: if you have any difficulty in getting my cheque cashed let me know and I'll send you two five pound notes instead –

In all it took poor Henry more than six hundred words to argue his case for returning the money to his determined daughter. He is so anxious not to appear as if he is rejecting her gift, yet is acutely aware of how important that hard-earned money is to her and what a very real and loving sacrifice she had made in sending it to him. The poignant image of Henry 'rather saving' the tie she had knitted him for a previous birthday would, if his words can be taken literally, seem to show something more than genteel poverty. In an even more extraordinary passage, he tells her of his indecision over whether or not he can afford to replace his old bicycle, now so shaky he hardly dare ride it. A new one would cost twelve guineas, but he thinks he now has sufficient 'funds' as the bicycle dealer has promised 'to allow me five pounds or guineas for the old one, so I would only have to pay seven pounds or guineas and I think I can manage that'.

This is an astonishing image, of the sixty-three-year-old Major Henry Bate agonizing over whether or not he could afford to replace his elderly bicycle. This was the man who gave lavish Christmas parties for the children of Tivy-side, and who sent his son Thomas to the best schools, even though the fees were reduced. The family's difficulties must surely have been very great for Dorothea to have sent her father money out of her meagre resources.

His letter finishes on an even more unsettling note: 'We have no idea yet as to where we will go after leaving this, but we hope things will come right somehow.' How defeated and passive Henry sounds, like Wilkins Micawber, waiting for anything to turn up. This new house, which they had moved to after three years at Bicknor Court, was clearly only temporary shelter. Thanking Dorothea again for her birthday gift, he signs himself 'your very loving father, Henry R. Bate'. And his hesitancy continues to the very end: 'P.S. Have sent a note for £10 instead of cheque as you might have trouble with latter – also return the £5 – In haste your loving father H.R.B.'[36]

Henry's agonized guilt and embarrassment at his impoverished circumstances were to have a profound effect on Dorothea; a recurrent theme throughout her life is a determination always to have the means to pay her way and maintain her independence from others, even if it meant hardship and hunger. More immediately, however, she was now, unexpectedly, £15 better off.

Back with the English community in their summer retreat, Dorothea gave all her attention in the first days to Francie Wodehouse, and then threw herself once more into the social round. She 'lunched, teaed and dined out', played croquet, tennis, or badminton almost every day and of course resumed her collecting of recent fauna. She looked for shrikes, shot a cole tit and a nightjar and observed hedgehogs, 'pugnacious beasts . . . often fighting and then make a noise like the "miaoul" of cats fighting'.[37] She set traps for small mammals and wondered whether she might find another creature as curious as one she had collected from the Kyrenia hills. At the end of May at Dikomo, her men had caught a mouse she had not seen before. They told her it was 'very scarce' and was never found in houses. It was about the size of a house mouse, 'but is grey on the back instead of yellowy brown and its hair on the hinder part of its back is quite stiff and spiny. My men described it to me as being like a hedgehog on the back.'[38] She procured in all five of these spiny-backed mice, 'a genus which', she wrote later, 'had not previously been recorded from this island'.[39] On her return to England she named this new species *Acomys nesiotes* Bate. The Troödos mountains, for all her efforts, offered nothing to compete with this, but she obtained a hare and butterflies and, probably because of her own susceptibility, became interested in the malaria mosquito, *Anopheles.* Major Girvin, who had recently acquired a new microscope, joined her in the hunt for its larvae. Local villagers, she noted with some scepticism, 'told us there were no mosquitoes in the houses but they occasionally got bitten outside the village – there is always a good deal of fever'.[40] On her own or accompanied by Major Girvin she explored streams, stagnant pools, and water tanks, closely observing and noting when the larvae changed to pupae and then to insects. Some larvae proved to be a non-malarial species, but the great mass were *Anopheles.* In spite of this evidence of its undisturbed breeding grounds and the virulence of the disease, with well over ten thousand cases a year, it was not until 1913 that the administration tackled the problem of malaria with any vigour.

On 9 August came the long-awaited coronation of Edward VII. In Troödos a service was held at the military camp after a review on the parade ground of the Cameron Highlanders. They sang the coronation hymn and the 'Te Deum'; Sir William Haynes Smith, the High Commissioner, made a speech; and after singing the National Anthem, they all went off to lunch. That night Dorothea went 'to the Dance at Govt. House – very jolly', she noted, 'grounds nicely illuminated'.[41] Over the next two weeks she spent every available hour in one social engagement or another, watched the 'very good' Highland Games at the military camp, and played in the celebratory tennis and croquet tournaments, passing successfully from one tie to the next, until on 23 August at the tennis club, 'played v. badly,' she wrote, 'and am out of it'. A day or so later she developed 'rather a bad headache, probably a touch of sun',[42] she thought, but it refused to go away. Nonetheless, almost every night she dined with friends, went to a reception at Government House or to 'the subscription dance on the hill at the Ordnance office. Rode up – very jolly – got back soon after 3am.'[43] She notes when Jack arrived for a couple of days at a time, and when he departed on his duties as a police inspector.

At the beginning of August she had received a letter from Dr Arthur Smith Woodward, telling her that they now had enough hippo remains. It was not until after the coronation that she found time to reply and decided this was the opportunity to tell Smith Woodward about her momentous find of the small elephant at Imboähry.

> I am writing to ask if you consider that it would be very important to get some more remains of this beast? I have only got about two dozen cheek teeth and one very broken tusk.
>
> Even if the Museum will buy any more that I might get I could hardly afford to go on as a lot of work is necessary for very little result. They are in a very hard bed of stuff and I fear I might not get any limb bones or many perfect pieces – except isolated cheek teeth.

And then she came to the essence of the matter, money.

> If you can tell me that I can for certain depend on getting anyway £10 I should be willing to risk that if you think it worth it. I should very much like to try and get some more of this in the hope that it might turn out to be a new species – but do not say anything about it yet. I am most anxious nothing should be known of it here and I have told no one beside yourself. Please write soon as I either go home or start work as soon as ever it is a bit cooler.'[44]

After all those months of searching for the second 'beast', Dorothea found it an unbearable prospect to have to return home with so few specimens. Smith Woodward acted swiftly. On 25 August, shortly after receiving her letter, he noted on it, 'Glad to give £10 for elephant remains.'[45] If she was careful, with the unexpected £15 from her father, she had enough to excavate for perhaps another month once the weather was cooler.

In the first week of October, the exodus from Troödos began. The

Wodehouses returned to Paphos and Dorothea left for Nicosia on 7 October by herself to resume excavating at Imboähry. Her head was 'bad again so could not hurry'.[46] She took a wrong path and by sunset she was lost. A group of men she met refused to show her to the right road and, fearing she might be left alone on the hillside all night, she had to follow them to their village. In the grip once again of malaria, when she continued her journey she felt so bad that she 'lay under a tree for a bit and drank a lot of chalky water – beastly hot wind blowing all day'.[47] Too ill to attempt again to reach Nicosia, she turned back to the mountains and spent the night at Troödos 'in a tent for the first time'.[48] By the morning she felt strong enough to travel, although when she had to dismount to retrieve some letters she had dropped, the mule trotted off and she 'had to walk 2 or 3 miles before catching her again'.[49] Quite exhausted, she finally arrived in Nicosia. She booked herself into the Armenian Hotel and found Jack there, 'in on duty'.[50]

In the morning Dorothea was still unwell, but simply carried on. She went to the village of Dikomo to organize men to start work at Imboähry and then, with some trepidation, went on to the site itself. 'Found nothing touched,'[51] she wrote with enormous relief. This second beast remained a secret between herself and the Museum. The digging was no easier than it had been in July. 'Stuff very hard and did not come across a sign of E[lephant] remains,' she wrote despairingly; 'fear my collection from here will only be Elephantine in one sense of the word.'[52] Nonetheless over the next few days she found one good tooth in a jaw and a broken piece, and when they reached softer ground, discovered several teeth and a 'tiny tusk only 5¼ inches round outer curve'.[53]

Torrential rain then brought everything to a halt, with two inches of rain falling in less than twenty-four hours. Unable to work, she accepted an invitation to the wedding of a daughter of Dervish Pasha, one of the leaders of the Turkish community. It was held in his splendid house, now restored as the Ethnographic Museum in the Turkish quarter of Nicosia. Dorothea had never been to a wedding like it and recorded every aspect of the experience in her diary: the 'nice looking' bride with crimped hair and 'complexion very much done up', her white satin gown and tinsel crown with long gold streamers hanging down; her bed covered with gold-embroidered white satin with 'pillow, nightgown cases and everything all to match – fourpost bed draped with same' and mosquito curtains of pale pink chiffon. 'We shook hands,' Dorothea noted; 'while doing so it is not etiquette for her to look up.' In the evening, 'the bridegroom had a show, all the guests (men) were given a lighted candle, also sweets. After a bit they had a procession from his house to that of bride all carrying these candles and fireworks being let off – left him,' she ended, 'at brides house.'[54] The care with which Dorothea describes this is striking. Perhaps it was simply of cultural interest to her, or perhaps, with her twenty-fourth birthday approaching, something more can be seen in the contrast between her description of this lavish wedding and her own highly unusual world.

When the rain stopped she returned to Imboähry, finding a few more broken elephant teeth and 'a very broken tusk'.[55] On some days she found nothing at all. 'Am altogether getting,' she wrote, 'a frightfully meagre collection.' So much so that when Jack asked her to go with him to the foothills of the Kyrenia mountains where he had a police enquiry to make, she was more than happy to go, leaving her workmen to dig on their own. A day or so later in Nicosia they went together to see one of the famous sights of the Ottoman Empire, the dancing dervishes, 'but unluckily', Dorothea wrote, 'they had just finished'. Again in great detail, she described in her diary the large square room where the dancing took place with its great railed-in circle for the dancers. Leading from the room was a series of domed chambers, containing the tombs of forty dervishes mostly in pairs, a large grave for an important dervish, a smaller one for his servant. 'On one end of nearly every one was placed one of the high conical khaki coloured hats distinctive of the sect, these again were covered in large handkerchiefs or cloths and over everything, walls, floors and tombs alike was a heavy pall of dust.'[56] They left the *tekke* (an Islamic monastery for dervishes) in such a rainstorm she decided there was no point in her going to Imboähry the following day. 'J. B. W. [Jack]', she noted, went back to Morphou the following afternoon.

Over the next four days, the bone deposit revealed a few more teeth, a small, slender tusk just three-quarters of an inch in circumference and broken in three pieces, and two pieces of a much larger one, with a 5¾-inch circumference, but 'Frightfully brittle – fear shall never get it home safe.' All this time she had been finding hippo bones as well, but these she left in the ground. On 8 November, her twenty-fourth birthday, she went to Imboähry for the last time, 'as I cannot afford to go on any longer', she wrote. There were no more options; it was time to go home. On this last day of excavating, she dug out 'a large tusk in several pieces and a small tooth'.[57] She measured the site, packed her finds and returned to Nicosia, to find that Jack had returned from Morphou.

Dorothea had intended to leave Cyprus as soon as possible, but Major Girvin told her of a most intriguing discovery: a large tusk, apparently like a mastodon's, had just been found in a quarry in the Troödos mountains near Peripedhia. It took a few days to pack and make all her farewells, but on 15 November she set off for the quarry, stopping for the night at the village of Lefka. 'JBW,' she wrote, 'arrived in evening.' In the morning they left at eight o'clock. 'J. came with me,' she noted, 'as far as Kalopanagiotis,' a picturesque village where they stopped for a 'couple of hours'.[58] Then they parted for the last time; Jack returned to Morphou and Dorothea continued on her way alone. Were they in love? Perhaps it is facile to read that into Dorothea's constant references to him, yet why else should she have recorded all his arrivals and departures? That is just what people in love do, although most are not constrained by the knowledge that their diaries might also be consulted for their scientific content. Nowhere, in any of Dorothea's letters and papers that have survived, is Jack ever mentioned again.

At the quarry Dorothea examined the tusk. It was about twelve inches in circumference at its larger end, she thought, but in several pieces, which seemed to fit together. Near the narrow end it had a roughened patch as if a branch had been there; 'also the "tusk" swells a little just below this.' In addition she found what she thought might be some long cylindrical pieces of iron pyrites. The 'tusk' appeared to be solid limestone and it worried her. 'Do not feel sure that it is a tusk,' she wrote, 'but can't think of what else it can be.'[59] Nor, for all her searching, could she find any site in the quarry that might have been a bone deposit. The detour had cost her dearly. She had hardly any money left even for food, and that in any event was scarce in outlying villages in November. The owner of the quarry sent her a partridge 'which a cat carried off', she wrote, 'and have only been able to get two eggs'.[60] The day after she noted 'rather short commons today', and the day after that she had 'only a plate of some vile native soup'.[61] She abandoned the search and, after a brief stop in Limassol for further farewells, headed for Larnaca. Dorothea appears not to have pursued the 'tusk' further, but it seems to have ended up in the Cyprus Natural History Society's museum in 1910, with a description completely inaccurate in every regard except presumably its size, 37½ inches by 15 inches in circumference. Its discovery was attributed to Dorothea; the place it was found was stated to be Kyrenia; and although the tusk was nearly three times larger than anything Dorothea had actually discovered, it was said to be of the same species of elephant.[62] In all probability Dorothea was quite right in doubting that it was a tusk, although, as its whereabouts are unknown, it is not possible to be certain; however, no elephant remains have ever been found so high in the Troödos mountains.*

Dorothea drove the 45 miles to Larnaca in the pouring rain. After eighteen months in Cyprus, during which she had made so many friends, she found the farewells of the last few weeks hard. She had just over a day in Larnaca to deal with a vast amount of packing, not only her fossils, but also all her mammal and bird skins and insects. After a few hours' sleep, on her last day in Cyprus, she 'packed violently all morning and went to see Mr Cobham'. The District Commissioner who had told her so much about the island on her arrival would be the one to escort her from it. At 1 p.m. she boarded the SS *Dundee*, a mail boat bound for Egypt; there was of course no direct passage to England. She never set foot on Cyprus again.

With her need to economize, Dorothea sailed 'as a deck passenger! Don't know what I should have done but the Chief Officer gave me up his cabin – Even that bad enough close by aft were lots of sheep and forward pigs etc some of which got loose and visited my cabin – made a fiendish row all night. Sixty sheep died in night and chucked overboard this morning.' It was only in the morning that she discovered that not only the first mate, but also the captain, came from near Tivy-side. 'Unfortunately the Capt did not know this at once or he would have given me his cabin – rough all day, boat rolls dreadfully. Only ill once in morning! But had to lie down all day.'[63]

* I consulted the palaeontologist Dr Noel Morris, of the Natural History Museum, who suggested that the most likely explanation is that it may have been a remnant from the floor of the Tethys ocean, from which the Troödos mountains are formed. Both the shape of the 'tusk' and the presence of iron pyrites could indicate that it might be the remains of the chimney or pipe of a 'hot-smoker', a hydrothermal vent, which sixty million or more years ago spewed out minerals, metals and water at enormously high temperatures. Around these vents lived (as in the oceans of today) a great variety of creatures, one of which, in a fossilized state, might also fit the description. None of these has ever been known to be of limestone.

At Port Said she had time to view the Suez canal, but was driven back by the myriads of mosquitoes; 'the fresh water canal seems alive with them,' she wrote; 'is an ideal place I should think from their point of view.'[64] The ship she took to England on 2 December was Japanese, the *Maba Maru*, and she managed to get a four-berth cabin to herself. There were few people to talk to and nothing to do. 'Sat on deck and loafed up and down,' she wrote, 'very bored.' At Marseilles, where the ship stopped for two days, she played the tourist. She explored the town and went to the Zoological Gardens, where she observed there was 'Not very much in the way of animals', although she noted a Corsican moufflon with 'average large horns but it is knock kneed'.[65] On the second day, after visiting the Marseilles Natural History Museum, she returned to the ship and found a letter from her mother. Leila, as Dorothea recorded in her diary, 'was dangerously ill with brain fever. I wired at once but got no reply before 4pm when the Maba Maru was leaving – so got out light luggage and left her. Mr Berkeley – 2nd Officer very civil to me.' She caught the overnight train and reached Paris about 10.30 the next morning. It was 'bitterly cold', she noted; 'spent most of my time in the Louvre standing over the hot air gratings!' Her train left Paris at 9 p.m. and reached Dieppe about midnight. She arrived at Victoria the next morning 'where Mother and Father met me – Li is a little better of course cannot see her yet.'[66]

Of all the endings that there might have been to her Cyprus expedition, the illness of her sister was not one she could have imagined, or, for all sorts of reasons, have wanted. Dorothea had spent eighteen months away from her family, and in spite of her very real concern for Leila, once again it was the eldest daughter who was the centre of attention. Dorothea's remarkable achievements would have to wait their turn, at least as far as her family was concerned. The Museum regarded her quite differently. During those months in Cyprus, Dorothea had collected around two hundred birds and mammals, butterflies, mosquitoes, beetles, and other insects – even she was not sure of the exact number. The range of her interests and observations is astounding. As for the extinct mammals, the Museum knew already of the importance of the little hippo. What the fossil elephant would reveal was keenly awaited. Within two or three days, as soon as Leila was out of danger, Dorothea made arrangements to return to the Museum to begin to examine her finds. She had also decided where she would excavate next, as soon as she had worked out the Cyprus remains and had the money to travel; it was to be that other large island of the eastern Mediterranean, the unsettled island of Crete.

CHAPTER 6

A Bit of a Shock to Me!

Ten days before Christmas 1902, Dorothea returned to the Museum. There was of course no triumphal entrance (as a man might have made), although her exploits and discoveries surely deserved one, but I like to think that, at the very least, the Keeper and his wife entertained her to tea. No longer was her status that of 'student'. She was now an acknowledged scientist and explorer whose work was recognized by the Royal Society as well as the Natural History Museum.

So demanding and exhilarating was the examination of her discoveries that, in spite of her long absence from her family, Dorothea barely acknowledged Christmas. The house her parents had moved to while she was in Cyprus, Horsley Court, meant nothing to her and she took just three days' holiday, returning on Boxing Day to the eerie stillness of the workrooms. There she examined her finds, comparing the specimens with others in the collections, refining her skills in precisely measuring fossil bones and teeth, and learning the complexities of the formulae used for identifying and comparing the molars of the different species. She may have been a woman in a male preserve, but she had qualities that ensured generous help from the department. She was intelligent and enthusiastic, she had proved herself as both explorer and excavator, but she had in addition a gift that she retained all her life: that of making the person she was talking to feel that only they mattered.[1]

The two scientists most important to her were Dr Charles Forsyth Major, who had identified *H. minutus* as a known species, and Dr Charles Andrews; both of them had helped with her Wye valley finds. Andrews was one of the most brilliant vertebrate palaeontologists of his generation, a former schoolmaster who in 1892, aged twenty-six, had joined the Geology Department as an assistant and subsequently worked extensively in Egypt. He was a spare, neatly bearded man, whose quiet good humour masked years of poor health. His knowledge and approach were indispensable to Dorothea and it was in long conversations with him that she began to develop her own formidable skills in the interpretation of fossil mammal finds.[2] Forsyth Major was nearly sixty, a striking-looking man with an exuberant silver beard and wide pale eyes. He never joined the staff of the Museum, remaining an unofficial scientific worker, a volunteer, whose knowledge was invaluable to the department. The son of a Scottish clergyman, he had been born in Switzerland and spent most of his life abroad. He had originally trained as a doctor and practised in Florence, but his interest in fossil mammals gradually took over. He spent some years exploring and excavating recent fauna and flora as well as fossil vertebrates in Italy, Corsica, Sardinia, and the Greek islands, and his fascination with the natural history of the Mediterranean islands had a determining impact on Dorothea.

She could not have had two more able, perceptive, and influential mentors. With their guidance, she discovered that she had found not one unique extinct species, but two. At Dikomo Mandra she had unearthed small bones as well as hippo remains, and these, her studies revealed, belonged to an extinct species of genet, a small catlike carnivore previously unknown in Cyprus, although its modern relations still live, she wrote, 'on the opposite shores of Palestine'.[3] It is most unlikely to have swum from the mainland to Cyprus, but may well have been carried there like other small mammals, drifting with the currents on an accidental raft of matted grasses. As for the pygmy elephant she had worked so hard to find, she determined, after comparing it with known species including fossil dwarf elephants from Malta and Sicily, that it too was unique. She named it *Elephas cypriotes* Bate. Smaller than the other known species, it stood no more than three or four feet high at shoulder level. It had arrived on Cyprus (like the hippo, it would have swum) as a full-sized, straight-tusked elephant from Europe and, over thousands of years, had lost its bulk as it adapted to this new, confined, and mountainous environment. Also like the hippo, it died out about 10,000 years ago, again for reasons unknown. As fossilized hippo bones came to be regarded as those of martyred saints, so the elephant skulls may have acquired their own kind of immortality as the origin of the Homeric Cyclops, the nasal opening so understandably imagined as a single, central, great eye socket.

The workrooms where Dorothea examined her 'Cyprus spoil' were hardly ideal. Although it was just twenty-one years since the Natural History Museum had opened, despite Richard Owen's great plans, scientists worked in cramped and overcrowded surroundings, as the Keeper of Geology, Dr Arthur Smith Woodward, constantly complained to the Trustees. The description by the entomologist Norman Denbigh Riley of his department applied throughout the Museum. 'The only heating,' he wrote, '. . . was from open fires which were locked behind screens at 4pm, artificial lighting was by naked gas jets and oil lamps which were so few and so unsafe that the staff was generally sent home in foggy weather or as soon as it got dark.' Even the exhibition galleries had no artificial light or heat, and electricity was not installed until 1906. A speaking tube connected the Keeper of each department with the Assistant Keeper, but, as Riley records, 'if either wished to call on the other he would don his silk hat before leaving his study, even if it only meant crossing the corridor.'[4] The hours of the Museum meant that Dorothea could not start work before 10 a.m. and, by three in the afternoon, the gloom of late December made close examination of specimens almost impossible. No one was allowed to stay after 5 p.m. and an evening patrol would insist on immediate departure, no matter how important or urgent the work in hand.

By the spring of 1903, Dorothea had produced reports on the little elephant and the genet, and Henry Woodward, the man who had shown such faith in her and supported her case for a Royal Society grant, presented them both to learned societies. On 7 May 1903, at a meeting of the Royal Society in Burlington House in Piccadilly, Woodward read to the assembled Fellows Dorothea's report, 'A preliminary note on the discovery of a pigmy elephant in the Pleistocene of Cyprus'. It was an extraordinary achievement for a twenty-four-year-old girl. She was not permitted, of course, to present her own paper; that could only be done by a Fellow of the Royal Society.* The Royal Society's records make no mention of those present at the meeting other than the President, but it is quite possible that Dorothea would have been invited to attend. A few days later Woodward read her report on the genet to a meeting of the Zoological Society of London, of which he was Vice-President.

Woodward himself heaped praise on Dorothea in the *Geological Magazine*. In a splendidly florid article in which he swoops over the 'exalted position' of the hunter of wild animals and the popularity of cave-hunting for fossilized creatures, he quotes at length Dorothea's own Royal Society report on the pygmy elephant and ends thus: 'It is to be hoped that this is but the commencement of a very successful scientific career for the author, who has evidently given her best energies to this most interesting and attractive line of investigation.'[5] Altogether, in 1903 Dorothea wrote and published five reports on her faunal and fossil finds in Cyprus, a quite remarkable concentration of work. The Museum also put on display specimens of the little hippo and elephant bones, tusks, and teeth. A hundred years later I looked for them in the galleries in vain. In one of the many reorganizations of the Museum they were removed from display and now can only be viewed in the collections by appointment.

It was not just in the Department of Geology that Dorothea's contribution was significant. She is listed in the Museum's *History of the Collections* as one of the more important contributors to the mammal collection in the Department of Zoology, donating and selling fifty-one mammals from Cyprus, including the new species *Acomys nesiotes* Bate, the spiny-backed mouse.[6] The Museum paid her £20 (about £1,000 at today's values) for sixty of her mammals and one hundred of her birds,[7] £2. 2*s.* 6*d.* for her beetles (which offset the amount she owed them for her collecting materials in Cyprus), and an unknown amount for 'other Orders of Insects which were a nice collection containing several new species'.[8] The money was essential to her, for she had somehow to pay for her keep in London. If there were no friends who could put her up, she would stay at a Young Women's Christian Association hostel for 'Business Girls', spartan and clean with its dormitories and shared washrooms. Bed and

* Although it was not uncommon for papers by women to be accepted by the Royal Society, it was not until 1945 that a woman was elected as a Fellow, the crystallographer Dame Kathleen Lonsdale.

breakfast cost her between 3*s.* 6*d.* and 6*s.* (about £11 to £16 in today's values) and she had to provide her own knife, fork, spoon and cup.[9]

She must also have received some sort of allowance from her father. Henry's unutterably weary 'we hope things will come right somehow' that he had written to Dorothea in Cyprus seems to have been answered. In early summer the family moved from Horsley Court in Gloucestershire, where they had lived for barely a year, to another large rented house. However, Wyseby, as it was called, was hundreds of miles north in Dumfriesshire, southwest Scotland. At least the family's moves around Wales and the West of England had some kind of logic, but I could find none to explain this move so far from everyone and everything they knew. Had Elizabeth's behaviour become so bad that Henry felt he had to remove her from some scandal? Or did Henry perhaps have debts that made the move advisable and Wyseby could be rented more cheaply than another house in Gloucestershire? There are no records to ease my speculation, but for Dorothea, at least, Wyseby, perched on a hill with a river not far away, seems to have been a pleasant place to live.

In August 1903, she went up to this new family home for the first time. The money she had left after living expenses was to go towards her trip to Crete and she hoped to add to it by collecting mammals for the Zoology Department from the countryside round Wyseby. The Curator of Mammals, Michael Oldfield Thomas, had been systematically expanding the collection since his appointment.* He persuaded and cajoled travellers, sportsmen, serving officers and colonial administrators, indeed, everyone he met at home and abroad, to contribute to the Museum's collections. He had a private income in addition to his not particularly generous Museum salary and financed needy collectors himself. Dorothea's attempts to collect small mammals in Scotland, however, did not prove very successful, as she wrote to Oldfield Thomas in September in a letter which reveals her extraordinary ability to talk to anyone as an equal and a friend. 'I won't attempt any excuse for the very small number of skins I am sending you! . . . so far I seem able to catch hardly anything but these voles and shrews which is a bit monotonous – I tried skinning

* Under his regime, between 1890 and 1900 the number of acquisitions had increased nearly threefold; between 1900 and 1910 it more than doubled, from 13,277 specimens to 33,458. J. E. Hill, 'A memoir and bibliography of Michael Rogers Oldfield Thomas, FRS', *Bulletin of the British Museum Natural History (Historical Series)*, 18:1 (May 1990), 25–113.

a stoat [its stench is notoriously dreadful] but have decided not to attempt anything of the sort again unless I believe I have something new or rare.' In addition, she was continually distracted: 'I find it very difficult to get anything done at this time of year,' she admitted to Oldfield Thomas, 'as I am almost always out, in spite of the dreadful weather – we have a rather nice trout stream which occupies a lot of my time.'[10]

In between the beguiling trout and monotonous shrews and voles, Dorothea was also finding out as much as she could about Crete. At the Museum she had come to know Charles Davies Sherborn, the natural history bibliographer who was compiling his great *Index Animalium* – a complete list of every known genus and species of living and extinct animals from 1750 to 1858, giving the date and place of publication of each name. There are more than 415,000 references in the finished work and Sherborn spent most of his time in the Museum's libraries. There was no better person to ask about references to Crete, and Sherborn obliged; despite his possibly undeserved reputation as a misogynist, the bachelor Sherborn was a kindly and generous man, referring her to works of earlier travellers, which she admitted were 'all new to me'.[11]

Like Forsyth Major, Sherborn was one of the many experts employed by the Museum on a temporary basis; in fact they often outnumbered the permanent staff. The size of the collections in all departments was far too large for the staff alone to manage. Furthermore, students, visiting scientists and members of the public who brought in potentially valuable fossils 'expect and need' the advice of the scientific staff, as Dr Smith Woodward constantly reminded the Trustees, and that left them little time to examine, catalogue or describe recent acquisitions.[12] Eminent academics, enthusiastic and knowledgeable amateurs, even retired members of staff, had to be employed as unofficial and temporary scientific workers to cope. Every specimen had to be identified, prepared, labelled, and then arranged in its correct place in the collections or exhibition galleries; in many cases a report, often for publication, had to be written about it. The Geology Department employed eleven permanent members of staff, of whom six were scientists, including the Keeper and Assistant Keeper; the rest were attendants whose work was unskilled. In 1901 a scientific post had been 'temporarily' transferred to the Zoology Department and, despite Dr Smith Woodward's constant complaint, ten years later it still had not been returned.[13] Without its unofficial workers, the productivity of the department would have suffered badly, with collections left unexamined for years in the storerooms.

Just how highly Dorothea was regarded by the Museum now becomes very clear: a few months after her return from Cyprus, she was asked to join this distinguished group. Unofficial workers with private incomes often worked for the love of their subject. Others were paid by the hour or by piecework. Henry Woodward, eminent but retired, was paid at the rate of one guinea a day, but he was a Fellow of the Royal Society and a former Keeper.* In December 1903, Dr Charles Forsyth Major, internationally renowned as a vertebrate palaeontologist, was paid £44 (approximately £2,400 in today's values) for 'preparing, labelling and arranging' about two thousand specimens of extinct carnivores in the exhibition cases and collections between October and December 1903. At the same time Dorothea, self-educated, no degree, just twenty-five years old and female, was paid £33. 10*s.* (about £1,800 today) for preparing, labelling, and arranging a staggering 6,000 specimens of 'Palaeotherridae, Equidae and hippopotamidae' (which are very large extinct tapir-like animals, extinct horses and extinct hippos) in the drawers of the table-cases in the southeast gallery.[14] There is nothing to indicate how long she worked on these; it may have been a few months, but it may equally well have taken her most of the year. It was vital experience for her, working on this vast quantity of material of different species and eras; the earliest known tapirs, for example, date back to the Eocene, more than fifty million years ago, and their relatives still thrive. She must also have learned more about mammal anatomy in these months than any university student on a formal undergraduate course. As for the money she earned, it meant, with a little help, she could now embark on her expedition to Crete.

Dr Smith Woodward had put on display at the entrance to the Geological Galleries a large piece of breccia he had excavated at Pikermi. It contains a wonderful assortment of mammal and bird bones, including the skull and foreleg of *Hipparion*, the extinct three-toed horse. Dorothea would have passed it every day. Like her Cyprus fossils, it is no longer on display but I came across it in the Museum's collections. The richness of

* He was also working on a specific project, the collection to illustrate 'dynamical geology' – the study of the events that lead to the formation, alteration and disturbance of rocks. Woodward received £200 on completion of the project. (18 March 1904, DF 102/9.)

this fauna from the mainland presents such a contrast with the ad hoc, limited and specialized species to be found on islands that it seemed to me to raise fundamental questions – questions that Dorothea must have asked too. How, for example, had the creatures arrived there; what global changes had made species now known only from Asia and Africa migrate; why should some animals swim to islands and be stranded there and not others; why should just some survive or evolve in pygmy or giant forms; and why then should they become extinct? These questions would occupy Dorothea's life and, most immediately, be the focus of her exploration of the island of Crete. Her quest again was to discover the extinct and recent fauna of the island, and in particular, as she wrote, 'to throw fresh light on the subject of the origin and development of pigmy forms'.[15] Once again this was a pioneering expedition; no one else had attempted a systematic investigation of the limestone caves of the island, although the existence there of animal bone deposits had been known for a considerable time.

Hundreds of millions of years ago, like Cyprus, Crete had been part of the seabed. With the movement of the continents, the seabed was uplifted and, for a time, Crete was part of the land mass that linked Europe to Asia. As the geological upheavals continued, all but the highest peaks were submerged, and where Crete is now lay a series of mountainous islands, separated by shallow water. After the cycles of rising and falling sea levels, and the drying up of the Mediterranean, Crete began to assume its current shape in the Early Pleistocene, about one and a half million years ago, although in historic times, violent earthquakes and volcanic eruptions have raised and submerged parts of the island. It is not impossible that, unlike Cyprus, during the global sea-level fluctuations Crete was at times linked to the mainland, allowing faunal migrations across land bridges. The mountainous islands now form a high central limestone ridge through Crete, divided in three places by fertile plains. In the mountains and along the coast there are, as in Cyprus, literally thousands of caves. As Dorothea discovered from the books Sherborn had recommended, a few remains of extinct pygmy hippos had been found in Crete and an extinct full-sized elephant, but not, as yet, an extinct dwarf elephant.

After she had paid her living expenses in London, all the money Dorothea earned went straight into her Crete fund. Even so, she needed to borrow and it was to her sister and father that she turned. Leila loaned her £5, while Henry, and perhaps this is an indication that things were beginning to come right for him, advanced a very generous £36. 10*s*. It was a more complicated trip than before; this time she had no friends to travel and stay with, or to introduce her to those who might be of help. The Museum itself had no formal department that dealt with expeditions, but William Ogilvie-Grant, who had succeeded Richard Bowdler Sharpe as head of the Bird Room, had travelled extensively and colleagues wanting to travel would turn to him for advice. So did Dorothea, writing to him in December 1903: 'I am very sorry to be troubling you so soon but I find now I may have to write to the Consul in Crete before I make my final arrangements – So I shall be most grateful if you could get me a letter of introduction to him some time soon. Do you know anything about Mrs Howard and if she lives out there too?'[16]

The Consul was Esmé Howard who, as consul-general to Crete, held the most difficult posting in the Mediterranean.* Turkey had ruled Crete for two hundred turbulent years. By the 1890s rebellion, violence and disorder were such that the four great powers of Britain, France, Russia and Italy intervened. Turkish rule ended and in 1898, Crete became nominally autonomous with Prince George of Greece as High Commissioner. In practice the Prince had to work closely with the consuls-general of the four powers to control the island. This was no easy task as there was frequent unrest in the majority Greek population who, like some Cypriots, wanted union, *enosis*, with Greece. Memories were still vivid of the barbaric violence between the Greek Christian and Turkish Muslim communities that had ended just a few years earlier. By 1904, however, according to statistics collected by Esmé Howard for the Foreign Office, the most prolific crime was theft of animals. In a population of just over 300,000,† there were just 150 instances 'of wounds inflicted on private

* Howard had joined the diplomatic service at the age of twenty-two and resigned seven years later in 1892 to fight the general election of that year as a Liberal. He was unsuccessful and spent the next few years travelling round the world, including a spell fighting in the Boer war in South Africa where he was captured but managed to escape. By the time he was forty, his interest in an adventurous life had waned; in 1903 he rejoined the diplomatic service and was sent to Crete. His career subsequently flourished, ending with six years as ambassador to Washington. He was raised to the peerage in 1930 as Baron Howard of Penrith. (See his autobiography, *Theatre of Life*, London: Hodder & Stoughton, 1935.)

† In 1900, 269,319 were Christian, 33,496 'Musselmen', 728 Jews, and 141 were British. (TNA (PRO), FO 421/207, Despatch 80.)

persons', four cases of rape, twenty-six of smuggling and fraud, and thirty of vandalism and 'motives of revenge'.[17] Good relations, nevertheless, between the consuls-general and with the Prince were imperative, and in this Howard excelled. In 1898 he had married the daughter of an Italian prince, Lady Maria Isabella Giovanna Teresa Gioachina, and she did indeed live out there with him.

On 23 February 1904, Dorothea left Wyseby for London, spent the next two days 'rushing about', and on the 25th caught the 9 p.m. train from Victoria, bound for Paris, where she bought a hat. From there she took the overnight train to Marseilles where she boarded the Black Sea steamer, *Guadiana*, and suffered for the entire five-day voyage. The sea was so rough at night that she 'hardly slept as was constantly shoved up and down my bunk',[18] while waves broke in through the porthole, which she had left open. By day it was 'not much fun' either, cold on deck and stuffy in the saloon.[19]

By the time the ship arrived at Piraeus on the morning of 3 March, she was unwell, quite apart from seasickness, but went straight into Athens, booking herself into the Hotel Victoria. She had two visits to make that afternoon, however poorly she felt. The first was to the Acropolis, which she had been unable to see when on her way to Cyprus three years previously. Now she went directly there, exploring and wandering through its wonderful ruins with a freedom that mass tourism, corrosion, and pollution have for decades denied to the rest of us.

The second visit was an appointment with Robert Carr Bosanquet, the director of the British School of Archaeology at Athens, and his wife Ellen. Bosanquet was just thirty-three, had been a brilliant scholar at both Eton and Cambridge, and had first started excavating in Greece as a student with the British School at Athens in 1892. In 1898 he excavated the Roman fort of Housesteads on Hadrian's Wall in Northumberland, as a result of which he was offered the post of assistant director at the British School at Athens in 1899, becoming director the following year. He had been excavating sites in Crete since then and was currently working at Palaikastro, a Minoan town in eastern Crete.[20] Bosanquet was an important contact for Dorothea, but the meeting was not quite what she had expected. 'They took me very calmly and was somewhat disappointed with my visit – however Mr B. seemed inclined to be very civil and to give me any information he could . . . Returned to the hotel very tired and somewhat depressed – bad cold and feel somewhat groggy.'[21] In Cyprus she had been of such interest to everyone she met, with her hunt for live and fossil mammals held in considerable regard. To an archaeologist, however, no matter how well disposed he was to Dorothea, that sort of work simply could not compete with the momentous discoveries then taking place in Crete of the sophisticated and wonderful Bronze Age Minoan civilization, or indeed with the antiquities of classical Greece. The day after her meeting with the Bosanquets she notes crisply, 'Went to see the Museum – some fine things there even to the uninitiated.'[22]

As a child I'd loved stories about Crete, the legends of the labyrinth and the Minotaur, and accounts of how Sir Arthur Evans had excavated and reconstructed the magnificent palace of Knossos; then it all came to life when successive family holidays were spent exploring those places I had so avidly read about. Now, reading Dorothea's Crete diary, I was transported to 1904, a marvellous year in the history of archaeological activity in Crete. At least four major archaeological teams were working there, all turning myth and legend into history. Bosanquet and the British School were working on Minoan cities at Praesos and Palaikastro, Arthur Evans was excavating Knossos; the Italians under Dr Federico Halbherr were excavating the palaces of Phaistos and Aghia Triada, while an American team of archaeologists led by Harriet Boyd* was excavating the Minoan town of Gournia and the village of Vasiliki.† A great shift had taken place in archaeology over the previous fifty years. Excavators were no longer treasure seekers; what they were after was an understanding of those past societies whose remains and ruins tantalized the modern world. Precious artefacts were still superb works of art with an intrinsic value, but for the archaeologists, of equal, if not more, importance was what they revealed of the society that had created them. There was also now an awareness of the importance of stratigraphy, the recording and understanding of the order in which events occurred.

Dorothea must have been made to feel keenly a perceived gulf between her work as a palaeontologist, looking for fossilized creatures to under-

* She became known as Harriet Boyd Hawes following her marriage in 1906 to the English anthropologist Charles Henry Hawes.

† In 1904 the term Mycenaean was still used for this civilization. The term Minoan was proposed by Sir Arthur Evans at the first International Archaeology Congress in Athens in 1905.

stand the history of the earth, and that of the archaeologists, revealing ancient civilizations and some of the finest works of mankind. What she may not have been aware of was the antipathy to female archaeologists that then prevailed. Harriet Boyd found that her requests to take part in the excavations of the American School of Classical Studies in Athens were initially rejected.[23] Even the very presence of women archaeologists on a dig with men was felt by some to be positively detrimental. The archaeologist J. P. Droop, a former student of the British School at Athens, sounded a famous blast against women archaeologists, at least those taking part in mixed digs. His chief objections were personal ones, although, of course, as he wrote, they had nothing to do with 'the particular ladies with whom I was associated; should these lines meet their eyes I hope they will believe me when I say that before and after the excavation I thought them charming; during it however because they, or we, were in the wrong place their charm was not seen'. Mixed digging, he believed, 'must add to all the strains of an excavation, and they are many, the further strains of politeness and self-restraint in moments of stress, moments that will occur on the best regulated dig, when you want to say just what you think without translation, which before ladies, whatever their feelings about it, cannot be done'.[24] There were of course exceptions, but in the early part of the twentieth century women archaeologists often either had all-female teams, or, if like Harriet Boyd they were in charge of the dig, men were employed in a junior, not an equal, capacity.

On 4 March, Dorothea boarded the overnight steamer from Piraeus to Crete. She arrived at Chania,* then the Cretan capital, early on a still, pale morning. It was a study in blue and pale golden-beige, topped by the terracotta shades of jumbled roofs. 'Came ashore as soon as could get out my baggage,' she wrote, 'at Customs began examining each box surrounded by crowd of men but when they got to about the 5th got tired of it and passed the rest!'[25] It was a ritual that everyone involved in excavating had to endure. The young Canadian archaeologist, Charles Trick Currelly, who was working with Bosanquet at Palaikastro, was well used to the customs officers. 'When any of us left it took about an hour for the customs officials, surrounded by admiring loafers, to go through

* Dorothea spells it Kanea – most of the time – but she uses a variety of transliterations for almost every Cretan name.

a suitcase. Pillboxes were emptied, hair-brushes were combed. This was a bit of comedy as surely, if we had wanted to smuggle out small things, we would have had them in our pockets. But all Cretans wanted government jobs, and so the customs officers demonstrated their efficiency as dramatically as possible.'[26]

There was reason behind this ritual. Although the wealth of archaeological sites had long been known to exist in Crete, there had until just a few years previously been a ban on excavating, not least because of fears that any treasures discovered would disappear to Turkey. Crete was also a highly dangerous place and few archaeologists would risk being caught up in the violence that was endemic in the island. When excavations began systematically after autonomy in 1898 it was under the strictest of rules. A petition had to be presented to the Cretan authorities, requesting permission to dig in a specific area. If this was obtained, a survey had to be carried out and then precise locations for excavations requested. When approval was granted, a representative of the Cretan government was appointed to make sure all agreements were kept.[27] In 1900, when Harriet Boyd discovered a site where she wanted to dig, permission was granted to her as a 'representative of the American School of Classical Studies to excavate in the name of the Cretan government'.[28] At that stage, nothing that she discovered was allowed to be exported. It all went to the museum in Herakleion.* As George Chacalli in Cyprus had complained, this was in great contrast to the casual attitude of the British towards Cypriot antiquities. It was an edict with which the archaeologists in Crete had some sympathy, as David Hogarth, Bosanquet's predecessor at the British School at Athens, wrote: 'All these precious things have been left either where they were found or in the Museum at Candia [Herakleion], which is now unique among the museums of the world. The island Government, still in its infancy, has made a praiseworthy effort to safeguard and house its treasures, a much more genuine effort than has been made, alas! by the British Administration in Cyprus.'[29]

Dorothea stayed at the Angleterre, a small hotel in Chania that looked out over the harbour. Behind, the stark snow-covered slopes of the White mountains, the Aspro Vouno, towered above the town. It was all very

* Dorothea often uses the old name Candia for what is now Herakleion, but she is not consistent, so I have used Herakleion, even where she does not.

beautiful, but just outside Chania, as Dorothea discovered to her distress when she took a wrong turning, was a leper colony. That afternoon she took a carriage to Khalepa, the elegant and secluded area of the town that was home to the High Commissioner Prince George, as well as the British Consul, and left her letters of introduction at Esmé Howard's house. The Consul responded immediately, inviting her to lunch the next day. 'Felt a bit nervous beforehand,' she wrote, 'but found them all very pleasant and easy to get on with . . . Lady Isabella H. is Italian but talks English quite well – Mr H. seems quite ready to help me which is an enormous relief.'[30] Exploring Cyprus had been difficult enough when she was surrounded by friends who respected and admired her work and who were only too willing to help. Alone in Crete, at least she now had Esmé Howard's support.

One of the books Charles Sherborn had recommended to her was by Thomas Spratt, the vice-admiral and archaeologist who had visited Crete in the mid-nineteenth century and written an account of his explorations in 1865, *Travels and Researches in Crete*. Spratt had included geological maps of the island in colour and, copying from these, Dorothea shaded in blue on her maps (one for the east, the other for the west of the island) all the limestone areas in which bone caves might be found. On an island that was about one hundred and fifty miles long, and at its narrowest point barely seven miles wide, there were over one thousand known caves. 'It is not surprising,' Dorothea wrote later, 'that caves are to be found of every imaginable size and shape: some single, others many-chambered, with tortuous passages and innumerable stalactites and stalagmites, which occasionally rise unbroken from the floor to a lofty roof which is but dimly discerned by the faint light of a lantern.'[31]

There were little more than twelve miles of serviceable roads, the rest being rough tracks or sheep and goat trails. She marked her route in black ink on the maps, a dark trail of courage, commitment and challenges greater even than in Cyprus. The west of Crete is the wildest and most rugged part of the whole mountainous island and today the least spoiled by tourism. In the far northwest three great limestone peninsulas riddled with caves jut into the Mediterranean: the bulbous Akrotiri Chania, which protects the great natural harbour of Suda Bay to the east with the town of Chania immediately to the west, and then Rodopolou and Gramvousa, like two misshapen bull's horns at the extreme end of the island.

With only her map as a guide, Dorothea spent her first fortnight in Crete riding and tramping round Chania. She began her cave-hunting on the nearest peninsula, the Akrotiri Chania, hiring 'rather a nice little pony – though an awful squealer'.[32] This was just the first of any number of animals whose distinctive, indeed almost anthropomorphic, character traits plagued Dorothea throughout her four months in Crete. The Akrotiri peninsula was edged with limestone cliffs, which at its far end plunged almost vertically from a considerable height to the sea, and was pierced in a few places by narrow winding gorges. In March it looked at its best: the purple, red and yellow of anemones, broom and a host of spring flowers clustered between low-growing shrubs, scented thyme, and the shaded grey of the rock. She noted a few butterflies, a red admiral, a clouded yellow, and a wall, a brown butterfly with orange-marked wings. Meeting 'several shepherds etc', she asked them whether they knew of any bone caves and was puzzled to hear from one of them that there used to be bones in some caves round there but 'they [the caves] were all gone now.' How caves had mysteriously disappeared she learned a few days later. 'They have a habit just here,' she noted, 'of cutting off the tops of caves to get the stone for burning lime which disguises and adds to the difficulties of finding these places.'[33] Her enquiries about a bone deposit on the peninsula near a ruined village called St George that had been recorded by the eighteenth-century English traveller Richard Pococke, were also 'met by the negative'.[34]

She continued the search the next morning, notwithstanding a contrary pony, which 'started off very well but turned out to be an utter slug – hardest work to get it out of a walk'. She had read of a possible bone cave found on the Akrotiri in the 1830s by the English traveller Robert Pashley. It had a little chapel built at its entrance and was known as the Cave of the Bear, after an old hermit who lived there, dressed in wild beasts' skins. 'This caused him,' Dorothea wrote, 'to be mistaken for an animal by a man who consequently shot him.' Wounded, the hermit crawled to the back of the cave where he died.[35] Her search for this cave was all too reminiscent of her experiences in Cyprus: an exhausting ride up hills and through gorges to the remote monastery of Aghios Jannos and beyond, almost to the farthermost reach of this rugged and desolate peninsula. She left her pony at the monastery and took as a guide a boy who said he knew the cave, but, as had happened to her so often before, he led her up horrendously steep paths to quite another, just above the ruins of the eleventh-century monastery of Katholiko. The cave had a roof so low in places that she had to wriggle along lying almost flat. Although when it opened up it was visually dramatic by the light of her flickering candle lantern, there was no sign of a bone deposit. In a crevice near the cave entrance, however, there was a pile of human bones, but whether they indicated that the cave had once been a mortuary for the monasteries or were the more recent and shocking relics of some 'barbaric act' between Greek and Turk, she was unable to discover.[36] When she did by chance find the real Cave of the Bear as they were clambering back to the monastery of Aghios Jannos, there was no sign of fossil mammal remains there either.

Exhausted and feeling unwell, she rested for a few minutes in the monastery. The boy returned to his village and she began the trek back to Chania, the pony 'so horribly slow the sun set long before I got back', leaving her riding in the growing darkness in a completely deserted landscape. Then against the dim shapes of rocks and trees, she saw 'one man on horizon who suddenly disappeared'. For a few moments her fear must have been acute. 'However,' she wrote, 'on coming up to him found he was drunk.'[37] Dorothea reached her hotel at about 8 p.m. The difficulties of her exploration of the Kyrenia hills in Cyprus seemed little compared to this. Wherever she looked there were limestone hills, the majority riddled with caves, and she had no means of knowing 'where or where not to hunt round'.[38]

Thomas Spratt had described a small bone cavern between Suda Bay and Chania and she set off to find it, on foot this time, taking with her a small hammer, a notebook, and a net and boxes for collecting insects. On her way over stony paths in hills thick with spiny shrubs, she found 'the hole and lid of a trapdoor spider – the first one I have ever found,' noted a few stonechats and filled the boxes with 'a good many insects, lot of bees and humble bees, latter look very large'. About midday she found a site close to the path, which seemed to answer Spratt's description. Just discernible were some very fragmentary remains of a bone deposit in the cave floor and extending for about forty yards, possibly the remains of two separate caves. In two places she saw teeth, but noted it was 'too public a place for me to do any digging here yet'.[39]

It really isn't clear at this point how much she knew of the Cretan archaeology laws, or how far she thought they applied to her. Certainly no one, not the Natural History Museum, not the Director of the British School at Athens Robert Bosanquet, nor even the British Consul Esmé Howard, all of whom knew perfectly well that she planned to excavate for fossil finds, appears to have told her that she must obtain a permit from the Cretan government before starting to dig. The Natural History Museum may not even have been aware of the legislation, and it is possible that everyone else simply assumed that fossil bones would not be classed as antiquities in the same way as artefacts. Dorothea, after all, was the first palaeontologist to excavate in Crete since the laws were introduced.

Her resolve not to dig by the public path barely survived the night. In the morning she went out 'bug hunting on foot again today and of course went off eventually to C[ave] as I wanted to get a few specimens'. This she did, extracting from the bone breccia remains of 'small things' including a small rodent and a shrew.[40] That these scant remains, like everything else she would find, belonged to the Cretan government and could not be exported on pain of imprisonment, no one seems to have told her either.

In the next few days she explored farther, riding along the green foothills through olive groves and crossing gingerly the streams rushing down from the bare snow-covered mountains. She noted a handsome pale-grey hawk with black-tipped wings and tried to purchase for the Museum a Cretan wild goat complete with horns. 'Enquired all along if any bone caves known but heard nothing satisfactory,' she wrote; 'offered a reward for finding any.'[41] A man took her up to a group of caves, but all she found was a lesser horseshoe bat. In the bazaar in Chania, after several days of trying, she bought a polecat skin,* the first she had found that was in good condition and not too expensive. She set some traps in limestone rocks near Suda and in one she found, 'greatly to my delight', an *Acomys*. This was the spiny-backed mouse she had identified in Cyprus and, until she found it, *Acomys* had not been known in Crete. She continued her cave-hunting, walking over the hills between Chania and Suda or to the far side of the Chania plain, but in all these places she 'found nothing and [they] did not seem at all promising'.[42]

What she needed was a guide and Esmé Howard recommended a

* It is still in the collections of the Natural History Museum with the original label 'bought in bazaar, Khania, Crete [signed] DMA Bate'. When it reached the Museum, the skin was stuffed with straw and sewn up with strong thread.

Cretan called Nikola, who lived in the hills south of Chania. Dorothea hired a pony, but it 'objected to my spur and when I got to its stables it refused to go on or do anything but try and kick me off'. She found a more suitable animal and, after a hot and dusty journey, arrived at Nikola's village, only to be told that he was down near the coast.[43] When she finally found him, she seems as unimpressed with him as she had been with Mr Perks in Cyprus, but she knew of no one else and, in spite of her reservations, employed him.

Thomas Spratt had discovered a bone cave in a limestone cliff near the monastery of Gonia at the foot of the Rodopolou peninsula. He sent the bone breccia he found there to his friend, Dr Falconer, whose *Palaeontological Memoirs* Dorothea so relied on. Falconer identified the fossil remains as being a goat, a roebuck or stag and a small *Myoxus* (a species of rodent).[44] There was no mention of pygmy hippos or the little elephant that Dorothea hoped to discover, but it was an indication of a more varied fossil fauna than Cyprus had offered. On the miserable and wet morning of 21 March, Dorothea set off for Rodopolou with Nikola and yet more unsatisfactory animals: 'Am awfully disappointed with the mules,' she wrote, 'wretched little beasts go donkey's pace.' Arriving at the monastery in the evening, she was greeted by the abbot and a 'large party of papas and villagers'. As in Cyprus, she was of great curiosity to the locals, not only as a woman travelling alone with just a guide, but also as a potential source of revenue. The local doctor talked to her in French and 'they took me in', she wrote, 'as I was a stranger from abroad'. It was all she could do to deal with this welcome. The cold she had caught on her journey had worsened and she was suffering now with toothache as well as a 'beastly cotton woolly feel in my head that I have had more or less all the time I've been out here'.[45]

The bone cave discovered by Spratt was clearly visible in a limestone cliff overlooking the monastery. The view from it over Chania Bay was spectacular, but she was 'woefully disappointed' with the cave itself. Spratt had found remains in the floor of the cave, but Dorothea could see none. All she discovered were some traces of bone in the cliff face below the cave. She extracted a couple of teeth, which seemed to be 'chiefly of some sheepy description', and noted some small rodent incisors. She climbed round the cliffs to a number of other caves, but 'without success'.[46]

The next morning brought more heavy showers, but she set off along the Rodopolou peninsula with Nikola and a guide called Stelianos who said he knew of a big cave. The peninsula, which is mainly limestone, was virtually uninhabited then as it is now, a barren, rocky place, with the ubiquitous spiny shrubs and sudden gullies. The cave Stelianos knew was named Kimido Spilia on her Admiralty chart and was not much above sea level. It was vast and labyrinthine, twisting and dividing so much that, in a splendid mythological echo, they had to hire a boy with a bag of leaves 'to mark our track'.[47] Even with her feverish cold, Dorothea felt a 'spirit of adventure' as she explored it.[48] Despite the interest of the cave itself, she found nothing there, nor in any other of the many caves she investigated that day. By the time they reached Aphrata where the guide Stelianos lived, it was raining so hard they stopped for a meal at his home. 'They eat snails etc,' Dorothea noted, and she opted for a couple of eggs instead. On the way back to Gonia she was again caught in a heavy and bitterly cold downpour; by morning the rain had turned to snow on the lower slopes. She went back to Spratt's cave with Nikola and Stelianos 'to blast away some of the rock'. This revealed bone breccia but, in spite of her efforts, no good specimens of teeth or bones.[49]

By now her activities had attracted considerable interest, blasting was hardly discreet, and it caused her much concern: 'Feel a bit depressed not having better luck and the fear of being interfered with – had rather an uncomfortable day as nearly the whole time had people up there watching me.' A police officer arrived first, then a crowd of villagers and peasants, and, most worryingly, the headman of Kalymvari, a village near the monastery. 'He,' Dorothea wrote, 'was evidently talking of writing to Chania about my digging here but hope my being vouched for by Mr Howard will stop that.'[50] Nevertheless, she continued to work, the men blasting the rock outside while she minutely examined the inside of the cave, finding a few bone traces in the walls and then in a small crevice, a 'sort of "pocket" I think', which was cramped and difficult to work. The following day brought further unwelcome attention. 'About mid day again,' she wrote, 'had an awful crowd about 20 men at one time.' Many, perhaps even most, people would have found this intimidating and would have stopped work. Not Dorothea. She and her two men continued to dig, surrounded now by all those men who had climbed up to the cave, watching, perhaps shouting, perhaps again threatening to report her. All she remarks in her diary with quite breathtaking understatement (although perhaps it does not quite conceal her unease) is that it was 'a bit trying'.[51]*

In fact she continued to dig at Spratt's cave, which she discovered was called Tripiti, for two more days. She thought she had found there two species, the goat and stag mentioned by Spratt, but the bone breccia was so hard she could obtain few good specimens. At the end of nearly a week of laborious and frustrating work, she decided to leave. On her last day here, 27 March, a Sunday, it was made clear to her that as a guest of the monastery she was expected to go to church, light a candle, and make a donation in front of the large congregation. After her recent exploits in the cave, she was, she felt, 'quite a rival attraction to the service!' As she left, a boy brought her a couple of hedgehogs. Unlike the Cypriot species, they were large and short-eared and, she noted ruefully, 'also unlike the Cypriote species they swarm with enormous fleas'.[52]

With Nikola she left Gonia and rode westwards towards the third and last of the peninsulas, Gramvousa. It was 'still pretty cold', she wrote, 'and I long for a flannel shirt! had rather a headache all aft[ernoon]'.[53] After a journey of nearly four hours, they stopped for the night at Kisamo Kastelli, a village on the shores of Kisamo Bay, thinking it was at the foot of the peninsula. At daybreak they discovered they were still three miles away and although she explored a few small caves, she hardly set foot on the peninsula itself. It was a bleak, inhospitable area, with little shelter and scarcely a track. With so many caves and no information about bones, it was an impossible task.

As if in answer to her need for guidance, the next morning a man brought in a lump of bone breccia, which he said had been found (although not by him) at Ennahoria, 'which', Dorothea remarked after looking at her map, 'is a big word and vague'.[54] Ennahoria was a huge valley about twenty miles away in the mountains to the south. Without knowing who had found the breccia and precisely where, Dorothea would have had a futile journey. 'So,' she wrote, 'went off in quest of the finder of this block.' He apparently lived in Vulgaros, a village in the mountains

* Very early on in my research I was told a story (although its source was unknown) that Dorothea had had to escape from a riot at a monastery in Crete. There is no evidence whatsoever for any such incident, except perhaps an embroidered version of this particular day. But far from it being either a riot or Dorothea escaping, there is instead this amazing image of her determinedly continuing to dig in front of this 'awful crowd'.

south of Kisamo Kastelli. The man himself was not there and the reason quite stunned her. 'Disappointed to find that he is at present in prison at Chania for trying to smuggle antiques (A bit of a shock to me!).' She knew now about the restrictions on excavating (how could she not after her experiences at Gonia), although she may not have known the details. However, that the penalty for attempting to take antiquities out of the island might be imprisonment had plainly not occurred to her. She met the man's brother, but he claimed that neither he nor anyone else knew where the bone breccia had come from. Without this information, it was useless to go on to Ennahoria, and she turned southeast, skirting plunging valleys and precipitous slopes to the village of Malatharios, which Nikola had told her had caves near by. 'The village,' she wrote in her diary, 'seemed to contain only relations of Nikola it is a tiny place but I got a decent newly built room. Bitterly cold and wet sat and shivered.' It was not just the weather that chilled her. The reality that confronted her as she sat there alone was that she too might find herself in prison, or at the very least banned from working, unless she obtained from the proper authorities permission to dig.

By morning she decided she had no choice except to return at once to Chania. In pelting rain and freezing wind, she rode back across the mountains, first to the monastery at Gonia to collect her fossil bones from the Tripiti cave and then on to Chania. She arrived in the late afternoon to find a letter from home that only added to her woes. Her brother Thomas, who was in India with the Royal Artillery and had been due to return home on leave, 'is not coming home after all – it is an awful disappointment to us all as we had been looking forward to seeing him for some months and this puts it off for a year anyway'.[55]

There were heavy showers again in the morning. Nikola came to see her and, with spectacular audacity, she sent him off to the gaol 'to get information' from the antique smuggler and finder of the bone breccia whose name, she thought, was Pappa Demetrios. It cost her a sovereign.[56] In the meantime, she wrote at length to William Ogilvie-Grant at the Museum, telling him of the two bone caves she had found, the difficulty in knowing where to look, and that she had not found

> anything very exciting yet – Nothing in the way of elephants or hippopotami!

> I have *the* collecting tin full of stuff that I am anxious to get off but don't yet know how I shall set about it. Am dining with the Howards tonight and will hear if there is any actual law against exporting fossils or not . . .
>
> I like the Howards so much and they are very kind to me – he has been a great help to me about getting a man and so on. Unlike in Cyprus, travelling about in the country is not the custom here, particularly in this part of the island, which makes it more difficult to arrange trips at first . . .
>
> I hope that all the family are very fit – Hope you will write and tell me all your news but don't use a B.M. envelope please![57]

To receive a letter in an official British Museum envelope, she felt, would draw even more attention to her activities.

That night as arranged she dined with the Howards. The news was everything she had feared. 'Find that the exportation of fossils is illegal so now suppose I must go off and try my luck with the authorities at Candia [Herakleion] – an awful sell if my trip is doomed to complete failure.'[58]

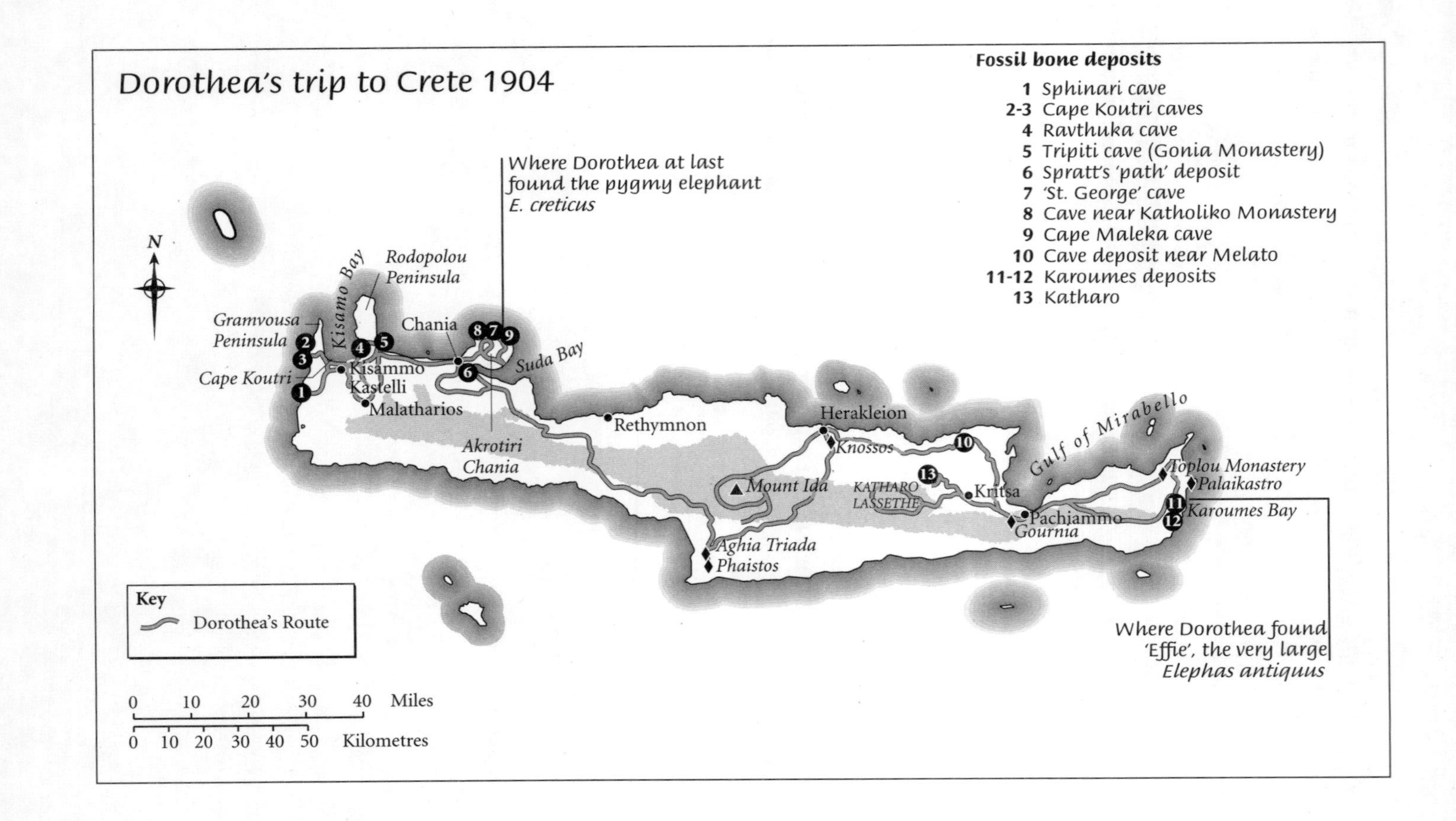
Dorothea's trip to Crete 1904
Fossil bone deposits
1 Sphinari cave
2-3 Cape Koutri caves
4 Ravthuka cave
5 Tripiti cave (Gonia Monastery)
6 Spratt's 'path' deposit
7 'St. George' cave
8 Cave near Katholiko Monastery
9 Cape Maleka cave
10 Cave deposit near Melato
11-12 Karoumes deposits
13 Katharo
Where Dorothea at last found the pygmy elephant E. creticus
N
Kisamo Bay
Rodopolou Peninsula
Gramvousa Peninsula
Cape Koutri
Chania
Suda Bay
Kisammo Kastelli
Malatharios
Akrotiri
Chania
Rethymnon
Herakleion
Knossos
Mount Ida
KATHARO
LASSETHE
Kritsa
Gulf of Mirabello
Toplou Monastery
Palaikastro
Karoumes Bay
Pachiammo
Gournia
Aghia Triada
Phaistos
Key
Dorothea's Route
Where Dorothea found 'Effie', the very large Elephas antiquus
0 10 20 30 40 Miles
0 10 20 30 40 50 Kilometres

CHAPTER 7

Courageous as Ever

Dorothea arrived in Herakleion on the evening of 6 April. Although not the capital, this was the island's business and intellectual centre. Here was the Museum and the all-powerful committee, which alone could grant permission to dig. It was led by the Ephor (overseer) of Antiquities, the archaeologist Dr Joseph Hazzidakis. Dorothea needed an introduction to him, and the man she approached to provide it was Dr Arthur Evans,* the archaeologist who had had the foresight (and the means) to purchase the land in which the ruins of the palace of Knossos were buried. With Bosanquet's predecessor at the British School, David Hogarth, Evans had set up the Cretan Exploration Fund to raise money for excavations throughout the island and with that, as well as his own considerable fortune, he had since 1900 excavated and reconstructed – or as he termed it, reconstituted – Knossos.

Dorothea rode the few miles south to Knossos to meet Arthur Evans, accompanied by (as she called him in her diary) 'Mr Corelli, most amusing Canadian'. This was Charles Trick Currelly, the young Canadian archaeologist who was excavating with the British School at Palaikastro and was staying for a few days with Evans. With Currelly as her guide, Dorothea explored the great site and watched in amazement the constant activity of Evans' army of 200 workmen. Evans was renowned for the

* Later Sir Arthur Evans.

delicious afternoon tea he lavished on his visitors, and it was nearly dusk when they returned to Herakleion. On the boat from Chania she had met another young archaeologist, and that evening she saw him again. This was 'Mr Seager, an American', she wrote, 'who is digging in the east with an American party'.[1] Richard Berry Seager was blond and blue-eyed, very charming, very rich, and twenty-two. She was to meet him again in May.

With an introduction from Arthur Evans and in the company of Charles Currelly, Dorothea confronted the Museum authorities the following morning. 'I saw Mr Hatzidakis [*sic*],' she wrote indignantly, 'but got no satisfaction as he only referred me to a man at Kanea [Chania]!'[2] which of course was where she had just come from. She felt that her time and money had been wasted in a fruitless venture and she left Herakleion the next day, returning to Chania in cold and windy weather. 'A Mr Dillon on board leaving Crete for Malta,'[3] she noted, and two days later recorded in her diary as innocently as if she were noting a horseshoe bat, 'Sent a small box of stuff with Mr Dillon to post at Malta'. Then she added, 'Went up hill to set traps and catch insects but rather windy.'[4] Perhaps smuggling had been in her mind all along from the moment she heard of the restrictions on excavating and exporting antiquities. Meeting Mr Dillon on the boat fortuitously gave her the means of doing it. He would simply have thought he was doing a favour for a charming young lady, not that he was being compromised as a partner in crime. It was a highly reckless venture, but an indication of her desperation. She was in Crete to discover fossil mammals for the Natural History Museum and could not tolerate the thought that her trip might be 'doomed to complete failure'. She may also have believed that, if she were caught, the British Consul would protect her from prosecution, but also, given what she was about to do next, it is very likely that she simply did not take the strictures of the Cretan government seriously.

On 13 April she packed and prepared to set off for Ennahoria, in search of the deposit of bone breccia. In return for the sovereign Nikola had given him, Pappa Demetrios, the man in gaol for smuggling antiquities, had at last revealed the name of the village nearest the site. With everything ready for her departure, she 'Went to see Mr Wilkinson etc', a fleeting acquaintance whom Dorothea had met through the Howards and who was shortly to leave Crete. She had succeeded in smuggling out a box with the innocent Mr Dillon; now she was trying again with the equally unaware Mr Wilkinson, this time involving Nikola, her by now loyal workman: 'Nikola came down about 8 and I got him to fix up that box of stuff and take it over to Mr Wilkinson!'[5] Clearly aware of the irony of the situation, she then notes, in writing visibly more scrawled than usual, 'Mr Howard had got me a letter of recommendation from the Chief of the Police and after starting found I had been provided with an escort of two zaptiehs [policemen] but not needing them sent them off with a letter to Mr H.' An escort of police was the last thing Dorothea wanted to travel with her. She must have given Mr Howard a very economical idea of her proposed activities on this trip. It was midday before she left for Ennahoria, heading through the mountains for Malatharios where she had spent such a cold wretched night just two weeks previously. They lost their way in the gathering night and arrived late, only for Dorothea to discover that the room she had rented on her previous visit was occupied by her landlady, 'laid up with a broken leg so I had to share her room!'

Any lingering hopes Dorothea had that this trip might now improve were dashed in the morning when she discovered she was in quite the wrong place for the bone cave. The village that Pappa Demetrios had named was not in the Ennahoria at all but across the mountains in the northwest of the island, near the foot of the Gramvousa peninsula. She spent that night near the peninsula in 'a miserable little house', she wrote wearily, 'and the fleas beyond description'. Pappa Demetrios had given Nikola the names of two men who knew the bone cave and this time the information was right. Both men, Dorothea wrote, 'have been in prison for 20 days for the same job. As far as I can make out it was [attempting to smuggle] a small statue they found in the Cave.'[6] After his experience in prison, one refused to help, but the other said he would, at a price, show them the cave. The crime statistics collected by Esmé Howard for this period showed that there were twenty-six cases of smuggling and fraud. In the space of just two weeks, Dorothea had consorted with three of those smugglers.

The cave that had taken her nearly three weeks to find was in the cliffs of Cape Koutri, a small promontory at the western base of the Gramvousa peninsula and near the site of the ancient port of Phalasarna. The port had been built by the sea, but the geological upheavals that had tilted the west of Crete upwards in early historical times had left the ruins marooned inland. The ex-prisoner, whose name was Manoli, led them past the remains of the ancient town and straight to the cave he knew so well from his attempts at smuggling. It was in low cliffs near the northern end of the site and was only visible from the sea or nearby cliffs, although relatively easy to reach – as long as you knew where it was. Along the cliff face was a distinct line that indicated the old sea level, about twenty feet higher than it is today. Dorothea saw at once that the effect of higher sea levels on the bone deposit in the cave had been devastating. On the floor of the cave were scattered fragments of bone breccia and one good-sized mass, containing, she thought, goat and roebuck; the invading seas had destroyed much of what had once clearly been a very large deposit. At the Tripiti cave, Dorothea had blasted the rock with gunpowder. Here at Koutri, she used dynamite for the first time, but with scant success. Bones and fragments of horns were plentiful but were so mixed up with hard breccia that she could extract few really good specimens. The effect on the cave of her efforts was, unfortunately but perhaps not surprisingly, to make it for later excavators 'unsuitable for scientific investigation'.[7]

In her pursuit of fossil mammals on this expedition, Dorothea so far had resorted to smuggling, bribery, deceit and high explosives, and had endured awful living conditions. Now she was told that bones had been seen in a cave about fifty yards along the cliff from where she was working. Nikola and Manoli took one look at the rocks and the sea below and refused to move. Dorothea attempted to climb along the cliffs to this new cave, but it was impossibly dangerous. There was only one thing to be done. Leaving the men behind, she removed her boots and stockings, climbed fifteen feet or so down the cliff and 'went for a swim in that direction instead – had a most painful climb barefooted up the rocks'.[8] If she removed anything else apart from her footwear she does not say, nor does she mention the discomfort of wet, clinging, and salt-encrusted clothes. Her reward when she reached the cave was to find the remnant of a sea-destroyed bone deposit much the same as where she was working, with similar remains. And then she swam back. The distance each way, as she discovered, was nearer one hundred yards than fifty.

For all her bravery, compared to the splendid deposits she had found in Cyprus this seemed disappointing stuff. Many of the specimens were just fragments and there was no sign at all of the hippo or elephant remains that had brought her to Crete. She left Koutri and went south, exploring mountains and arduous sea cliffs, pursuing over the next few days every possible lead, yet finding very little at all. The guide Stelianos had been following her around to tell her of a cave he had found at the foot of the Rodopolou peninsula close to Ravthuka, but even that produced little except a few remains similar to those at Koutri. She acquired a young eagle from a man who otherwise said he would eat it, and a boy brought her a hedgehog, which, she noted, 'eats oats all night'. Anxious still to avoid people, she spent the night outside a village, in a small house on a hillside. 'Fleas,' she observed, 'terrible here.'[9]

It had been a frustrating, miserable and intensely uncomfortable twelve days. She had few good specimens to show for her considerable labours: the goat and roebuck, which she later refined to antelope and deer, and some fossil rodent bones. In her anxiety to avoid discovery, she had endured sleepless nights in hovels alive with fleas that infested her clothes and belongings, tormenting her dreadfully. As she rode back to Chania on 27 April, it was 'very hot and close in morning thunder storm not far off'.[10] If she believed in such things, the oppressive weather was an undoubted omen.

The following day she had a meeting with Esmé Howard that must have been deeply unpleasant for both of them – so awful in fact that she has written in the margin of her diary, 'Fiasco'. The entry reads: 'Find that Mr Howard prevented Mr Wilkinson from taking that box of mine. Also that all my specimens will have to go up to the Museum here and I may be allowed to have some duplicates! Also I have to apply for Govt permit to dig.'[11] How Mr Wilkinson came to reveal his mission for Dorothea to Mr Howard she does not say, although it was most probable that he was as ignorant of the law as the unsuspecting Mr Dillon and simply made a chance remark to the Consul. However it had occurred, Dorothea now was in a situation that she had no idea how to handle. These were all her worst fears; she had put the British Consul in an exceedingly embarrassing position, and she had no friend to advise her. She was mortified at a predicament that seemed incapable of solution. Wretchedly she packed her specimens and the next morning in heavy rain 'took my things up to Museum don't know when I shall hear verdict'.[12] Her diary entry for 30 April is two angry lines slashed across the page. That for the following day, Sunday, 1 May, originally read 'Lunched with Howards', but it has been forcefully scribbled out. How she could ever rebuild her relationship with Esmé Howard after this debacle, she could not, for the moment, contemplate.

At the front of this diary, Dorothea has written four maxims:

1 Blessed are those who expect nothing for they won't be disappointed.
2 The Lord helps those who help themselves.
3 Never refuse a letter of introduction.
4 If you want a thing done well – do it yourself.

These were maxims she held to throughout her life, but they were never so appropriate as now. The first particularly could be applied to the awful twelve days she had just endured, while she was about to put maxims two and four into operation. Far from being cowed by the experience of her confrontation with Esmé Howard and having to take her specimens up to the Museum, her resolve now to do what she believes to be right is absolute. Just four days later she wrote in her diary, 'Believe I have got a box off!'[13] This time she did not rely on anyone else to get the box out of Crete. She simply used the shipping agents. In June she wrote to the Keeper of Geology at the Natural History Museum, Dr Arthur Smith Woodward: 'Just a line to tell you that you will probably soon receive a box from me through Messrs Stahlschmidt and Co – theirs is the only name on it but I have already communicated with them about it. There ought also to be a small parcel addressed to me from Malta [posted by the unsuspecting Mr Dillon] – I shall be so glad if you would have both of them put on one side till my return.' Then, as if her audacity were the most natural thing in the world, she added:

> and of course nothing must be said about them as they have been got out pretty much on the quiet. I just failed to get away a much larger box [the one that Mr Wilkinson revealed to Esmé Howard] but may succeed later though of course these attempts have to be concealed from the authorities – as I have to take all my finds to the museum where they are to be examined by a commission who may allow me to take away what they consider worthless for their museum![14]

From Dorothea's point of view, the Cretans had absolutely no rights in the matter; they were 'my finds', discovered through her own tenacity and expertise and with her own money. The hardships she endured for her science were certainly not on behalf of the Cretans and their museum. She had excavated the fossil bones on behalf of the Natural History Museum and, in her mind, that simply was where they had to go. As far as the British Museum (Natural History) was concerned, in common with most similar institutions in 1904, not too many questions were asked.

No crusader had greater zeal now than Dorothea. Without even attempting to apply for a government permit to dig, she packed her belongings and set off to go cave-hunting in the east of the island, where she knew fossilized hippo bones had been obtained in the 1860s by the French traveller Felix Victor Raulin. They came, he had been told, from Katharo, a remote plateau in the Dikti (or Lassethe) mountains above the village of Kritsa. To get there she took a circuitous route through the south of the island, a journey that would take her over a week.

Her travels took her to the palaces of Phaistos and Aghia Triada, which for the past four years had been excavated by an Italian team of archaeologists led by Professor Federico Halbherr. On a hot and thundery day she reached the ruins of the summer palace of Aghia Triada on a hillside overlooking the grey-green sea of olive and pine trees of the plain of Dibaki. It was here that Halbherr had found some of the most glorious of the Minoan remains, including frescoes and pottery painted in the Kamares style, so named after the cave on Mount Ida in which this exquisitely patterned ware had first been discovered. Halbherr, who was excavating a cemetery, showed Dorothea a curious circular tomb filled with a jumbled mass of human bones and told her of the superb painted sarcophagus he had found, which depicted in marvellous detail the funeral rites of the Minoans. The sarcophagus, in accordance with the law, was – and still is – in the Archaeological Museum in Herakleion.

For four more days Dorothea travelled through this towering, beautiful island, wandering from her path to explore caves and archaeological sites, observing and collecting bats, birds and insects, and frequently being soaked by heavy, thundery showers. On 15 May they reached the Dikti mountains and began the climb to the village of Kritsa. Then, as now, countless windmills dotted the landscape. Spratt describes Kritsa as being 'picturesquely situated under some cliffy steeps',[15] which scarcely conveys the almost perpendicular nature of this large village of narrow, winding streets, perched high above the Gulf of Mirabello. On my visit there, when negotiating a particularly sharp and seemingly vertical bend we found ourselves in the path of a funeral procession. With clutch slipping and smoking, somehow we reversed unscathed, but had we known the difficulties of the road ahead, we might have driven no farther.

Dorothea made enquiries about the bone deposit at Katharo and 'Had town crier out',[16] but this curious call for news of ancient bones produced little result, just one villager who brought her some pieces of limb bones, while another tried to sell her a 'perfect' hippo tooth, but she considered the price for this extortionate. In the end it was Nikola who found a man who claimed to know where the bones had come from and so, Dorothea wrote, 'went off up to Katharo'. If only we had found it that easy. The road from Kritsa to Katharo was first tarmacked in the 1990s and the views are spectacular, but its death-defying bends and lack of any barrier between road and a thousand feet of nothing make driving truly horrendous.* We could not decide which was worse: to meet another vehicle coming the other way, or to meet none at all, which meant no witness (so our imaginations ran) to our car plunging over a precipice. It took Dorothea three hours to ride up what was then a narrow, precipitous, and treacherous track; sharp spiny shrubs caught at her clothes, and branches from stunted, wind-wizened pines had constantly to be avoided.

The plateau was nearly four thousand feet above sea level, high enough for delicate pink cyclamen and single white peonies still to be in full bloom. As Dorothea rounded a final cliff, before her spread a great green shallow basin, about two and a half miles long by three-quarters of a mile wide, between the peaks of the limestone mountains. It was used as a remote market garden by the villagers of Kritsa, who pastured their flocks here and cultivated fruit trees, corn and vegetables. Water ran freely between low rounded hillocks in the cooler months, although in midwinter the plateau could be snow-covered and uninhabitable. At the western end where the hills were lowest, she noted that streams had worn away a

* My beloved partner, who was driving, discovered (a little late) that he had a problem with precipitous mountain roads. My offer to drive was declined with a terse 'that would make it even worse'.

deep and narrow gorge, allowing water to flow down to the extensive upland basin of Lassethe, nearly a thousand feet below.[17]

It is here at Katharo, after two frustrating and exhausting months, that she discovered part at least of what she had come to Crete to find. In an area near two hillocks and just lying on the surface of the ground were pieces of bone and teeth of what she was sure was *Hippopotamus pentlandi*. In her diary she has written the 'H' with a definite and delighted flourish. This was a Pleistocene species of extinct dwarf hippopotamus, larger than the little hippo she had discovered in Cyprus and already known to have existed in Malta and Sardinia.* But her excitement – and relief – at this find were qualified. The specimens were fragmentary and scattered. According to her guide, many years ago 'a Turk' had plundered the site, carrying off two mule-loads of hippo bones. 'This,' she wrote, 'seems likely and probably what is scattered around are his rejected fragments,' and she adds with feeling, 'the brute.'[18] She would have tolerated almost any accommodation to stay at Katharo rather than make that precipitous trek twice a day, but all that was available were 'rudely built and indescribably filthy hovels',[19] which served to shelter man *and* his beasts.

The following day she was told of a monastery at which she could stay. It was on the great fertile plateau of Lassethe, which lies below Katharo, a patchwork of tiny fields stretching to the enclosing mountains beyond, with little villages dotted around its edges. She found it a haunting place, strange and shut off from the world. They rode up a small hill at the eastern end near the monastery and surveyed the plain lying beneath them. In the gathering dusk, it looked to her like 'some vast and irregular draught board',[20] the fallow and cultivated land forming squares of dark and light.

For nine days Dorothea made a twice-daily journey of two and a half hours up an 'awful road' that climbed the thousand feet of mountain between Lassethe and Katharo, as she explored and excavated the plateau. Even in late May, at that altitude it was so bitterly cold her hands and face became red and raw, while a strong icy wind brought with it a damp

* The creature she found has subsequently been shown to be smaller than *H. pentlandi* and in 1966 was renamed *H. creutzburgi* Boekschoten and Sondaar. See G. J. Boekschoten and P. Y. Sondaar, *The Pleistocene of the Katharo Basin (Crete) and its Hippopotamus* (Amsterdam: University of Amsterdam, 1966).

and all-enveloping fog. But she was finding deposits that produced good hippo bones, and while the men worked on these she tramped over the plateau, hunting for more. The best deposit was in a channel cut by a stream. The bones were six to eight feet below the surface, and although many had rotted in the wet, she extracted some 'goodish' specimens, although a skull, she feared, was out of the question.[21]

By 26 May, after such intense work, she was finding few good specimens, and with two workmen and Nikola to pay she could not afford to continue for little return. Early the next morning she packed up the hippo bones and left Katharo. What puzzled her most, as she rode down the mountain through cloud and driving mist, was how hippos, even dwarf ones, had come to live 4,000 feet up in a mountain basin. The plateau had some vegetation and some water, but it was covered during the winter with snow, conditions impossible for the hippo. Not even one adapted to an island environment would have been agile enough to migrate seasonally down precipitous slopes to the warmer coastal plains. The climate, she thought, must once have been warmer than now with much more vegetation. Before the channel had worn through the barrier of surrounding hills, water would have been trapped in the great shallow basin, transforming it into a lake or swamp, according to the season of the year.[22] What Dorothea wanted next to discover was whether the little hippo had occupied any other region of Crete.

By the time Dorothea reached Kritsa the cloud had given way to hot sunshine and she continued down the mountains to resume her journey eastwards round the Gulf of Mirabello. As she rode through the little fishing village of Pachiammo near where the Americans were excavating, she came across Richard Seager, the young, good-looking archaeologist she had met on the boat to Herakleion. The Americans had a house here and they asked her to stay. 'Such a treat,' Dorothea wrote in her diary, 'to be with kind English speaking people – Their party consists of Miss Boyd, Miss Hall and Mr Seager – all charming to me.'[23]

Dorothea's weeks of travel through remote mountains with only the faithful Nikola and workmen for company had clearly taken their toll. Her Greek was passable but not fluent, and it was a huge relief not only to be able to converse in her own language but with people near her own age. They had in common the delights and disappointments of digging for concealed riches, but how much the archaeologists and the palaeontologist actually understood of each other's work is another matter. Harriet Boyd, who led the party, was thirty-three, a graduate and lecturer in Greek and Greek archaeology at Smith College in Massachusetts.* She was described by the *New York Times* as 'a little woman, dark-haired and slim. She has a bright, attractive, pleasant face, and winning manner. The masculine work she has been doing in the world has had the effect of giving her a business-like briskness in conversation, but it does not destroy the charms of her manner, but rather adds piquancy to it.'[24] According to Richard Seager, however, that briskness could often turn into a quick temper and sharp tongue.[25] This was her third season excavating at Gournia, the Minoan town she had discovered on a hillside overlooking the Gulf of Mirabello. Employing a hundred workmen, Harriet had unearthed thousands of tools, pottery pieces and figurines, all of which she sent to the Museum at Herakleion, carefully wrapped in seaweed, straw, and newspaper.[26]

Harriet Boyd had two archaeology students working for her, Edith Hall and Richard Seager. Seager, who had spent just a year at Harvard, had worked with Harriet at Gournia in 1903. Now she put him in charge of the excavations at the nearby Minoan settlement of Vasiliki where he was discovering unique and beautiful pottery. Edith Hall, who held a fellowship from Bryn Mawr College in Pennsylvania, had met Seager some months earlier at the American School of Classical Studies in Athens. Edith's first impression of Seager, as she wrote to her parents, was that he was 'good-looking and carefully dressed, but maybe nice'.[27] In March 1904, before resuming excavations, Harriet and Edith had ridden around western Crete together, hunting for new sites and carrying with them their bedding, a hamper of food and, indispensable for American women abroad, a rubber tub.[28]

One of the delights of Dorothea's time in Crete is the joy of recognition. I already knew something of the major Minoan sites that Dorothea visited, but now I discovered that, of the many octopus-decorated pots that Harriet found, one was the splendid stirrup cup (prominently displayed in the Archaeological Museum in Herakleion) that I had been fascinated by as a child. She had found it in eighty-six well-preserved pieces in

* Just seventeen days before Dorothea's arrival, on 10 May 1904, Charles Henry Hawes, the English anthropologist who was visiting Crete to measure heads, met Harriet Boyd for the first time. They married in 1906.

Gournia.[29] The astonishing richness of this and neighbouring sites, and Harriet's considerable achievement in discovering them, are evident from Dorothea's first day with the Americans. On a hillside near Pachiammo, Dorothea stood and watched as two sarcophagi were discovered, one decorated with 'conventional octopuses', the other with cows suckling calves. In recesses in the soft limestone of a nearby ledge, Dorothea also saw thirty almost perfect vases emerging from the crumbling rock.[30] From here she went on to the excavations at Gournia where she witnessed even more pottery being unearthed.

To Dorothea's untrained eye, these just seemed to be 'a lot of pots'.[31] But how could she have understood their significance any more than archaeologists could understand her passion for digging up bones? Dorothea's expertise lay in the history of the earth, not of civilizations. As for how some archaeologists viewed her work, that is probably reflected in a letter that Harriet Boyd received from the archaeologist Sir John Linton Myres. Writing to Boyd in August 1904 from the meeting in Cambridge of the British Association for the Advancement of Science, he remarks with evident disdain, 'The rumour here is that Miss Bate has dug up the original Cacotherium!' which loosely translates as Cretan monster.[32] Classical archaeologists of the time tended to view with disgust animal bones excavated from their sites and would throw them away, little realizing that they were destroying significant evidence in interpreting the lives of ancient peoples.*

In the two days she spent with them, Dorothea made a notable impression on the Americans. She searched for bone caves with Richard Seager, explaining to him the type of rock that might be bone-bearing. She swam in the early morning with Edith Hall, and all four of them, the three academic women and Seager, walked in moonlight on the long beach, the surf breaking at their feet. Edith described Dorothea to her parents, providing a valuable and all too rare image of her: 'The Miss Bates [*sic*] I spoke of is one of the jolliest, most capable, and fearless girls I ever knew. She is interested in palaeontology, and is digging caves to get bones . . . She goes about by herself with one native to guide and help manage her luggage. And the beauty of it is that she is entirely unconscious

* Not so Palaeolithic archaeologists and prehistorians, who already understood the importance of animal bones found with Stone Age cultures.

and girlish, she dresses well, and is altogether a most companionable person.'[33]*

On 29 May, Dorothea resumed her journey eastwards, heading for the Toplou monastery at the foot of the Sideros peninsula. It is a massive fort-like building on a bare stony plateau high in the hills, with a windmill by the front gates. Centuries of fighting in Crete had left the building in a poor state (it has now been lavishly restored); Spratt thought the monastery had 'nothing to recommend it to the attention of the traveller'.[34] However, the monks treated Dorothea well. 'Am in room,' she wrote, 'which was built in anticipation of a visit from Prince George.'[35]

In the morning she explored a nearby gorge, but there was no sign of a bone deposit, nor even of any caves. Money again was becoming an issue and she pressed on, riding to Palaikastro, the Minoan town on a plain near the sea in the far east of the island where the British School were excavating. Robert Bosanquet had come out to Crete earlier in May and working with him were Richard McGillivray Dawkins, who was in charge of the excavations in Bosanquet's absence; the Canadian Charles Currelly, and C. H. Heaton Comyn, an architect and architectural student of the British School who was working on the plans of Palaikastro. Dorothea arrived at a marvellous moment, just as the archaeologists were celebrating one of the finest discoveries made at Palaikastro, two exquisitely carved ivory statuettes of boys.[36] 'All most kind to me,' she wrote, 'and I stayed to supper.'[37] The talk of course turned to Dorothea's work, and as she was describing the type of deposit she was looking for, one of the Palaikastro foremen, a man called Michael, 'said he knew of some bones along the coast' at the Bay of Karoumes. With great generosity, Robert Bosanquet immediately proposed that Michael show Dorothea the site and work with her there. Dorothea does not mention his surname, but he was probably Michael Katsarakes, whom Bosanquet compares with another foreman thus: 'Michael is a younger man, tall, good-looking and better trained in the art of digging but has less force of character and less

* I was fascinated at the idea of Dorothea dressing well in the middle of Crete, given the primitive nature of her accommodation. Edith's letter also serves to dispose of another erroneous story, which has had some currency. According to this one, Dorothea rode around Crete dressed as a man. There is not even a hint in her diary of anything that could even remotely account for this, nor in any known letters. It is a splendidly romantic notion but, like the riot at a monastery, without foundation.

go. Both have perfect control of the men and of their own tempers.'[38] With such an agreeable workman and guide promised to her for the morning, Dorothea left Palaikastro after supper, enduring a 'very cold wind riding back to Toplou'.[39]

In the morning in the muddy aftermath of a torrential downpour, she returned to Palaikastro and, with Nikola and Michael, set off for the Bay of Karoumes. At the village of Chochlakes, they left the mules and baggage and entered a deep, narrow gorge leading to the sea. For more than a mile it twists and turns through a range of hills. Huge boulders frequently block the way, thistles tear at legs and clothes, and loose rocks threaten sprained ankles or worse. It is oppressively still, the silence only broken by the sudden sound of a falling stone or the lone cry of a bird. The temperature beneath the great, looming cliffs easily reaches over 100° Fahrenheit at midday.

It took more than an hour to reach the sea. They turned south, clambering along the shoreline at the foot of limestone cliffs and, as Michael had said, there in the rocks were faunal remains. It had once been an enormous deposit, destroyed, as at Koutri, by the sea. Dorothea found patches of bone breccia over a distance of some one hundred and fifty yards, and in a cave. It seemed similar to the deposit at Koutri: deer, goat-like bones and a species of rodent. And then, in what she thought was the site of a separate cave, now long ago destroyed, the bones she found were those of the hippo. One of her questions at least was answered; the little hippo had lived on the coast as well as high up in Katharo. Just a few minutes later she found a small hole to the southeast of this cave where the upper surface of the rock had broken away, leaving exposed the bone breccia underneath. 'I stooped down to examine it,' she wrote triumphantly, 'and inside saw part of a jaw bone and tooth of an elephant! – Could not get it out today as had no tools.'[40]

She sent a message to Palaikastro to ask for the loan of some tools and was back at the deposit early in the morning, impatient to begin excavating. The tools, specially sharpened, arrived in the afternoon accompanied by a letter from Bosanquet and another workman, who, Bosanquet wrote to Dorothea, 'you will find strong and intelligent. We all wish you good luck and wish we could pay you a visit, but there is rather a press of work – We have plenty of stores; let us know what we can supply.' He also sent Dorothea some gunpowder, which they had found in the village and, with even greater generosity, he told her that the Cretan excavation fund would pay both men's wages for ten days.[41]

The elephant bones, to her great surprise, were 'Very large', she wrote; 'should think it is a full sized E[lephant]'.[42] She knew that in 1893 remains of a large fossil elephant had been discovered by the Italian explorer, Vittorio Simonelli, but that had been near Rethymnon on the north coast of Crete.* The financial help from Bosanquet should have enabled Dorothea to work out the deposit as fully as possible, but she lacked basic equipment. She could not properly explore another cave that Michael found as she was short of candles, and she also needed geologists' tools, as the ones loaned to her by Bosanquet could not do the job. For a morning she worked away to 'get out the jaw bone but find it impossible to do so (intact) with the tools I have'. There was another even greater problem. In this remote corner of Crete there was nowhere to buy food. 'Not expecting to stay,' she wrote in her diary, 'am short of provisions and have come down to bad black barley bread – seem to have come to an end of eggs in village and Nikola says he cannot work having nothing to eat!! so regretfully decided to leave it and return later if possible.'[43] For some unexplained reason, possibly her pride, she did not take up Bosanquet's offer of stores. She covered the elephant with earth and stones and made her way back with Nikola through that oppressive gorge. Wearily she rode across the hot bare limestone hills towards Pachiammo.

Dorothea arrived at the Americans' house in the late afternoon to find that Seager was ill with a fever. Harriet Boyd and Edith Hall were (in some measure) looking after him. He was, Edith told her parents, 'staying in bed where we pay him occasional visits, and mix up his malted milk for him'.[44] But there was real concern. Seager had heart disease and in 1902 had been given a very precise eleven years to live by a doctor in Germany, although by 1904 his prognosis was rather more hopeful. Nonetheless, when he went swimming, Edith observed that one leg was purple with varicose veins.[45] After her initial cool reaction to Seager, Edith had by then decided that he was 'one of the cleverest and most considerate people I ever knew'.[46] Witty and a good friend, he protected her on more than one occasion from Harriet Boyd's acerbic tongue. Edith refers to his

* Dorothea's own copy of Simonelli's *Candia* is now in the NHM's Earth Sciences Library. A faint pencil line highlights just two paragraphs, those which mention on p. 171 his most interesting discovery of 'magnifici avanzi di elefanti' – magnificent remains of elephants.

'handsome high forehead' and writes to her sister that he is 'a love'. She was evidently very fond of him, and probably more than that; Seager was young, good-looking and charming and Dorothea may well have been captivated too.

Dorothea had intended to travel straight to Herakleion when she left Pachiammo but, as Edith Hall wrote to her parents, she did not leave alone:

> Richard Seager has gone this morning. Luckily enough Miss Bate, the English girl who digs bones, came last night and can keep an eye on him if he gets ill on the journey to Candia [Herakleion]. I am sorry to have him go, for he is amusing and always talkative and cheerful and has fielded over many a trying time with Miss Boyd. He is going as straight as he can to Nauheim [in Germany] where he takes a heart cure and where his mother is waiting for him.[47]

Dorothea's diary reveals rather more – that they spent the greater part of the next ten days together. She records only where they went; of her feelings, once again, there is nothing. But this was of course a work diary and she could hardly write anything she would not wish to be seen. She sent Nikola up to Lassethe to collect the Katharo hippo bones and arranged to meet him at Herakleion. She herself left for Aghios Nikolaos after breakfast, noting, 'Mr S. came on just after . . . We stayed at little hotel.' After supper, Dorothea and Seager went down 'to see a small lake near the harbour which has an inlet to the sea – It is most extraordinarily phosphorescent lot of fish in it which dart about looking as if made of fire.'[48]* At 5 a.m. the next day, Dorothea and Seager caught the steamer *Enosis* bound for Herakleion, which they reached that evening. At their hotel, they met an American, a Miss Stone, and a (nameless) Russian countess. In the morning they all visited the Museum and after lunch Dorothea and Seager went off on their own to the bazaar. She looked at some embroideries to buy but 'nothing very nice'.[49]

They spent the next day together again, riding out to view the excavations at Knossos in the afternoon. Evans was enjoying a wonderful season. Fragments of wall paintings had been found, some illustrating the famous bullring and crowds of spectators, others a pillar shrine with

* This is Lake Voulismeni, some sixty-five metres deep.

double axes stuck into the columns. In the northeast of the site *pithoi*, great storage jars, had been restored so that they now reached their original height of over two metres. He had also discovered the remains of what he thought must be the royal arsenal, and a hundred tombs containing jewellery and gems, 'magnificently' painted pottery, bronze vessels and swords.[50]

All this was destined for the island's Archaeology Museum; watching Evans at work may well have finally persuaded Dorothea to play by the rules. On 10 June, accompanied by Seager who was well known to the authorities, she went to the Museum to ask for a permit to dig. This time Dr Hazzidakis (the Ephor of Antiquities) seems to have made every effort to be helpful and straightaway took her to see a bone deposit ('a marine deposit', she noted), which had been found about five miles east of Herakleion.[51] Dorothea was well aware of how to put her considerable charm to good use and it was now apparent to her that being granted a permit was much less trouble than attempting to avoid the authorities.

Seager was due to leave Herakleion in five days but he was unwell again. So was Dorothea, who had been suffering from a 'bad head' for some time with a touch of fever every afternoon.[52] With his weak heart, Seager had to rest, but Dorothea just kept going. On 12 June she went for a sail with the Danish artist Halvor Bagge, who was working at Knossos. 'Got fairly wet,' she wrote, 'as good breeze on.'[53] That night she 'dined at Mr Evans', who lived in appropriate style in Herakleion in a house which, according to Charles Currelly, was more of a palace, built around three sides of a courtyard.[54]* On 15 June, Richard Seager, still unwell – as, indeed, was Dorothea – was due to leave Herakleion. That evening she went with him to the harbour for the boat to Athens and 'saw him on board with my big coat on!!'[55] What her feelings were at his departure we cannot know, and it is unwise to read too much into two exclamation marks: they could refer as much to wearing a big coat in the Mediterranean summer because of her fever as to lingering romantic notions. Even if Dorothea did feel some attachment (which I suspect), she was too focused to allow herself to be more than pleasantly diverted; her work had to come first. She had still to excavate the elephant deposit at Karoumes Bay and this time she had permission.

* Arthur Evans did not build the famous Villa Ariadne at Knossos until 1906.

Nikola had returned from collecting the hippo bones from Lassethe that morning 'ill and dejected', and she had doubts as to whether either of them would be well enough to travel. Still feverish in the morning, they nonetheless set off, but it was a 'horrible ride', she wrote; 'N and mules slept practically the whole way.' They stopped to rest for most of the afternoon, 'and then crept on to Melato which reached after 8pm'. She felt wretched and her temperature at bedtime was '103 and 4/5'.[56] Far from calling a doctor, after a dreadful night she was up before dawn and went out to see a bone deposit she had been told of several weeks previously. She was too ill to do much digging, but observed that there were several fragments of hippo molars and canines and excavated 'a young imperfect lower jaw'. Two hours later she left Melato, driving the mules as Nikola was still too ill. She did not stop till she arrived at Pachiammo in the late afternoon, only to find that both Harriet and Edith were away. 'Waited there an hour,' she wrote, 'and then crawled to Kavousi.'[57] Here she stayed the night, too unwell even to send a message to the Americans. At dawn she dragged herself out of bed and, with Nikola still unwell, again drove the mules. She had hoped to reach Toplou but, arriving at Sitia in the early afternoon, found she could go no farther. Even Dorothea's indomitable spirit could take only so much.

Word that she was at Sitia reached the British School archaeologists and both Bosanquet and Currelly came to see her. For Bosanquet, the visit was to say goodbye, as he wrote to his wife Ellen: 'Miss Bate, courageous as ever, has turned up again, to have another go at the Caroumais [Karoumes] Cave near P.K. I'm sorry we are going and can't help her.'[58] What he did arrange was for a man to help her dig. Disregarding her own fever and with Nikola 'still very sorry for himself',[59] they rode off the next day for Chochlakes where they stayed. Nikola did not appear at all in the morning and she went off with just Bosanquet's man and worked 'away down at Karoumis blasting etc but got no specimens'. Dorothea's temperature that afternoon, after working in the hot sun and being soaked every so often by a wave, was '102 and 3/5 F cannot get thermom. below 102'.[60] The next day her perseverance paid off and they finally excavated the elephant jaw and tooth, although it was still encased in a big block of matrix. Nikola, she noted, was 'still invisible'.[61] She battled on for a few more days, feverish and utterly miserable, while every fragment was a struggle to excavate. 'Beastly place this is to work – so hard and can't find another sign of a tooth working all day and hardly any result to show for it don't believe there is very much more of Effie [as she fondly calls this huge beast] probably been washed away when the sea was over it.'[62] On 25 June, with a supreme effort, she managed to extricate the thigh bone of the elephant, but it 'is *very* fragmentary worse luck – found one other limb bone part of which I got out, in pieces, but can see nothing else . . . so perhaps as well I am stopping work cannot afford to dig largely on the off chance of there being more of Effie'.[63]

Her money was exhausted. She sent Nikola (at last recovered) to Herakleion with the elephant remains, while she rode to Sitia to follow him there by steamer. She was sitting quietly in the hotel after dinner when she received an unexpected and most unwelcome visitor. 'After N[ikola] and bones had safely gone,' she wrote, 'in evening a "Govt Spy" sent by Mr Hadzidakis and on his way to see what I was getting at Kharoumes arrived – We drank coffee which I provided and tried to make polite conversation without giving any interesting information!'[64] Perhaps she talked about Bosanquet's excavations or the brilliant pottery finds she had seen Richard Seager make at Vasiliki. Most likely she did not talk about Crete at all; how much safer to chatter, in her passable Greek, about the Welsh or Scottish countryside, anything except a subject that could bring the conversation around to elephant or hippo bones. According to Edith Hall, Dorothea was an 'ideal conversationalist'.[65] Never had those skills been so needed.

With no further interference, she boarded the steamer the next night. It was so hot and 'disgustingly airless' that she slept on deck.[66] At 4 a.m. they reached Aghios Nikolaos in the Gulf of Mirabello, where Edith Hall came on board. Edith had planned to leave Pachiammo the previous week but, as she wrote to her father, Harriet needed help packing the finds, 'so I waited for the next boat . . . Miss Bate, the English girl who digs bones is to be on the boat . . . and will probably stay in Candia [Herakleion] as long as I do. It will be very nice to have company at the hotel.'[67] Like two girls at a sleep-over, Edith put up her camp bed and Dorothea spread her blanket on the deck; 'and there we lay all day,' wrote Edith to her mother and sister, 'talking and sleeping and eating the most delicious cherries I ever tasted'.[68]

In Herakleion, Edith was revived by 'clean clothes and the first hot bath I have had since April',[69] while Dorothea received a letter that revitalized her even more. Three weeks earlier, on 9 June, Henry Woodward had read to the Royal Society her substantial and detailed report of the Cypriot pygmy elephant which she had completed shortly before leaving for Crete. Her preliminary paper on these finds he had read to the Royal Society the previous year. In the morning with great energy, Dorothea repacked her belongings and sent Nikola off to Chania 'with several boxes of stuff'.[70]

Free for a day or so to relax in Herakleion, it is easy to imagine Dorothea and Edith, both now in excellent form, bumping into Charles Currelly and Richard Dawkins (who had been working at Palaikastro) and agreeing to what, had they been at their respective homes in Scotland and America, would have been a most unusual scheme. The two men were about to climb Mount Ida, the highest mountain in Crete. Robert Bosanquet had made this famous trek to watch the sunrise from the summit the previous year. As Edith told her parents (after the event): 'They asked Miss Bate and me to go, and we, though we had not planned it before, seized time by the forelock and said "yes".'[71] It was, to say the least, an interesting foursome. Edith, a passionate archaeologist, in 1911 was appointed Curator of Grecian Antiquities at the University Museum of Pennsylvania University and in 1915, when she finally recognized that nothing would ever come of her friendship with Richard Seager, she married Joseph Dohan, continuing to work after her marriage. The Canadian Charles Currelly, who was twenty-eight, was 'silly and fat', according to Edith, and 'most amusing', according to Dorothea, who was much nearer the mark. Currelly had excavated with that doyen of British archaeology, Professor Flinders Petrie. In 1914 he founded and became Director of the Royal Ontario Museum of Archaeology. The fourth member of the group, Richard McGillivray Dawkins, who was thirty-three, became Director of the British School at Athens in 1906. After the First World War (when he was an intelligence officer in Crete), he was appointed to a chair in Byzantine and Modern Greek at Oxford University and became a Fellow of Exeter College. He never married and had something of a reputation as a misogynist, and yet there he was, on 1 July 1904, about to climb the highest mountain in Crete with two young women.[72]*

* Dawkins was something of a late starter. After school he had become an electrical engineer, which he hated, and in lonely evenings after work read Greek and Latin classics, and taught himself Sanskrit, Italian, German and some Icelandic, Irish, and Finnish. So successful was his

At 6 a.m. they set off, passing through some of the most beautiful scenery in Crete. Dorothea noted every bird while Edith, in the hazy morning heat, 'was so sleepy I nearly fell off my horse'.[73] They reached the Nidha plateau where they were to make camp shortly after seven in the evening. It reminded Dorothea of Lassethe on a very small scale. To the west of the plain was a little chapel where the women were to sleep; 'the others,' Dorothea noted, 'slept out.' They lit a fire and dined off oxtail soup and sausages, bread and tomatoes. 'Rather absurd,' remarked Dorothea, 'how many necessaries such as spoons etc we have come without.'[74] Then domestic nuisances became irrelevant as the snow-tipped mountains became shadows and the huge black sky filled with stars.

Dorothea was first up and went 'bug-hunting' before they all set off for the Kamares cave, 1,000 feet above the plateau and a three-hour trudge up a rough track. It had been a shelter during Neolithic times and used by the Minoans for religious rites. The eponymous pottery had first been found here, wonderfully painted with abstract swirls, delicate and detailed flowers, or creatures in white and subtle shades of deepest red to orange, against a background of black or terracotta. The floor of the cave was covered in sherds. 'I got one or two nice pieces of a white daisy on a black ground,' wrote Dorothea. 'Mr Dawkins got an agrimi [wild goat] head on a piece of pottery.'[75] They arrived back at their camp in time for tea, which was, Edith told her parents, 'a late tea which ran imperceptibly into dinner. Altogether counting the time we waited for things to cook, we were busy eating, as Mr Dawkins said, for four hours. We ended up with a canned plum pudding.'[76] After which, wrote Dorothea, 'Others went to bed about 8 but I had bugs to pin.'[77]

Dorothea, after her bug-pinning, had little more than an hour's sleep before they began the climb to the summit by 'splendid moonlight'.[78] Guiding them were two shepherds, with a little mule to carry rugs and food. Edith did not feel at her best, 'and I must say,' she told her family, 'that when I started off for Ida with only one raw egg to hold me up, I was a little doubtful of the wisdom of proceeding . . . But Mr Dawkins gave me his hand for many of the highest steps, and after an hour I was

---

self-education that aged twenty-six he went up to Cambridge to read classics. J. H. Droop dedicated his book *Archaeological Excavation* to Dawkins who had been Director at the BSA when Droop was a student: 'all-sagacious in our art, Breeder in me of what poor skill I boast', Cambridge: CUP, 1915.

thoroughly enjoying myself.'[79] They reached the first patches of snow about 2 a.m. and two weary hours later arrived at the summit. It was bitterly cold with a gale-force wind and just a faint tinge of red in the sky. 'Lay down,' Dorothea wrote, 'but was too frozen to sleep.' Less than an hour later the sun appeared while there were 'flocks of the most beautiful fleecy white clouds rushing about at a great rate round the hills below us'.[80] She took out her camera to photograph the clouds, but it was so cold, Edith noted, 'she couldn't turn the screws, and we were obliged to call on Mr Dawkins who had kept his hands covered'.[81] The summit was at a height of 8,000 feet, with spectacular views over the whole island and, hazily merging with the sky, the steel-shiny sea of early morning. While the others napped or admired the landscape below, Dorothea continued her bug-hunting, although in the high wind she found only ladybirds and one or two wasps and flies. She noted white- and purple-shaded crocuses along the edge of the melting snow, 'a dear little pink flower which the shepherds call stone almond', and a small yellow buttercup.

The descent was difficult, sliding on loose shale, and even more unpleasant was the sun, burning remorselessly in the clear mountain air. 'It got pretty hot coming down,' Dorothea wrote, 'and what with sun and wind my face and hands were a sight, the latter also swollen and all so painful.' They reached Nidha in time for lunch, after which 'the others slept and I went after a few more bugs'.[82] Her energy seems boundless; while the others collapsed, she continued to collect and observe everything around her. At five o'clock the others woke up and, with no great enthusiasm from any of them, they climbed up to the Iddaean cave, celebrated as a place of worship by the Minoans and in Greek mythology as the place where Zeus spent his childhood. In spite of the wealth of artefacts that 3,000 years of worship had left there and which Dr Halbherr and the Italian School had recently excavated, exhaustion dulled their curiosity. They returned to the chapel and 'spent the evening talking round the campfire', wrote Edith to her family. 'Mr Dawkins and Miss Bate are two of the best read and keenest people I ever heard talk.'[83] Edith went even further than that in her admiration of Dorothea in a letter to her sister Anne, her words remarkably similar to those used to me by scientists who knew Dorothea in her later years: 'I still sing the praises of Miss Bate. Such quickness of decision, such a masculine grip on her work, such politeness and kindness to everybody I have never seen . . . Entirely unconsciously she interests herself in everybody she talks to, and draws them on and on making them feel as if they were talking to their own mother. She has "bucked me up" a lot – to use English slang.'[84] This is high praise indeed; Edith seems to have been very close to her own mother, although it is an interesting comparison; Edith was actually a few months older than Dorothea.

After a slow, difficult descent, which included struggling up steep ridges and then leading their mules down the other side, the four reached Herakleion the next night. For all of them it was almost the end of the archaeological season and time to pack and go home.

Dorothea spent two days 'busy packing up Effie' and her other finds. She then went to see Dr Hazzidakis at the Museum to establish just what she could take with her. He would not give her an answer. Instead he told her that, with his entire committee, he would come to Chania and decide there which of her fossils must stay and which the Natural History Museum could have. As for when this visitation would take place, that had not yet been decided. On the evening of 6 July, Dorothea and Edith saw Richard Dawkins and Charles Currelly off on the steamer to Athens. Edith was remaining in Herakleion for another week, while Dorothea returned to Chania at dawn by boat.*

As soon as she arrived at Chania, Dorothea lost no time in calling on the Howards and re-establishing, with the charm the Americans had found so attractive, their relationship that had been fractured after the 'fiasco'. The Consul, she was delighted to hear, promised to help her in any negotiations with Dr Hazzidakis. For nine days Dorothea waited for the committee, spending the time bug-hunting and attending social functions with the Howards. She lunched with Nikola in his village of Malaxa, and in Chania she went up to the 'Dervishes place just outside the town where they had a service on with dancing'. In Cyprus she had visited the dervishes' *tekke*, but had arrived too late to see the dancing. Now she watched, fascinated, as rhythmically they whirled themselves to

* Dorothea's boat trip to Chania is notable for just one wonderfully human admission. At Rethymnon, she wrote in her diary on 7 July, 'I went ashore with an old Turkish woman who took me with her to a friends house where I smoked cigs. [*sic*] and ate various messes [*mezes*].' A cigarette-smoking Dorothea is one I had not quite imagined, but I suspect this episode had more to do with novelty than any indication of a lifelong (and still unladylike) habit.

a higher spiritual plain, one palm upraised, the other pointing down. Nine men including the priest and two little boys danced. 'In the gallery,' she wrote, 'men played weird long reed pipes and sort of tomtoms, also chanted nasally – After allowed me to photo[graph] them, something unusual.'[85] The strain of waiting for Hazzidakis was unbearable. She was determined to have one last attempt at finding the bone deposit on the Akrotiri mentioned by Richard Pococke and which she had failed to find in March, yet she dared not leave Chania until Hazzidakis had made his fateful decision.

At last, on the evening of 16 July, Dr Hazzidakis arrived, and he was alone. 'The rest of the committee to examine my things,' she wrote, 'were afraid of the rough weather,' and Hazzidakis refused to examine her fossils without the committee.[86] They finally arrived on 19 July and their verdict, she notes bitterly, could not have been worse: 'They have not been modest – and taken all the best looking things besides *all* the hippo limb bones, including many duplicates, and nearly all the isolated teeth of same.'[87] It was devastating. The whole purpose of her trip had been to 'throw light on the subject of the origin and development of pigmy forms' on the Mediterranean islands and the committee had, by their decision, deprived her of a whole mass of evidence. With even greater determination to find Pococke's bone deposit on the Akrotiri, Dorothea prepared for one last expedition. She had found dwarf hippos and a full-sized elephant, and she was convinced that dwarf elephants must also have lived on the island.

Pococke's account was impossibly vague. It began: 'We went two miles to the west among the mountains and I saw a ruined village called St George, and a church in a grotto, under which there is another grotto, where I was informed there were petrifyed bones of a larger size than ordinary, and I actually found some bones in the softer part of the rock, but not petrified.'[88] All her enquiries about the ruined village of St George when she had first looked for it in March had met with ignorance or misdirection. This time was different. After one or two false starts, she found a man in the west of the Akrotiri who not only knew of the village, but also exactly where the bone cave was, well concealed among the hills at the end of the peninsula.[89] The cave did contain fossil remains, but they were similar to those at Koutri; of an elephant there was no sign at all. Wearily she returned to Chania and had just reached Khalepa on the outskirts when, to her astonishment, a man who she refers to as 'Mr

Zitelli' showed her a fossilized tooth.[90] She bought it from him at once. It was from a small elephant, but apart from telling her it came from the Akrotiri, 'he could, or would', she wrote, 'give no information'.[91]

That evening, with the departure of the steamer for Athens imminent, she packed into two boxes those fossils she had been allowed to keep and wrote to the Keeper of Geology, Dr Arthur Smith Woodward, her resentment very evident: 'These two boxes contain all that the Museum were kind enough to allow me to keep! I had a very good lot of limb bones (many duplicates) of the hippo. I think they kept all except one damaged piece besides retaining pretty well everything else they thought worth having.' But she had sent off rather more than that, as she revealed to Dr Smith Woodward: 'I believe I got another case off today addressed Stahlschmidt and Co and if there is a bill of lading I will send it to them.'[92] The anonymity of the shipping agents had worked successfully once and she saw no reason not to use them again. She felt she had little to lose. But in her anger at the committee's work she failed to realize the significance of what they had done: they had actually given her permission to keep two boxes of material, even if they were poor specimens, and export them from Crete. Dorothea told Smith Woodward she would be leaving Crete in a few days, but not until she had made one last effort to find the elusive extinct pygmy elephant.

At 4 a.m. on 22 July she rode right across the Akrotiri to meet a man 'who said he knew of some bones'. Leaving the ponies near the Cave of the Bear, they walked over hills and down a stream bed to the sea. A short distance along the cliffs she found a few scattered fragments of bones in the rocks, a 'sort of "last faint memory" of a bone cave', but that was all. Unsure which way to go next, she met a shepherd who said he knew of another bone deposit in the opposite direction. It was, she wrote, 'beastly hot and rough walking', but in the blazing sunshine and with the heat of the stones coming through her boots, she struggled on. And here in the sea-battered cliffs of the Akrotiri she was finally 'well rewarded for this at last was the place I have been looking for'. It was the very place where the elephant tooth she had been given at Khalepa had come from and, incredibly, in the rock she 'saw the imprint from where it had been broken off'. It had taken her four and a half months to discover this place. There was no trace left of the cave walls and the deposit itself had largely disappeared, although she thought it must once have been very large. Embedded in 'frightfully hard' matrix there were about half a dozen elephant teeth showing and some pieces of tusks. She managed to extract one fragment, but needed her good tools to extract more.[93]

The following morning, this time with the right tools, she managed to extract a section of the tusk and several portions of molars, but the matrix was impossibly hard. 'Did not get any very good specs.,' she wrote, 'and fear they will be almost impossible to clean.' She returned to Chania at 8 p.m., 'dog tired'.[94] This deposit, she felt, typified her whole Cretan experience: the promise of so much but in the end relatively few specimens for enormous endeavour. Her exhaustion made her undervalue her achievement. She had, in an island with an overwhelming number of uncharted caves, found the fossil remains of Pleistocene deer, rodents, the pygmy hippos and elephants, and a full-sized elephant, which may have been the ancestor of the dwarf species. Her finds were, in the words of a later palaeontologist, 'a treasure of fossils'.[95] In both Crete and Cyprus, her solitary, pioneering work had revealed the main extinct species of the islands.

She made sure at least that Dr Hazzidakis did not hear of the elephant and poured out her frustration at the Cretan antiquity laws in a letter to Edith Hall.

> The Museum people (Mr Hatzidakis being one of them) were just beastly about my bones – took everything that they thought looked any good – I had a large collection of very good specs. from [Katharo] and they took practically the lot – there were 14 perfect specimens of one bone and I wasn't given one! – If I hadn't taken steps to help myself I should certainly have made a complaint through Mr Howard who was quite willing to do this – but under the circumstances [that is, smuggling] I thought it wiser to leave things alone – I nearly killed myself with work the last week or ten days I was at Chania and besides had such a hot room at the hotel, couldn't go in for a minute without beginning to drip.'[96]

In addition to this, she received a letter from home. Cryptically, in faint letters she has written in her diary, 'Heard from L about Mr L.'[97] Her elder sister Leila, now twenty-seven, had become engaged to Henry

Tansley Luddington, a wealthy landowner and farmer with an estate in Cambridgeshire. Just as Dorothea was preparing to return home after her second pioneering expedition, once again Leila moved into centre stage. This time, however, the consequences for Dorothea would be appalling. While she may have been happy for Leila, the engagement meant that her parents would assume that their unmarried daughter would end her travels and devote her time to them. Henry and Elizabeth appear to have had no understanding whatsoever that Dorothea, at the age of twenty-five, was an explorer and palaeontologist of note.

Dorothea booked a passage on a ship leaving Chania on 25 July for Alexandria, where she hoped to find a ship bound for England. She had been given three live baby rabbits from the island of Dhia, which lies just off Herakleion, and these she put into a portable hutch. 'All Leightons,' she wrote, referring to a family she had met in Chania, 'and Nikola came down to see me off and help with Effie.'[98] She was smuggling to the end. At one o'clock in a rough and rolling sea, her ship left Chania. 'Stayed up all night in a deck chair,' she noted unhappily and suffered all the next day too. At Alexandria* she booked herself into a modest little hotel and discovered at Cooks the following morning that a boat was due to sail almost at once for Marseilles, 'but must go to Cairo,' she wrote, 'so will have to wait week or ten days!'[99] Leaving Egypt without visiting the Pyramids was inconceivable, particularly as she had no idea when she might be able to travel again.

In Cairo she stayed in 'an awfully nice flat' owned by the YWCA and for the next week became a tourist. She went round Gezirah island by moonlight and tasted her first mango. She visited museums during the day and at night 'went by train to see the Gizah Pyramids and Sphinx – went half way up Cheop Pyramid . . . no time to go all the way – awfully easy, can't understand people wanting 2 or 3 men to help them. Nice moonlight and no-one else there.'[100] Most people, however, had not spent the last four months clambering up cliffs, climbing Mount Ida and blasting bone out of rock. She explored old Cairo and visited mosques and churches. On 3 August she 'Got up at 3am' and by moonlight with a party of 'missionaries!' rode on a donkey out into the desert. 'It was wonderfully

* The steamer reached Alexandria in the evening after Customs had shut but, Dorothea noted in her diary, 'being English was allowed to bring a few things through'.

nice out there as it happened to be cloudy in the early morning. Saw the sun rise in the desert.'[101]

She returned to Alexandria on 5 August. 'Went to Hotel for luggage,' she noted, 'and found that two of my three rabbits had died (or been eaten).'[102] Her ship to Marseilles was comfortable and the company congenial; it was, she wrote, a 'very good passage'. At Marseilles she caught the train for Paris. 'Six of us, including two midshipmen from HMS Irressistible [*sic*] . . . had dinner at Lyon – At Dijon which we reached at midnight we got out and stayed too long only catching train by jumping on to the footboard on wrong side as it was moving off!!'[103]

Dorothea arrived in England on 10 August, just as the Cretan government was granting permission to Harriet Boyd under a new law to export objects 'without any scientific value or interest whatever for Cretan Museums'.[104] Nine days later, Harriet left Crete with her permitted antiquities from Gournia and Vasiliki. On her arrival in the United States, she told the *New York Times* that these were 'the first antiquities that have ever left Crete for any foreign country with permission from the Cretan government'.[105] It does appear, though, from the evidence of Dorothea's diary entry for 20 July and her letter to Dr Arthur Smith Woodward the following day,[106] that this honour was actually hers, although, of course, her antiquities were fossil bones, not works of art from the Minoan civilization. That anything was (legally) allowed to be exported at all however, was thanks to the tireless efforts of Sir Arthur Evans, and in fact his antiquities agreed for export were just a few days behind Harriet's. Almost from the moment he had begun to excavate at Knossos in 1900, Evans (who was not above a little illegal exporting of his own) had petitioned the Cretan government to amend the archaeological laws to permit the export of duplicate finds. He received the support of Hazzidakis and other senior officials, and the new Act giving assent to this was published in June 1903[107] with Dorothea seemingly its first beneficiary in July the following year.

Back in London, Dorothea spent just two days at the Museum and then went home to Wyseby. In marked contrast to the weeks she had spent poring over her finds on her return from Cyprus, she did not reappear at the Museum until the end of the year. At Wyseby she began writing for the *Geological Magazine* an account of her travels, based on her diary entries rather than a palaeontological description of what she found.[108] She did not report on her fossil material from Crete (depleted as it was) for another two and a half years. For most of the remaining months of 1904, there was, for the entire Bate family, whether they wished it or not, just one preoccupation: the marriage of Leila to Henry Tansley Luddington.

CHAPTER 8

Exile in Ecclefechan

On 5 September 1904, Dorothea wrote to Edith Hall when she had been back at Wyseby for just three weeks.

> I have only had a glimpse of my future brother-in law but think I quite approve – not that that would make any difference! – Of course I am very glad for her sake – but (don't think me a beast) can't help feeling a bit low on my own account as I fear my travelling and nat. hist. collecting will have to cease – of course I don't say anything about this so am relieving my mind by grumbling to you! –
>
> At present I cannot feel reconciled to the idea of just living an absolutely aimless life for I don't seem able to make things for myself to do at home. Even now I miss most awfully not having lots of work to do as I have almost continuously had for the last three years – however being an adaptable beast I daresay I shall settle down quite comfortably, if not excitingly, in a very short time – so forgive and forget this grumbling.

And then she adds yearningly, 'When you do return to Athens, write and tell me all the gossip of everyone I know – when you do.'[1]

This letter is unique. It is the *only* purely personal letter from Dorothea that appears to have survived anywhere, and it makes almost unbearable reading. Quite early on in my research, when I discovered how little time she spent at the Museum after her return from Crete, I wrote how bereft and miserable she must have felt, prevented from pursuing her career. The subsequent discovery of this letter to Edith Hall in the archives of the Pennsylvania University Museum library came as a shock, partly because of how closely I had judged her character, but mainly because of the letter's content, the desperate hurt she was feeling, and her valiant attempt to try to minimize her distress, even to her friend. Like so many women of her age and class, she was a casualty of the conflict between her desire for an independent career and her family's expectations. That she could not even talk about it to her family appears today almost incomprehensible, yet she was expected without a murmur to abandon a career and achievements for which a man would have received considerable acclaim. She had worked so hard to establish herself as a palaeontologist and to be accepted by the scientists at the Natural History Museum; her reports had been read to the Royal Society and published in learned journals; while her exploits abroad were, as Robert Bosanquet wrote, truly courageous. Instead of a life as a respected scientist, all she could see before her was an endless, dull existence of quiet gentility, her parents' companion, no longer the pioneering explorer. Much as she loved her family, it seemed to her that whatever she did, her needs and wishes would always be secondary to theirs. It had been a revelation for her to discover that her love of natural history could be channelled to a purpose, that instead of being simply the awkward middle child, she could establish herself even more exceptionally and noticeably than Leila, and find that her gift for science was taken very seriously by one of the great scientific institutions of the world. As Leila prepared to embark on her new life as Mrs Henry Tansley Luddington, the expectations of their parents seemed to Dorothea to bring to an end everything that made her own worth while.

As for the man who inadvertently was the cause of all this, he had met Leila at a wedding in Harrow in north London in 1900. At Cambridge, Tansley, as Leila called him, had represented the university in athletics, rowing, cricket and football in 1876 and 1877, the year Leila was born. He was just five years younger than her mother. His career was preordained, running the family estates in Cambridgeshire and Norfolk. In marrying Tansley, the beautiful, talented and vivacious Leila was exchanging the rootless wanderings of her parents to live, for the first time in her life, in a house that was owned and not rented.

In spite of these almost overwhelming domestic pressures, Dorothea would not give up her scientific work quietly. She had brought home with her some of her mammal finds from Crete (what her parents made of this is unrecorded), and in November 1904 she wrote to Michael Oldfield Thomas, the Curator of Mammals: 'Very many apologies for the dreadfully smelly parcel – hope you won't mind! – It contains the few skulls got in Crete and I shall be so much obliged if you will have them cleaned for me.'[2] Small mammals were cleaned at the Museum in a room in which, wrote a reporter from *The Pictorial Magazine*, 'the smell was a good deal more powerful than agreeable.'[3] Skulls and skeletons were placed in running water until all traces of flesh had been thoroughly removed, a process that could take from a few days to several weeks. She told Oldfield Thomas that she would be in London, 'Sometime before the end of the month and want very much to work at the mammals'. Even to him she sadly confides, 'In future my time won't be so much my own as my only sister is going to be married early in December.'[4]

To my surprise, the marriage, on 15 December 1904, did not take place in Dumfriesshire or Cambridgeshire, but in the grandeur of the Chapel Royal at Hampton Court Palace 'by special permission of His Majesty'. The King's chaplain officiated and the reception was held in the Orangery. All this, I discovered, had been arranged through Charlotte, the widow of Major-General Sir Charles Metcalfe MacGregor, who had made his reputation in India and Afghanistan. On his death, Lady MacGregor had been granted a grace-and-favour house at Hampton Court Palace[5] and Leila, whether through family connections or one she had made herself, often stayed with her. According to newspaper reports, Leila's gown was of 'rich white satin, draped with flounces of very beautiful Limerick lace . . . The Court train was of chiffon and silver, with antique Limerick lace presented by Lady MacGregor.' Her veil was tulle and her bouquet included white orchids and lilies. Among Leila's jewels were an antique pendant of rubies and diamonds and a string of pearls.[6] Dorothea was one of five bridesmaids, there were two little train-bearers, and the honeymoon was spent in Cairo, India and Japan. Not one photograph of the wedding appears to have been either published or to have survived the fire at Leila's house.

At home after the wedding, Dorothea finished the account of her trip to Crete for the *Geological Magazine*, and heard from the shipping agents,

The magnificent Natural History Museum about 1880, the year before it opened to the public.

The panelled ceiling of the Museum's Central Hall.

Henry Bate encouraged Dorothea's fascination with natural history and taught her to fish and become an excellent shot. It is only perhaps in his far-focussed eyes in this photograph that there is any hint of his anxiety over the family's financial difficulties.

Elizabeth Bate was wonderfully energetic and an accomplished musician. She contributed much to the society around her and ensured her children did as well. She became, however, 'a bit of a handful', and her behaviour may well have been behind the family's sudden move from Wales to Gloucestershire.

Gellidywyll on Tivy-side where the Bates lived from 1888–1898. This photograph from 1973 shows its sad state after decades of neglect. It was demolished a few months later.

From Gellidywyll the Bates moved to Bicknor Court in Gloucestershire. It was from here that Dorothea explored the Wye Valley, discovering the bone cave that launched her career as a palaeontologist.

The hand-ferry at Symond's Yat. It is more than a hundred years since this photograph was taken, but the means of crossing the river Wye at this point is exactly the same today. There has been a crossing here since Roman times.

This is the earliest likeness of Dorothea, but apart from that splendidly determined chin, it is hard to equate this image with the vital explorer of Crete and Cyprus. Dorothea may well have sat for this during the interminable five years that intervened between her explorations of Crete and Majorca when her parents refused to allow her to travel abroad.

The curator of birds, Dr Richard Bowdler Sharpe, photographed here in about 1907, gave Dorothea her first opportunity to work in the Natural History Museum when she was just nineteen.

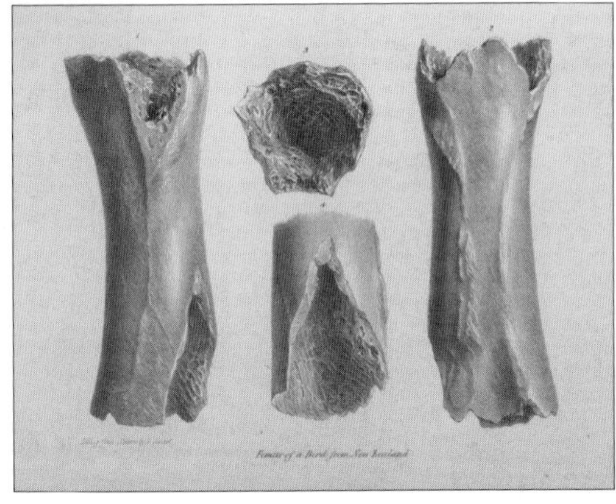

The Superintendent of the Natural History Museum, Sir Richard Owen. He is pictured here (*c.*1877) with the skeleton of the Giant Moa (*Dinornis*). It was from this bone, just six inches long (above), that Owen was able to describe a large, flightless bird that was 'a heavier and more sluggish species than an ostrich'. Four years later he was sent from New Zealand more complete bones which confirmed his identification of *Dinornis*.

Dr Henry Woodward, Keeper of Geology at the Natural History Museum from 1880–1901 painted *c.*1906 by Lucas. His encouragement and support were invaluable to Dorothea from her earliest days as explorer and palaeontologist.

His successor, Dr (later Sir) Arthur Smith Woodward, photographed here in 1906, was similarly supportive of Dorothea. A world authority on fossil fishes, his reputation never recovered from the Piltdown debacle.

Clarence Wodehouse, district commissioner for Papho in Cyprus and his wife Francie gave Dorothea the opportunity for her first fossil hunting expedition to the Mediterranean, when they invited her to stay with them in Cyprus.

Their son Jack, photographed here at Harrow School in 1890, formed a close friendship with Dorothea during her eighteen months in Cyprus.

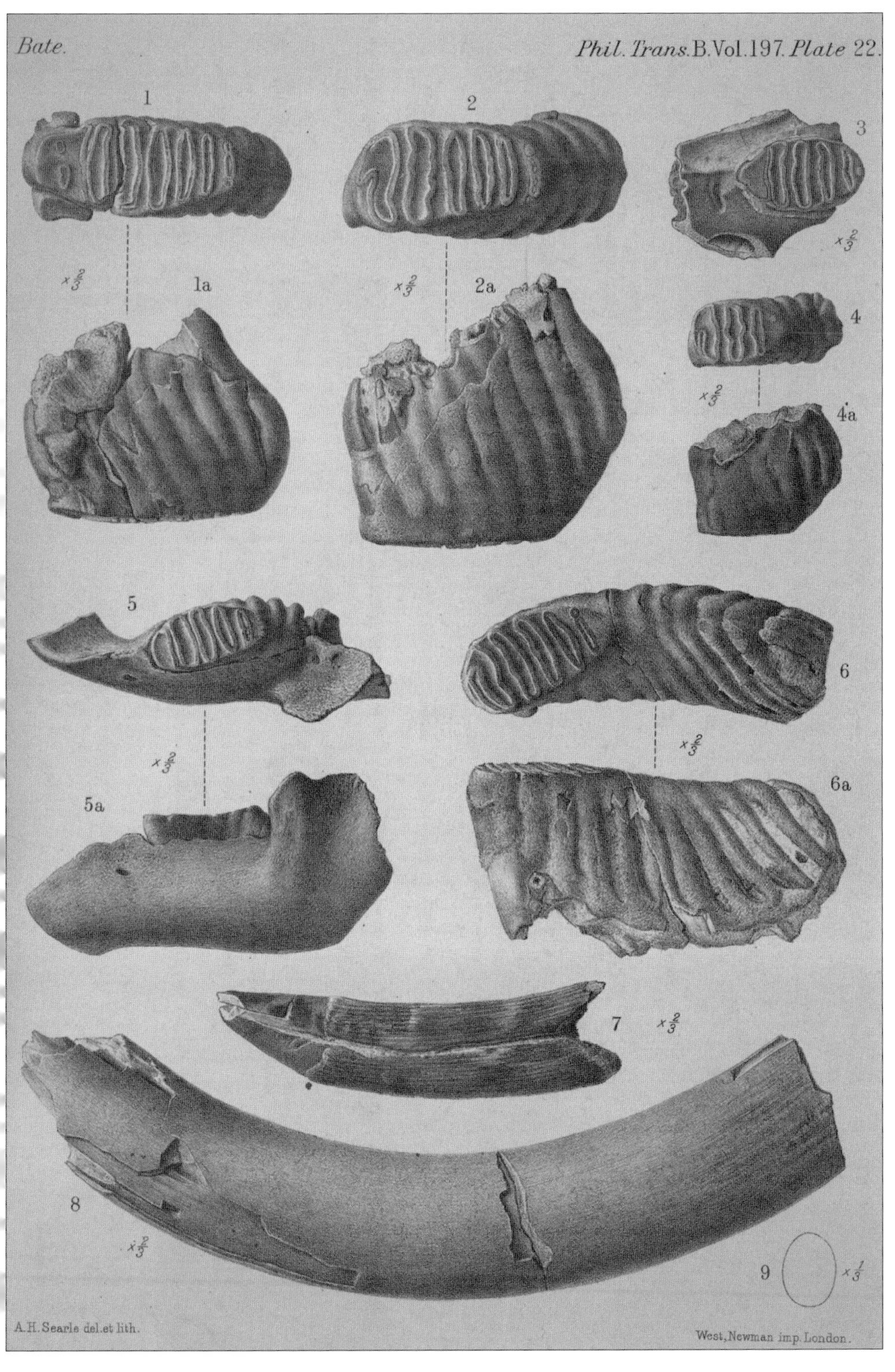

These drawings of Dorothea's discoveries of the fossil remains of the extinct pygmy elephant (*Elephas cypriotes* Bate) that she found at Imboähry, clearly show the characteristic ridge pattern of the teeth.

Two of the American archaeologists whom Dorothea met on Crete: Harriet Boyd Hawes, seen here in Herakleion in 1904 examining pottery sherds, and Richard Berry Seagar, who was, according to fellow archaeologist Edith Hall, 'good-looking and carefully dressed, but maybe nice.'

The Katharo plateau, high in the Dikti mountains where Dorothea found fossil remains of extinct pygmy hippos.

Stahlschmidt, that one of her (smuggled) boxes from Crete would be delivered to the Museum on 15 March. It was not until May that her parents allowed her to join it. She arrived the day after the unveiling of what is one of the most famous exhibits in the Museum, the enormous skeleton of a *Diplodocus*, which now dominates the Central Hall. *Diplodocus* was donated to the Museum by Andrew Carnegie, the wealthy Scottish-American industrialist, and is actually a cast made of plaster. The original skeleton had been discovered in Wyoming in the United States and Carnegie ordered the Director of the Carnegie Museum in Pittsburgh, Dr William Holland, to buy it. On a visit to Carnegie's Scottish seat of Skibo Castle, the King, Edward VII (who as Prince of Wales had been an enthusiastic and supportive Trustee of the British Museum), saw a picture of the skeleton and was so impressed that he asked Carnegie whether one could be obtained for the Natural History Museum. Even for a king, the chances of finding another *Diplodocus* skeleton immediately were clearly remote, but it was suggested instead that a cast be made of the original, an offer that was swiftly accepted. The cost was £2,000 (more than £100,000 in today's values), and Carnegie paid for it. At ninety feet long, the cast arrived at the Museum in thirty-six cases and its assembly in the Reptile Gallery was supervised by Dr Holland.

On 12 May 1905, *Diplodocus* was formally presented to the Museum by Carnegie in a ceremony attended by 300 guests. The newspapers loved it: 'New resident in South Kensington,'[7] trumpeted the *Westminster Gazette*, and 'Welcome colossal stranger',[8] headlined the *Daily Graphic*, while the cartoonists suggested, among other fancies, that *Diplodocus carnegii* might be used as a coat and umbrella stand. Many of the papers referred to the creature as an 'antediluvian monster', a popular though outdated reference to the long-held belief that fossils were the remains of animals and plants wiped out by the great deluge of biblical times. Until the eighteenth century, this had conveniently accounted for the presence of sea creatures found buried in rocks far inland and on the tops of mountains. However, as observations of the earth became more detailed, serious naturalists began to form very different ideas. By the eighteenth century, it was apparent that all these plants and creatures could not possibly have lived – and died – at the same time; it was also realized that the surface of the earth had been subjected to immense upheavals, causing frequent redistribution of oceans and continents, and therefore of the habitats of these ancient forms of life. The different strata of rock, it was recognized, represented different stages of the earth's development, and the relative age of rocks and fossils could be determined by where they came in successive layers. The age of the earth, until then, was thought to be around 6,000 years old, according to James Ussher, the Protestant Archbishop of Armagh who, in 1650, had calculated from the chronology of the Bible that the Creation took place in 4004 BC. As the new science of geology developed, it became apparent that the earth must in fact be many millions of years old.

In his address at the presentation of *Diplodocus carnegii*, Dr Holland proclaimed that 'forty thousand centuries are looking down upon us today from this expressionless skull'.[9] He was wrong by about 150 million years. It was only in these early years of the twentieth century that the means of establishing absolute dates for geological time was discovered, and that was the recognition that the radioactive decay of certain elements in rocks and minerals could be measured and the results used to calculate the age of the earth. Even so, it is only in the last thirty or forty years that it has been determined with any certainty that the earth is 4.6 billion years old.

The reconstruction of *Diplodocus carnegii* may have given Dorothea an idea, and that was to build a 'restoration', as she called it, of the little hippo from Cyprus, *Hippopotamus minutus*, that she had found in such quantity. Smith Woodward agreed and work to construct the skeleton began under the skilled hands of the Geology Department's chief preparator, Frank Barlow. Dorothea, meanwhile, began work on her finds from Crete, both extinct and recent. The skins she had sent to the Zoology Department had been cleaned and disinfected, and those that were too hard for stuffing into shape were hung up by the heels over a tank of running water in the taxidermists' room until sufficiently pliable. It was known rather engagingly as the relaxing tank. She wrote a report for the Zoological Society on the sixteen species she had collected in Crete, but by the time it was published in September 1905 she had been back at Wyseby for three months.[10]*

* These were four types of bat, a hedgehog, polecat, badger, beech marten, weasel, two types of rat, field mouse, Cretan spiny mouse, hare, Cretan wild goat, and a rabbit. This last was the only survivor of the three rabbits that had been given to her from the island of Dhia, the other two of which she suspected had been stolen and eaten in Alexandria while she had been sightseeing. It was also the only live creature she had brought back and it lived for some months after her return in the Zoological Society's gardens in Regent's Park. Gratifyingly for

There she remained until March 1906, apart from a few fleeting days at the Museum in the summer and late autumn. That she was listed as a major contributor to the Entomology Department when she presented to them 363 insects she had collected – and pinned – in Crete[11] only served to emphasize for her how remote the possibility of further travel seemed to be. She heard, probably through a letter from Edith Hall, of the engagement of Harriet Boyd and the English anthropologist Charles Henry Hawes, which had been greeted in the American press with such witticisms as 'Love among the Ruins' and 'Hunt Dead Cities and Find Love' to the considerable embarrassment of the couple.[12] She wrote to Arthur Evans, remarking on Harriet's engagement and expressing curiosity about the 'horns', as Evans called them, that he had found in a shrine at Knossos. Evans replied that he was 'most interested' to hear of the engagement and gave Dorothea permission to have a photograph taken of the horns. 'I shall be glad,' he wrote, 'to know what you make of them. The shrine belongs . . . to the first period of the palace, – not later, – I reckon, – than about 2000 B.C.'[13] The horns, as she was later to note, were the antlers of *Anaglochis*, the extinct Cretan small deer,[14] but with so many of her 'antelope or deer' remains retained by the Cretan government, even by the 1940s Dorothea had been unable to examine them. It is was not until eighty years after her discoveries that no fewer than eight species of deer from the Pleistocene of Crete were identified, which cover the range of island peculiarities from dwarf to gigantic.[15]

In February 1906, Dorothea received a letter from the American palaeontologist, Professor Henry Fairfield Osborn, of the American Museum of Natural History in New York, asking her whether it would be possible to obtain any of her pygmy hippo and elephant finds from, as Dorothea wickedly quotes him to Dr Arthur Smith Woodward, '"Cypress in Crete"!' As she told Smith Woodward, 'I am writing to tell him that there are some Cyprus hippo remains at the Museum and that I will see you about the matter. I think you once told me you thought there would be about £20 [about £1,000 in today's money] worth of stuff left when you had picked out what you required – so perhaps this might be arranged.' And she reveals something of her unhappiness as she adds, 'I might as well

---

her, four of these species were unique: two of the bats, the field mouse, and the Cretan spiny mouse: *Rhinolophus ferum-equinuum* and *R. hipposideros*, *Micromys sylvaticus hayi*, and *Acomys dimidiatus minoūs*.

get something for these specimens as I expect I shall want to wander again some day – I may be in London before long when I hope to get in a little work on the Cretan spoil.'[16]

Three weeks later she arrived at the Museum and worked not on the 'Cretan spoil', but on her Cyprus hippo finds. The 'restoration' of the skeleton was nearly complete and it was extraordinarily small, just 55½ inches long and 26¾ inches high 'at withers'. The hippo was exhibited in the Museum and it was photographed to accompany a new report on *H. minutus*, which Dorothea published in the *Geological Magazine* in June 1906. In this she described how the skeleton had been constructed from actual bone and plaster where necessary, and how Barlow had skilfully recreated the skull from fragments as she had not found one intact. She wrote of the new evidence that had emerged in the four years since Forsyth Major's initial identification of *H. minutus*, and discussed both the phenomenon of dwarfism in islands and possible reasons why the pygmy hippos and elephants had become extinct. Human activity she rules out as there was then no evidence that either animal had existed contemporaneously with man; she suggests instead that climate change and consequent alteration in vegetation may have been a cause, environmental factors whose impact she would develop throughout her career.* Drawing on her knowledge of Darwin (in 1906 his ideas were barely fifty years old) and the expertise of her friend and mentor Charles Andrews, she concluded that the hippos' extinction might have been because they belonged to 'an ancient and effete race' and were unable to evolve fast enough to survive their changing surroundings.[17]

Shortly after the publication of this report, there seems to have been a remarkable change in her parents' attitude and consequently in Dorothea's future. In August she sent the most surprising letter to Arthur Smith Woodward: 'I am just writing to tell you that I am very anxious to go abroad in the Autumn and would much like to spend a month in the Balearic Islands looking for cave deposits if I can overcome the various

* The question of human influence is still open. One interpretation of more recent finds, particularly at Aetokremnos in Cyprus, is that the bones seem to show signs of cooking and indicate that not only did humans and dwarf hippos coexist, but that man hunted these animals for food. See David S. Reese, 'Tracking the extinct pygmy hippopotamus of Cyprus', *Field Museum of Natural History Bulletin*, 60: 2 (1989), and David S. Reese, 'Cypriot hippo hunters no myth', *Journal of Mediterranean Archaeology*, 9: 1 (1996), 107–12.

difficulties which at present appear to be in the way. One of these is of course with regard to the financial part of the scheme.' Professor Osborn, she told him, no longer seemed interested in buying any of her *H. minutus* specimens as he wanted only complete skulls or skeletons.

> So now I am wondering if you would be so *very* kind as to arrange for the Museum to buy the elephants' remains from Crete? – I should like £40 for them if you think that a fair price. My trip to Crete cost me £150 [approximately £8,000 in today's values] and I should certainly wish to receive £100 for the remains I brought home which include those of five species of large mammals and at least three small ones (rodents etc.). So I thought I might have £40 for the elephants and the others could be arranged about later when I have worked at them.
>
> I do so very much hope that you will approve of this – and if so do you think you could possibly get it for me by the commencement of October?
>
> . . . With many apologies for the trouble I fear I may be giving you and hope to hear from you soon.[18]

How Dorothea had managed to persuade her parents to allow her to travel isn't clear at this point, but her plans seem astonishingly well advanced. She mentions here to Smith Woodward just one of the difficulties in her way, money. Smith Woodward replied by return that he was glad to learn that she proposed to make another expedition and made what seemed to be a very generous offer – although one that tellingly demonstrates the rigid financial constraints of the publicly funded Museum:

> There are only two difficulties about your proposal this morning. We think that £40 is too high a price for the elephant remains alone, and we are unable to pay any money until the end of October.
>
> What, then, do you think of this proposal? The Museum to take all it needs from the collection of fossil vertebrates from Crete for the total inclusive sum of £50. I personally to lend you half this sum at the beginning of October if you really need it so early.[19]

On a personal level this was extraordinarily generous, but it was not the price that Dorothea required and she had no inhibitions about haggling with the Keeper of Geology. While she found his own offer extremely kind and thoughtful,

> at the present moment I do not feel very inclined to fall in with your suggestion that I should let the Museum have for £50 all it requires of my fossil mammals from Crete which would practically be almost the entire collection which is more varied than bulky.
>
> I cannot help thinking that £50 is a very small return, even in a scientific undertaking, for the results of a successful expedition which meant six months hard work and an expenditure of £150 – also I have a dislike to parting unless necessary, with such portions of my collection as have not been worked out or described.

Smith Woodward knew Dorothea well enough by now to know that she would not meekly accept an inadequate offer, but even he may have been taken aback by her vehemence in defence of her work, no matter how circumspectly her rejection was phrased. There is, though, a note of desperation as she counters his offer: 'At any rate for the present could you make me an offer for the elephant remains? Are they worth £25? The pigmy species is new therefore ought to be of value to the Museum which rather specialises in elephant remains.' She ended the letter with, 'Hoping you will be able to meet me in this.'[20] And he did, by return agreeing to the £25 for the elephant remains.

It is a measure of how highly Smith Woodward regarded Dorothea and her achievements that he took further her tart reminder of the importance of the Museum's elephant collection. It is noted in the Minutes of the Trustees that her elephant collection from Crete was 'new to the Museum and to Science'.[21] Indeed, not only did Smith Woodward present her case to the Museum for a reasonable payment for her fossils, he also offered to try to sell some of her *H. minutus* remains to a dealer in Bonn who specialized in such things.[22] The Minutes note her collection of forty pieces to be of a pygmy elephant and of *Elephas meridionalis*, which were the bones she identified as *E. antiquus*. This animal had reached enormous size, standing around fifteen feet at the shoulder. Her attempts at smuggling the elephants from Crete had clearly been successful. She completed her report on the two species of elephants by the end of 1906 and it was published the following year.[23]

At the beginning of September 1906, Dorothea was still planning the trip and she wrote to Michael Oldfield Thomas who had explored the Balearics in 1900: 'I am afraid you will not feel very pleased with me when I tell you that I am trying to spend a month in the Balearic islands in the Autumn [she had promised to send him mammal skins] – but all the same I hope you will be very kind and help me by giving me some information about the place as it is rather difficult to find out much from the country – I want to explore as many caves as possible in the time, I suppose there would be no difficulty in doing so?' And then, from the woman who had achieved so much on her own in Cyprus and Crete, came an extraordinary question, which shows the true nature of the difficulties she was trying so hard to overcome: 'I wonder if by any chance you may know of someone who might care to join me? I am trying very hard to find a travelling companion as I don't think my people will allow me to go without!'[24] What family rows must have lain behind that request. Not only was she unable to research the Balearic islands as she would have wished in uninterrupted hours in the libraries of the Museum and in conversations with Oldfield Thomas, Andrews, Forsyth Major, and Sherborn, but she was mortified to find that, after all her adventures and triumphs, at the age of almost twenty-eight her parents would not allow her to travel without a chaperone.

September crept on and still she had not found a companion. She wrote to Mrs Smith Woodward who promised to try to help but by October still had found no one. Dorothea had wanted to go abroad in the autumn while it was still warm and light; that now was impossible and in the end she did not go. Life at Wyseby must have been horribly strained, and for the next two years there is little information as to how she spent her time. Occasionally she went to London and worked for a few days at the Museum; she undertook some archaeology locally, her mother could hardly prevent her from doing that,* but she was

* This included the excavation of a cairn at Mossknow on the Kirtle Water in Dumfriesshire. The archaeologist, Roger Mercer writes of 'This mutilated stony cairn' and 'The cairn is poorly preserved having been excavated in 1908 by Miss DMA Bate', although Dorothea was doing no more than following the accepted archaeological practice of the day. In Roger Mercer, with contributions by Richard Tipping *et al.*, *Kirkpatrick Fleming, Dumfriesshire. An Anatomy of a*

now approaching thirty and it was little compensation for the vanishing years.

In November 1908 she returned to the Museum, in time to say farewell to Charles Forsyth Major. For years he had struggled financially, trying to balance the needs of his family with a wholly inadequate income and continually badgering the Museum for prompt payment as he completed sections of the catalogue that he was writing on fossil rodents in its collections. In what seems an uncharacteristically sudden decision, and even more oddly for such a meticulous man, at the end of 1908, leaving his work on the rodent catalogue unfinished, Forsyth Major left not only the Museum but England. He retired to Corsica where he could devote himself to Pleistocene mammals of the Mediterranean, his obsession that he had imparted to Dorothea. For Dorothea herself, his departure meant the loss of a friend as well as a mentor. Within a few months, however, she received a letter that revived her dusty plans to visit the Balearics and gave her the energy to overcome all her parents' obstructions.

Parish in South West Scotland (Dumfriesshire and Galloway Natural History and Antiquarian Society, 1997), p. 39.

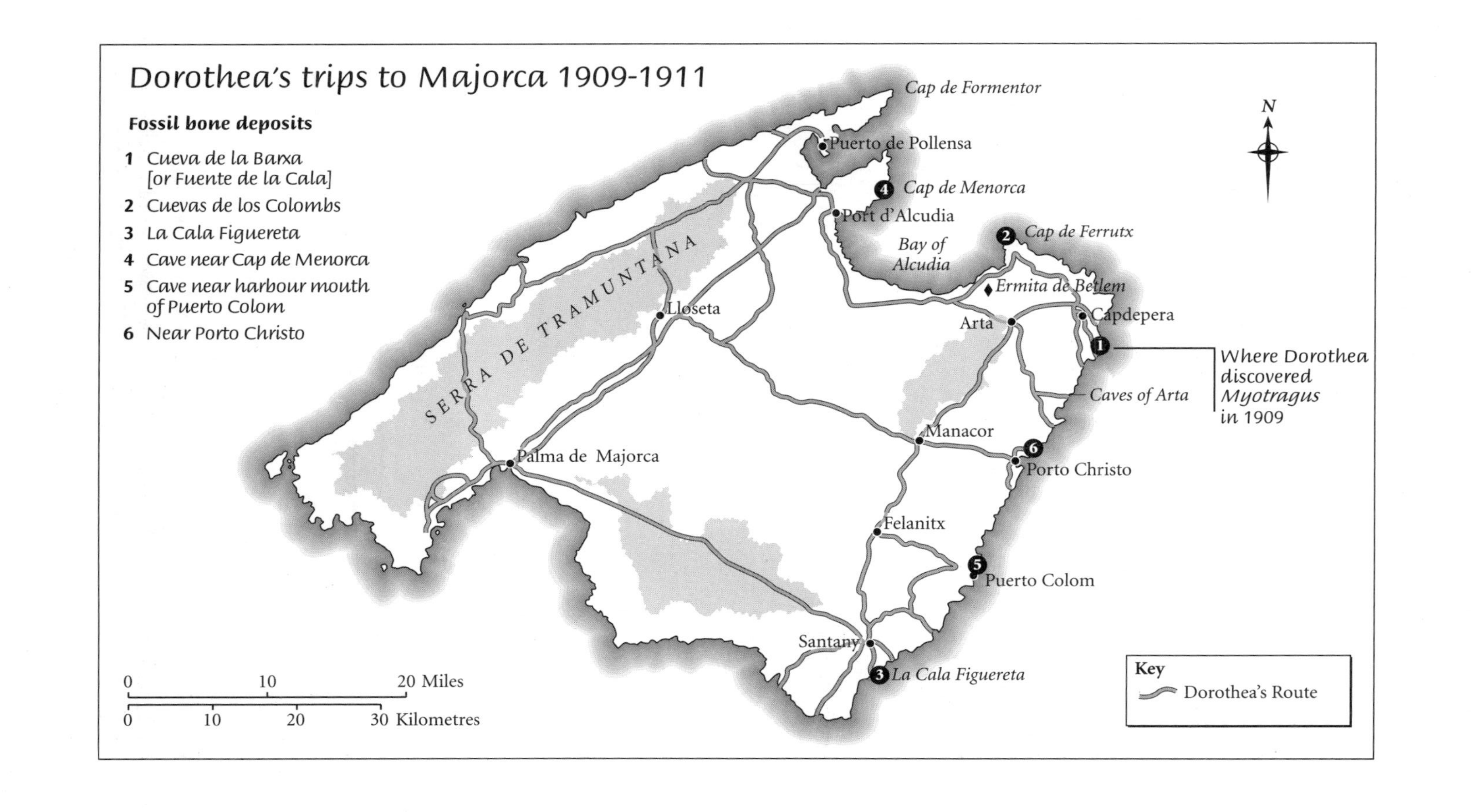
Dorothea's trips to Majorca 1909-1911
Fossil bone deposits
1 Cueva de la Barxa [or Fuente de la Cala]
2 Cuevas de los Colombs
3 La Cala Figuereta
4 Cave near Cap de Menorca
5 Cave near harbour mouth of Puerto Colom
6 Near Porto Christo
N
Cap de Formentor
Puerto de Pollensa
Cap de Menorca
Port d'Alcudia
Bay of Alcudia
Cap de Ferrutx
Ermita de Betlem
Capdepera
Arta
Where Dorothea discovered Myotragus in 1909
Caves of Arta
SERRA DE TRAMUNTANA
Lloseta
Manacor
Porto Christo
Palma de Majorca
Felanitx
Puerto Colom
Santany
La Cala Figuereta
0 10 20 Miles
0 10 20 30 Kilometres
Key
Dorothea's Route

CHAPTER 9

A Needle in a Bundle of Hay

The letter was from a keen amateur fossil collector with a passion for molluscs, the Reverend Robert Ashington Bullen, who was a frequent visitor to the Museum. In the early spring of 1909, he went to the Balearics and on the east coast of Majorca he found what appeared to be a deposit of bone breccia and wrote to Dorothea to tell her. It was the spur she needed.[1] Within a few weeks she had acquired both sufficient funds and a travelling companion to appease her parents and an interminable five years after her last exploration, set off on the long-deferred project to explore the Balearic islands. Once again, Dorothea was breaking new ground. Until Ashington Bullen's discovery, there was only one record of any Pleistocene mammalian remains in the Balearics and that had been made in 1855 by the renowned Sardinian archaeologist Count Alberto Ferrero de la Marmora, who had found what appeared to be a bone deposit on the hill below the Castell de Bellver, near Palma.

Lying about eighty miles off the Spanish coast, Majorca was a place of retreat for many of Spain's greatest families, who came to enjoy its beauty and tranquillity. From the late nineteenth century, the reputation of the Balearics began to spread across Europe, through the industry of the Austrian Archduke Ludwig Salvator who, at the age of nineteen had come to live in Majorca. He wrote extensively on the archaeology, culture and natural history of the islands and was *the* acknowledged authority.[2] In 1909, however, only a trickle of visitors came to admire the unspoiled islands.

The writer Mary Stuart Boyd, who spent six months in the Balearics with her family, was warned beforehand by a friend who had been there that 'you won't enjoy them. They are quite out of the world. There are no tourists. Not a soul understands a word of English and there's nothing whatever to do.'[3] Little changed in the islands until the 1960s when the development of package holidays brought tourists there by the million.

Dorothea's diary for this trip has vanished. There is no record now of who her travelling companion was or details of this first visit, although there is some information in her letters, her diaries from subsequent visits to the islands, and her scientific reports. On 12 April she wrote to Dr Smith Woodward from Palma de Majorca, 'I have arrived here somewhat to my surprise as our plans have been so uncertain. This seems a delightful place and I am sorry we shall not be able to remain more than two or three weeks as that means very little time for fossilising. I found your letter and the map [of all the Balearic islands] awaiting me for which very many thanks.'[4] In fact she remained for a month, searching a 'considerable tract' of the limestone country for more creatures, extinct and living, that were new to science.[5]

Dorothea's task was Herculean, more difficult even than either Cyprus or Crete had been, as so little was known. The Balearics had once formed the peaks of a long arm of land that extended from the Iberian peninsula into the Mediterranean. Over time as sea levels rose, the highest peaks were isolated as islands. They are divided into two groups, the Gymnesic islands of Majorca, Menorca and Cabrera, and, to the south, the Pityusic islands of Ibiza and Formentera. In Majorca, Dorothea was confronted with a largely limestone mountainous island of some sixty miles by forty miles, with 3,500 miles of indented coastline, countless caves and fissures, and absolutely no indication of the type of extinct fauna she might discover. As she knew to her cost from her previous explorations, a guide was essential and within a few days of her arrival, she had made a valuable contact: an amateur geologist and naturalist whom she calls Padre Caldentey.* He seemed to know everyone of importance or interest

* His identity remains unclear. Dorothea confusingly calls him Padre Caldentey, although he was evidently not a priest – she mentions those she met clearly and besides, he was married and had a grown-up son called Miguel. Given her often erratic spelling of proper names, his name may have been Pedro, though this only confuses matters further. There was a Pedro Caldentey in Majorca at the time, a geologist and speleologist who discovered the Cuevas dels Hams in 1905, another of the great 'show' caves. Nice though it would be if they were one and

throughout the Balearics and helped her to find reliable workmen and guides for all three of her visits.

What Dorothea discovered in Majorca in the four short weeks of her first visit is truly remarkable. In this unknown island, which had not even been comprehensively mapped, Dorothea discovered, among the 'innumerable caves' as she called them, the fossilized remains of an extinct creature, which is still described today as one of the strangest species to inhabit the planet.[6] Her discovery is even now the focus of academic study (and argument) on the scale of a small industry. The remains were in a cave in the cliffs to the north of the show caves of Arta, in a particularly remote and difficult place.* Known locally as the Cueva de la Barxa, its only entrance was through, as Dorothea wrote in her report, 'a rent in the roof and a drop of 7 or 8 feet'.[7] It was flooded periodically by the sea, but she found here the remains of a bone deposit that had originally been six and a half feet thick, but had been almost entirely demolished by the force of the waves. From what was left, Dorothea excavated a number of perfect examples of limb bones and 'a well-preserved skull of an old individual with the associated mandible [jaw]'.[8] It appeared to be a goat-like creature, but was unlike anything she had seen before.

Searching farther north, she discovered caves in the great limestone cliffs of Cap de Ferrutx† at the eastern point of the Bay of Alcudia. They were known as the Cuevas de los Colombs, a series of caves and crevices that stretch in a steeply ascending line for about seventy yards, the lowest

---

the same, this man was only 26 in 1909, he did not have a son called Miguel, and the family have no knowledge of Dorothea.

* From her reports it is evident that Dorothea had read extensively of the great show caves of Majorca, particularly the accounts by the French speleologist E. A. Martel who had explored them in the nineteenth century, and she almost certainly visited them on this trip. The most famous, the caves of Arta and of the Drach or Dragon in the east of the island, were renowned for their fantastically shaped stalactites and underground lakes. Martel was astonished by their vast grandeur, but he also noted that in places they were as black as the interior of a chimney, coated with soot from torches used by the guides. 'Le désastre,' he wrote, 'est irréparable.' (E. A. Martel, 'Les cavernes de Majorque', *Spelunca, Bulletin & Mémoires de la Société de Spéléologie*, 32 (February 1903).) None of them contained bone deposits. Dorothea was to write later of her dislike of the glib comments of the cave guides and how the natural magnificence of the caves was trivialized for visitors. (Bate, 'The Caves of Crete', in Aubyn Trevor-Battye, *Camping in Crete* (London: Witherby & Co., 1913).)

† Dorothea spells it Cap Farouch; apart from this, all names are as Dorothea spells them.

cave no more than fifteen feet above the water. In one of the lower caves she found a patch of rock that looked of very great promise. She organized tools and men to dig but, as suddenly as she had found herself in Majorca, she was summoned home; her father was unwell. Her missing diary would have reflected her real concern for her father as well as her response to this abrupt termination to her first expedition for five years. All she could do was mark the place in the cave and hope that the health of her father – and her parents' concerns, which had halted her career for so long – would not prevent her return.

Dorothea left Majorca immediately, but by the time she arrived back in England Henry was already making a good recovery. On 16 May she wrote to Smith Woodward, the frustrations of having to leave her work barely begun very evident. She is also curiously coy about the goat-like skull, not yet revealing how strange it appeared to be. 'I have only a few specimens and those I fear not of an interesting description. There is so much ground to be examined in Majorca that it is rather like searching for the proverbial needle in a bundle of hay.' The one box of fossils she had excavated was being sent directly to the Museum and she asked Smith Woodward to have it put on one side for her, 'as I should like to unpack it myself'.[9] The box was delivered to the Museum on 9 June and, by the middle of July, Henry had recovered sufficiently for her to be allowed to travel to London.

From the carefully packed crate, Dorothea lifted out the skull and short, straight, sharply pointed horns of this small, goat-like animal. Nothing quite like it had ever been seen before and she examined it with Charles Andrews. What made it so immediately remarkable were its teeth. Instead of the usual six lower incisors (lower front teeth) characteristic of ruminants, there were instead just two, which were enormously enlarged and ratlike, continuously growing, like rodents' teeth, from open roots.[10] It was the only known ruminant with just a single pair of lower front teeth. As they examined the creature, many more peculiarities became apparent. The jaw was massive, while the position of the anterior teeth suggested, Dorothea wrote, 'that the animal's nose must have been very short'. The skull she found to be characterized by a general 'shortening', and the eyes were placed very far forward. It was not only in the skull that a general shortening and thickening was so noticeable; it also 'appears to have been characteristic of the whole animal', she wrote, 'for this tendency is also very strongly marked in the limb bones'. It was, Dorothea could only conclude in her preliminary report on the creature, 'an ungulate [hoofed mammal] which appears to be without parallel'.[11]

She named this goaty antelope with ratlike teeth *Myotragus balearicus* Bate. It is so specialized that there is not even a common name for it. It is referred to variously as a ratlike goat, a cave goat, an antelope gazelle or, simply, the self-explanatory *Myotragus*, which translates from the Greek literally as mouse-goat. From its strange shortened and thickened shape, it had evidently adapted to climbing very steep rocks and crags 'to a degree', Dr Andrews stated, 'equalled in no other animal'.[12] The legs were wide apart and its normal gait would have been a walk; it is unlikely to have been able to run away from man, even if *Myotragus* recognized man as a predator.[13] On Majorca it was very small; the specimens she found stood about forty-five centimetres high at the shoulder, although some adult skeletons found in recent years are no more than twenty-five centimetres.[14]

How the creature had come to evolve in so specialized a fashion, and what caused its extinction after five million years or so of peaceful browsing with those great teeth, was to perplex not only Dorothea and Charles Andrews, but has still not been conclusively resolved. Current research suggests that about 5.7 million years ago, a species of goat crossed the promontory from the Iberian peninsula and, as goats do, climbed up crags and mountains. When the waters of the Mediterranean rose and these mountains, which formed the Balearics, became isolated, so was the creature. It was that isolation, over a period of just over two million years, that caused its specialization, but Dorothea's questions as to why and how this should have happened are still being pursued today.

Before Dorothea could even begin to enjoy the celebrity of this bizarre and curious creature, after just ten days' work at the Museum she had to return home and remain there. This time it was not illness that interrupted her work; after six years in one place her parents once more were moving house and they could not cope without her. As with all the previous upheavals, no record remains of why they left Dumfriesshire. They stayed in Scotland, however, moving to Gordon in Berwickshire, to Bassendean House, a seventeenth-century mansion with Victorian embellishments. Here Henry Bate held the tenancy not just of the main house but also of two lodges, extensive grounds, and shooting rights.

Dorothea's patience with her family is extraordinary. This was a woman aged thirty, whose achievements were manifold, whose reputation as a scientist and explorer was established and growing, and who had just discovered an astonishing and unique extinct creature, yet still her parents' demands in 1909 dictated her life. She could write her short preliminary report on *Myotragus* at home, but do little else. Meanwhile *Myotragus* was of such interest that the type specimen skull she had found at the Cueva de la Barxa had been put on display at the Royal Society.

In the early spring of 1910, Dorothea made a brief visit to London. The great Japan-British Exhibition was due to be held that year from May to the end of October at the White City Stadium in West London, with the aim of furthering trade and friendship between the two countries. Exhibits would range from sumptuous silks and ceramics (with demonstrations) to Sumo wrestlers, naval and military exhibits, Japanese gardens, and a section on science. The Natural History Museum had been asked to loan some fossils and a collection of minerals. Dorothea discussed with Dr Smith Woodward the possibility of lending casts of the *Myotragus* skull and the restored skull of the little Cyprus hippo, as well as 'a couple of pieces' of pygmy elephant from Crete or Cyprus. However, as she wrote to Dr Smith Woodward on her return to Scotland, 'the question is would you very kindly allow [the casts] to be made at the Museum? I cannot possibly afford to spend much over this as the "Majorcan Exploration Fund!" is in great need of all the support I can give it.'[15]

The Keeper promised to try, but a couple of days later, Dorothea received a letter from Dr Francis Bather, the Assistant Keeper, which infuriated her, as she wrote to Smith Woodward: 'Since writing to you I have had a letter from Dr Bather from which I gather he is not very anxious to have only casts – So, as I am away, perhaps you would find out from him if he really cares to have them or not.' And then, acidly, 'I would not wish to send an exhibit that he did not consider entirely satisfactory.' Bather, renowned for his fussiness, plaintively noted to Smith Woodward on the foot of her letter, 'I said that the public did not greatly appreciate casts, so that if she could supply nothing else the casts should be coloured and I expressed fear that there might be some difficulty in having this done in time. That however seemed to be a matter for your decision. Of course I should be grateful for anything she can do.'[16] The entire exhibition, visited by an astonishing 8.5 million people in the five months it was open, was a huge success, and so, as Smith Woodward told Dorothea, were her casts.[17]

Perhaps it was this interest in *Myotragus* that made Dorothea's parents at last appreciate the importance of her work. Their daughter was an exhibitor at both the Royal Society and a great international exhibition attended by millions of people; Henry and Elizabeth may even have made the journey to see Dorothea's work for themselves. She had published another report in a learned journal and she was invited to one of the great social events of the summer, the Royal Society's soirée, which took place in June. Her work in Cyprus had also recently been acknowledged by the Cyprus Natural History Society, which invited her to become one of only two honorary members.[18] Dorothea's impatience to return to the Balearic islands was surely matched by the Museum's, eager to receive further specimens of that unique extinct fauna.

In May she reached agreement with Smith Woodward over the sale of some of her hippo finds from Crete for the sum of £40 (about £2,200 in today's values). It was money essential for her trip and she wanted to be paid at once, nagging him mercilessly, if apologetically. 'I am sorry to be in such a hurry about it,' she wrote in one of three letters she sent him in June; 'my reason is that – unless anything unforeseen happens – I think of returning for a few weeks to Majorca and should wish to start about the 3rd of July. Do you think this can be managed?'[19] Astonishingly it could, and just three days later than she had planned, on the evening of 6 July, she set off from Victoria Station for Majorca. This time there was no nonsense from her parents about the need for a chaperone. Her diary for this expedition still exists, a record of endeavour, obstinacy, and perseverance, and a testimony to the hardship that an instinctive natural historian like Dorothea will endure in pursuit of knowledge.

The journey took three uncomfortable days with little sleep. The crossing by steamer from Newhaven to Dieppe was smooth, but any rest she had hoped for on the journey through France was interrupted by three changes of train, one in the middle of the night. She reached Barcelona early in the morning. As the ferry to Majorca did not sail until the evening, she spent the day at the Benedictine monastery at Montserrat, famous for its Black Madonna and cave-filled cliffs, although the cost of this meant she could only afford to travel second class on the ferry to Majorca. 'Two children their garlic eating mother and friend shared my cabin!' she wrote. 'So tired I got several hours sleep and provided food for two enormous fleas on whom I was later revenged.'[20]

It was dawn as the steamer approached the island, the sheer cliffs of the southwest coast opening up to the great Bay of Palma with the massive, circular Castell de Bellver emerging above its surrounding woods on the hilltop above the harbour. Dominating Palma was the majestic Gothic cathedral of La Seu, pale gold in the early light. A substantial building programme in the previous few years had demolished much of the ancient city walls, but among the new buildings were an opera house and a splendid hotel, the Grand. Dorothea had stayed here the year before and on her return was welcomed as an old friend. 'One of my chamber maid friends greeted me with effusion,' she wrote, while at the other end of the social scale, the British Vice-Consul, Señor Bosch, and her acquaintance from the previous year, Padre Caldentey, were 'both delighted to see me'.[21] The mosquitoes were, too, plaguing her so much that she woke the following day with 'eyes bunged up and face swollen'. Her first purchase in Palma for her expedition was a mosquito net. At lunchtime Señor Bosch arrived at the hotel and took her through the narrow wandering streets and steps of the town to an exhibition of Majorcan culture and art.

The following morning Dorothea caught a train eastwards to begin her cave-hunting. The major towns were linked by rail, but the road network was poor. With the exception of Palma, Majorcan towns were set back several miles from the sea, linked to small ports by roads that were often little better than mule tracks, potholed and rutted. Persistent raids by Barbary pirates had made coastal living hazardous until well into the nineteenth century. She left the train at the medieval town of Felanitx and had to continue her journey to Santany, the southernmost town in Majorca, by trap. She waited for it in a café where, over cake and lemonade, she watched the town pass by, noting as well a sand martin on its nest. Her wait in the end was for four hours. With visitors to the island a rarity, particularly in the more remote parts, finding transport was almost impossible. At Santany the following day, there was no carriage at all to be had and it was only the unexpected arrival in a pony-trap of an acquaintance from her previous visit that enabled Dorothea to reach the sea, by what she laconically describes as an 'awful road'.[22] The journey was in vain. Her plan had been to explore the cliffs for caves by boat, but the waves were so fierce a fisherman told her it would be dangerous to sail and impossible to land. It was a pattern of delays and reversals that was to haunt her for the next five weeks.

With a fisherman and a carpenter as guides, she explored the landward side of the cliffs in the southeast of the island. In the intense July heat, she drove or walked through a landscape of gold and dusty green, noting junipers, thistles, cistus and rosemary, flocks of goldfinches, blackbirds, and 'an eagle of some sort',[23] but she found not one cave worth examining. Her diary echoes with dismay at delays caused by the lack of transport, at the agonizing jolting she suffered from springless carts, and at seas too dangerous for cliff exploration by boat. All seemed to conspire against her and was made worse by sciatica (she goes so far as to call it 'annoying'), which she had been suffering from since her arrival, although she refuses to allow it to hinder her investigation. She had already lost five years of scientific exploration because of her parents; the pain of sciatica she simply endured.

A fisherman told her of a bone in the rocks by the sea near La Cala Figuereta. There she climbed down and across the cliffs, to find on the shoreline a limb bone, probably of *Myotragus*, 'and they were quite right in saying it was the only one – could see no trace of another anywhere about – had to wade in a little bit . . . V. disappointed not to find more.' However, it was, she thought, the last remnant of a cave deposit, and it 'encourages me to continue my search by the sea'.[24] Later that day the waves quietened and she found a fisherman to take her in a boat to the southernmost tip of the island, Cap des Salines. This took two and a half hours. Although the cliffs they passed were full of caves, 'they did not look very promising and anyhow,' she wrote, 'I had not the energy to do more than look at them with my glasses for what with the hot sun beating down on me and the fishy smells I was sea sick and felt wretched!'

She returned to Santany in a trap in the cool of evening, noting in her diary partridges, rooks, the use of seaweed as a fertilizer, and her driver, a 'Very nice "cochero" [coachman], Miguel and his little boy.'[25] In the morning she packed up and hired Miguel with his small son to drive her northwards to Puerta Petro, a little whitewashed fishing village about two hours away. 'Should like to have stayed here,' she wrote, 'as I dislike being in big villages and there is generally a breeze by the sea.' Miguel made enquiries about bones, but to no avail. Then with a fisherman Dorothea walked along the top of the cliffs to some caves round the point of the bay. It was very rough going, 'but good for climbing as lots of hold,'[26] but again there was no sign of a bone deposit and the fisherman knew of none.

Miguel and his son drove her farther north to Porto Colom, the port of Felanitx, where she 'Parted affectionately!' from them, the little boy presenting her with an antique silver coin from Santany. The following morning she was so weary she could scarcely drag herself out of bed. In her diary she admits to feeling 'beastly tired' from tramping and climbing in the heat, but there is an uncharacteristic note of despondency in her writing. She had been cave-hunting for less than a week, but every hour had seemed a struggle. She went for a walk, asking of any fishermen she met in her halting Mallorquin, her much-rehearsed questions about bones in rock. As she walked, with *Myotragus* and its extraordinary ratlike teeth in mind, she examined the vegetation, old-looking scrub, thistles, rosemary, junipers, and a few pines and remained gloomy. 'Can see nothing in vegetation,' she wrote, 'that I think can account for my beast's teeth.'[27] A few years later, as she worked on *Myotragus* with Charles Andrews, they noted that the teeth had modified to deal with very hard, tough vegetation, possibly 'lichens, the hard fibrous stems of heath-like plants, or, perhaps, tough bark and woody tissue'. What they could not establish was why the food should have been so restricted and could only suggest that conditions were so arid that vegetation had become stunted.[28] It is a problem that has still not been resolved.* Recent palaeontological

* In the north of the Serra de Tramuntana, the mountain chain that runs along the west side of Majorca, excavations in the Cova Estreta in the 1990s revealed *Myotragus* coprolites (fossilized dung). From these it has been possible to identify the pollen grains of the plants that *Myotragus* in this area ate and, surprisingly, the predominant plant was box, *Buxus balearica*, a leathery-leaved plant with a hard, woody stem that is highly toxic and also hallucinogenic. It was once rampant in Majorca but is now quite scarce. The consequences for a ruminant today of eating a similar species of box, range from gastroenteritis, to vertigo, convulsions or death, depending on the quantity ingested. All sorts of questions have been raised by this discovery. For example: it is not known whether box was eaten by *Myotragus* elsewhere in the island, whether it was an habitual part of its diet, or was eaten to avoid starvation, whether this diet was the cause of the animal dying out or whether it had adapted to the toxicity of the plant (the palaeontologists who discovered the coprolites noted from the shape that 'they are unlikely to belong to sick animals') or whether the clearing of box from the land by early settlers contributed to the animal's extinction. (Josep Antoni Alcover, Ramon Perez-Obiol, Errikarta-Imanol Yll and Pere Bover, 'The diet of *Myotragus balearicus* Bate 1909 (*Artiodactyla caprinae*), an extinct bovid from the Balearic islands: evidence from coprolites', *BJLS*, 66: 1 (January 1999), 57–74.)

discoveries have shown that *Myotragus*' ratlike teeth had evolved by about 2.3 million years ago, following a period when the climate changed from warm and moist to the relatively cooler, drier Mediterranean climate, with a consequent change in vegetation from fleshy-leaved subtropical plants to a harsher, woodier plant life. The current hypothesis is that this change in climate and vegetation and the evolution of *Myotragus*' teeth is cause and effect, but it has yet to be proved.[29] Nearly one hundred years after Dorothea first raised these questions, they still remain.

She returned to the *fonda* (a small inn) deeply disheartened. She had enquired of so many fishermen and farmers about bones in rocks and had received the same negative answer from them all. Exploring some caves across the bay, she found what had evidently been a cave deposit, 'but again an absolutely useless one as so little left . . . in one only saw a few land shells and in the other just enough bone to swear by'. Nothing, she thought, was to be gained by remaining in this part of the island, for 'it would be the work of years to examine the whole coastline and all caves.'[30]

On 17 July on a hot and oppressive morning, she went north by diligence, a public stagecoach, to Capdepera. This was the nearest town to the Cueva de la Barxa, the cave in which she had first discovered *Myotragus*. Here she met Padre Caldentey, who had helped her the previous year, and his son, Miguel. At lunch she told them of her plans and they found her a workman with a mule and cart, a man called Gabriel.

At dawn the next day, still feeling dreadfully tired, Dorothea set off with Gabriel to 'my old Cueva de la Barxa'. Although she now felt so unwell that she could not even eat, she began digging where she had found the skull the previous year, 'but not much good'. When she did find a few specimens at the back of the cave in a very awkward pocket of rock, she found it almost impossible to extract them, and she was too weak to persevere.[31] This was more than exhaustion. The next morning she could not walk. 'Supposed it malaria,' she wrote in her diary, 'so dosed myself and took nothing all day but boiled milk and water.' This may well be the cause of the despondent tone of her diary entries. Unable to sleep and aching everywhere, she 'did nothing but toss about'. She sent for the doctor, 'who opened blinds to examine me and found I was covered with a bright scarlet rash! He says it is Scarlatina!'[32] For three days, 'in varying degrees of discomfort', and with a raging fever, she lay in bed. One of the servants at the *fonda*, Margarita, nursed her, while nuns came in to help at night. By 25 July she was allowed to sit in a chair for a short while, 'but felt deadly tired after it'. One thing in particular perturbed her and it is a rare personal note: 'My hair never touched all the time I was in bed and as I tossed continually as long as the fever was on me it was in a nice state. When better I lay with my head on the nun's lap and she and Margarita combed it out with the aid of spirits (contraband!) – it took two days!'[33] The next day she had just enough strength to walk down the passage to the next room and to see Miguel Caldentey who visited her that evening. By 27 July, the rash had nearly gone and she was able to dress and go downstairs. She even managed to walk up to the convent in the evening, although it tired her out. 'However,' she wrote, clearly willing herself to recover, 'I get stronger every day.'

Ignoring the admonitions of the doctor, she persuaded Miguel Caldentey to drive her in his donkey cart to Cala Rajada on the coast to look for caves. They walked through pine trees to the lighthouse, but she was too weak to do more and was back at the *fonda* by mid-morning. It was just eight days since she had become ill. The next day, refusing to rest, she hired a carriage and, again accompanied by Miguel, drove south to Son Severa. There she made her usual enquiries about bones and received what had by now become the expected negative replies. She sent some boys to a cave near by, 'but they only brought back recent things'. As the local schoolmaster was supposed to know about fossils, she went to his house, only to find that he was away and the few fossils she could see were marine or invertebrates.[34] With Miguel she drove down to a little port, but the fishermen there knew of nothing, apart from the bone in the rock that she had already seen. Not only was she still weak, but she could scarcely have imagined how difficult her work in Majorca would prove to be.

With an extraordinary effort of will, she made arrangements to move on to the Cuevas de los Colombs at Cap de Ferrutx, at the eastern point of the Bay of Alcudia, which she had briefly visited the previous year. That the doctor had told her to convalesce for considerably longer she simply ignored. For six pesetas a day she engaged a workman, another Gabriel, together with his mule and country cart, which she describes as a wooden frame with sides and floor of grass matting, 'no springs though this is *slightly* modified by a swinging seat'. It was a journey of three and a half hours to Cap de Ferrutx, to the last farm on the peninsula where they were to stay.* It was not that far from the cape itself, 'but on the opposite side to my caves which I fear will take a long time to reach'. Exhausted by the journey, she had to rest as soon as they arrived, 'but could not sleep just dripped and waved off flies.'[35]

It was barely light as Dorothea, Gabriel, and a 'donkey boy' called Guiem started off for the caves, Dorothea riding a donkey 'as I did not wish to walk much so soon'. The peninsula was mountainous, unmapped, and largely uninhabited. They had only been going an hour or so when Gabriel said that the donkey could go no farther, which left her with no alternative but to proceed on foot. Neither of her guides knew of the caves she wanted to see, and Dorothea had only the vaguest directions. 'Awful nuisance they are not mapped,' she wrote with classic understatement, 'or I might have been saved a lot of trouble.' It took them an incredible six hours of arduous walking over harsh limestone hills in ever-increasing heat, before they finally reached the caves. 'They are *not* at C[ap] Faruch itself,' she wrote in dismay, 'but more Capdepera way.' It took all her reserves of energy to clamber down the cliffs to the cave that had seemed so promising a year ago. First indications could not have been worse. 'I felt dead and awfully disappointed,' she wrote, 'when they dug a bit into a place I had marked last year and there seems to be nothing in it.' This is something every palaeontologist knows, the feeling of meeting an insurmountable wall. Experience dictates that the stuff is there, but it won't be found. Yet just when giving up seemed like a sensible option, somehow Dorothea was driven to go on, to make that extra effort until the elusive finds were revealed. In a small, awkwardly positioned crevice, or 'pocket' as she calls it, she suddenly uncovered a few specimens of *Myotragus*. Then, sifting through some earth a little later, she found 'a very queer looking (to me) bone covered with tiny teeth and a second row apparently to come on – found it among *Myotragus* remains. A curse on my ignorance! It may be nothing interesting but I do wonder!' Always at the back of her mind was the thought of finding another species, possibly even as unique as *Myotragus*, which might have inhabited the

* She calls it Belem, but it may have been part of the Ermita de Betlem, or Bethlehem, a hermitage founded in 1800, which supposedly acquired its name from the requirement that its inmates had to sleep in a stall on their first night there.

island during the Pleistocene. 'Cannot help my hopes rising,' she added, 'temporarily anyway!'[36]

The cave was difficult to work. It was just above the sea and the bone deposits in the floor had, she noted, almost been destroyed by waves sweeping in with great violence during the winter or particularly stormy seas.[37] What fossils remained were mainly in holes and crevices protected from the full force of the breakers, and that meant they were in the least accessible parts of the cave. With another lengthy journey ahead, she had little time to work before beginning the trek back. Even so, it was dark before she reached the farm, managing with a touch of her old spirit to remark, 'a good long day considering I have not been out of bed a week yet!'[38] Although concerned at the small amount of remains that seemed to be in the cave, she knew 'of no other deposits', she wrote that night, 'and it seems quite likely I may find none'.

The sea at least was kind the following day, calm enough for her to get to the caves by boat. She scanned the cliffs for new caves as they passed, 'but saw no satisfactory looking ones – went into one but the sea went in also!' The journey took barely two hours. She arrived refreshed, not exhausted by a six-hour trek, and by ten o'clock she had started work, her humour quickly returning. She found tantalizing scraps of 'mousy' jaw bones, possibly one of a shrew, a small tooth she did not recognize, and a piece of jawbone 'that I believe must be ordinary goat . . . This is not quite the playfellow of *Myotragus* that I came all this way to look for!' With a full day in which to work, she also found a few scattered *Myotragus* bones and teeth, 'just enough', she wrote, sounding brighter than at almost any time since her arrival, 'to keep up my hopes of a good haul'.[39]

The weather held and on 4 August five of them set off for the caves, Dorothea, Gabriel, Guiem, a fisherman to help with the digging as well as another who owned the boat. It was, she wrote happily, 'the best day I have had'. The finds included several pairs of *Myotragus* horns and a block of breccia, 'with I believe the nose of a young *Myotragus*!' Lunch was a feast of freshly caught small fish cooked by the fishermen. They lit a fire of dried sticks in the cave, and when it had burned down, 'they put the fish, split, on the hot embers – then when cooked laid one on a slab of bread about 1½" thick which had first had some oil poured on it – to top up the piece of tomato was squeezed over all!'[40]

With a break of just one storm-filled day, the weather continued to hold and the men even worked on a Sunday, digging ever deeper into the cave. As they excavated two pockets in the inner wall, Gabriel in his found a good bit of earth with bones, then a hole that led into a tiny cave about two yards long, tapering to a point. He enlarged the hole and was able to 'creep in', Dorothea wrote, 'and work the earth which was soft and very wet, like clay, while lying nearly at full length'. One of the fishermen, meanwhile, was working in another hole. Dorothea, who by this stage was in 'a filthy mess', dissolved into laughter at the sight of the men, who looked, she wrote, 'like two gigantic rabbits burrowing!' It may have been messy, but it was an exceptional day. The pocket into which Gabriel had burrowed was filled almost to its roof by a deposit of fossil bones, most of them *Myotragus*. The second pocket had similar contents, although birds and rodents were more plentiful.[41] Amongst all this material there were several 'more or less' perfect skulls of *Myotragus*.[42] It was a most satisfactory haul. She dug a little in the lowest cave and found a pocket with a few more specimens, but as she admitted to herself that night, 'the bones before I commenced work appeared to be very scattered and few in number – so that had I known of other deposits I should probably not have come here particularly considering the difficulties in way.' As she had found with the elephant deposit at Imboähry in Cyprus, it was the least likely-looking sites that so often produced the most remarkable finds.

Before dawn the next morning, she left the farm with Gabriel to explore the north of the island. It was a long drive, partly on the main road, but mainly on tracks so awful she feared her fossils might be smashed. In the pine woods that separated the vast marshland area of S'Albufera from the sea she noted a great number of birds including striped hoopoes, turtledoves, and shrikes.* It was evening before she arrived at a *fonda* at the port of Alcudia and could scrawl in her diary, 'Oh the purgatory of all those hours in that springless cart and my sciatica leg pained horribly, could hardly walk when I eventually got out.'[43]

Gabriel returned to Capdepera, and Dorothea went cave-hunting by boat with the 'padrone' of the *fonda* and a fisherman. For three days, when wind and torrential rain allowed, she searched for caves round the

* Long renowned for their wildlife, these wetlands are now protected and have been designated a nature reserve and bird sanctuary.

northern tip of the Bay of Alcudia and past the Cap de Menorca. She found remains in cliffs on the Bay of Alcudia, in a cave known as the Cueva del Contrabanda, reached after a climb of twenty-five feet (with no mention of sciatica). It was only about three feet high and she spent nearly four hours crouched in here, finding some bones and teeth including *Myotragus*, but none of the other caves she explored revealed anything. Past the Cap de Menorca they came to an area of cliff peppered with caves and holes. These too were empty, one in particular to the annoyance of the *padrone*. It had been, he told her, 'a very good contraband hole . . . but it had been "given away" to the Carabineros and they lost a lot of tobacco'.[44] Smuggling in the Balearics had become something of an art form; the thousands of little bays and uninhabited cliffs made law enforcement impossible. As the traveller Frederick Chamberlin noted, 'Every owner of a row-boat, of a fishing-smack, of anything with a sail or a motor, was in the [smuggling] industry . . . The whole administration was local in birth. The soldiers were born in the islands, and not one of them had a relative who was not a smuggler. Every official protected his friends, until finally the highest officials had to be sent from the mainland, or there would have been no revenue at all.'[45]

With no further indications of bone deposits, Dorothea packed her fossils and journeyed south, forced to endure another springless cart as 'don't know how I should have kept bones safe in train'. She was in pursuit of 'black fossils', fossil bones found in coal, and she headed for the small town of Lloseta on the edge of the Serra de Tramuntana where there were lignite mines. On every level it was a futile and exhausting exercise. It even started badly when she discovered there was no *fonda* at Lloseta and the only café she could see had no rooms. 'Did not quite know what to do,' she wrote, 'when a disreputable looking man came up to me and said he had a room if I would come and see it – Walked to the edge of the village where came to a tiny Café of the poorest sort – however there was no choice so I decided to stay.'[46] Her room at the café was in a vermin-infested loft with just a flea-ridden mattress on the floor. She made many enquiries about black fossils and even resorted to her old trick of using the town crier, but 'heard nothing satisfactory'.

In the morning, revived by a cup of milky coffee, she set off with her host, the 'disreputable looking man . . . in a most awful old cart and antediluvian horse', to the lignite mines near the village. She found the dialect here so impenetrable that she hired a man who spoke French (learned as a fruit seller in France) to go with her. Of the six mines she visited over the course of that long day, bones had been found at three, but were there no longer, while at a fourth, a newly opened mine near Selva to the north of Lloseta, she was told they had found a jaw and some teeth, 'and other bits but had smashed them all'. There were just a few fragments left, which they gave to her. That night she wrote in her diary, 'Came back tired and disappointed at my utter want of success.'[47]

She awoke in the morning still 'not feeling particularly flourishing', and waited until midday to see if any fossils might be brought to her. When none came, she packed and left for Palma 'in horrid cart really too small to hold all my things – sciatica as usual – rather hot dusty road arrived at Grand Hotel about 5.30. My trip over.'[48] All she could think of for the moment was the exhaustion and awfulness of the last few days, not the crates of significant fossil bones that she had successfully found. In the morning she began her packing and shopped for mementoes, buying little chain silver purses for which the island was famous for her mother and Leila.

Two nights before she left Palma, Señor Bosch, the British Vice-Consul, who had greeted her so warmly on her return to the island, came to see her at her hotel. The diary that Dorothea kept so assiduously was a work diary; she might include details of her illnesses and such like, but that was because they impinged on her work. In the only entry of its kind in any of her papers, she scrawled on 16 August, 'I do hate old men who try to make love to one – and he ought not in his official position.'[49] She has then, at a later date, scribbled this out, although it is still easily legible. Bosch may have made amorous overtures to her and been curtly dismissed, but it can only be a matter of conjecture how she may have responded to other men who tried – or perhaps did – make love to her. The true nature of her relationships with Jack Wodehouse and Richard Seager, for example, with the absence of any personal evidence, can never be known. As for marriage, it would – between the demands of her parents and the expectations of a husband, no matter how understanding – most probably have ended her career. It is hard to see how she could even contemplate jeopardizing all that she had struggled so hard to achieve. That little scribbled-out sentence on 16 August 1910 is one of the very rare occasions where Dorothea the woman is briefly glimpsed without the mask of Miss Bate the palaeontologist, and even then, only questions are raised; she gives no answers.

On 18 August she left Palma by steamer for Barcelona on a 'lovely moonlight night', after a visit of five intense weeks. Her health had recovered, apart from the sciatica, and she had collected in very difficult circumstances some splendid specimens. All, however, had come from Majorca. Still to be explored were the two other substantial islands of the group, Menorca and the scarcely populated and primitive Ibiza.

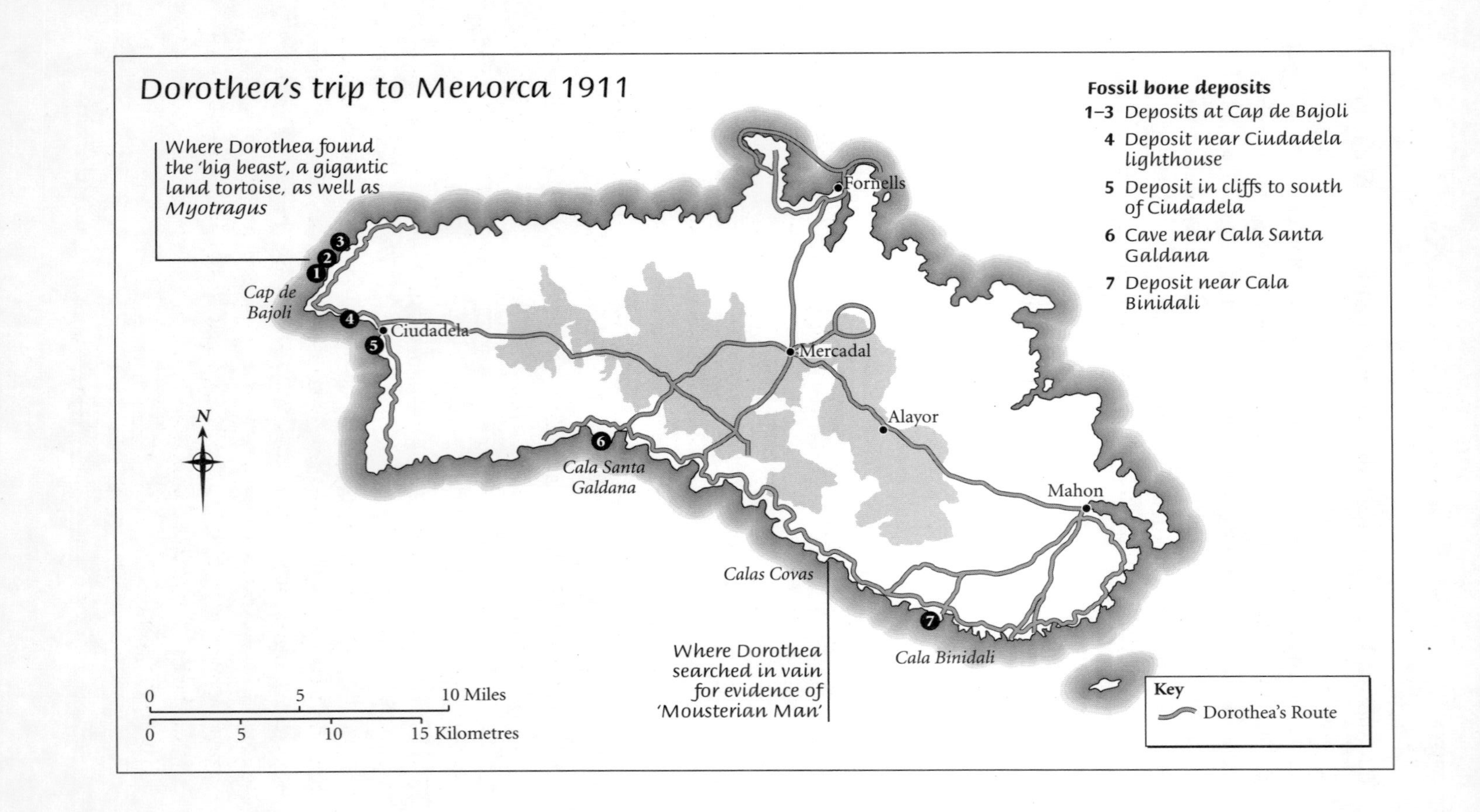
Dorothea's trip to Menorca 1911
Where Dorothea found the 'big beast', a gigantic land tortoise, as well as Myotragus
Fossil bone deposits
1–3 Deposits at Cap de Bajoli
4 Deposit near Ciudadela lighthouse
5 Deposit in cliffs to south of Ciudadela
6 Cave near Cala Santa Galdana
7 Deposit near Cala Binidali
Fornells
Cap de Bajoli
Ciudadela
Mercadal
Alayor
Mahon
N
Cala Santa Galdana
Calas Covas
Cala Binidali
Where Dorothea searched in vain for evidence of 'Mousterian Man'
0 5 10 Miles
0 5 10 15 Kilometres
Key
Dorothea's Route

CHAPTER 10

The Fates Seem against Me

Dorothea's departure from Spain was a flurry of missed trains and negotiations over the cost of transporting her boxes of fossils. She arrived back in Britain at the end of August 1910, the bones some weeks later. Her priority, as she wrote later to Dr Smith Woodward, was to arrange a third trip to the Balearics for the following spring. 'My idea was to explore Menorca this time – do you approve? It seems a pity to leave it untouched – there might be more wonders there!'[1]

Her time was still not her own. During the first three months of 1911 she was with her parents, first in South Wales and then, as she told Michael Oldfield Thomas, 'languishing' in Brighton.[2] It was April before she could return to London and work at her latest Majorcan haul. In the few days available to her, she had time only to examine the tiny teeth which she had found with *Myotragus* in the Cuevas de los Colombs, which had so tantalized her. Far from being a new species of small mammal, however, the teeth turned out to belong to a common and still-surviving species of *fish*, the wrasse, notable for its thick lips and strong teeth, which are designed for eating shelled animals such as barnacles.[3] Its presence in the caves was attributable simply to its being brought in by seabirds and had evidently been deposited there over a considerable period of time. Her latest specimens of *Myotragus* remained untouched. On her return home, she concentrated on planning her third visit to the islands.

What she needed was a grant. She applied to the Percy Sladen Memorial Fund, which funded research work, particularly expeditions, in natural science (and still does). Her application was 'To make an expedition to the Balearic Islands to search for and to collect the remains of extinct vertebrates from the caverns and other fossiliferous deposits'. She hoped to obtain further specimens of 'the peculiar ungulate *Myotragus*' that she had discovered in Majorca, to throw light on its origins, and to visit Menorca and Ibiza, whose extinct fauna was 'at present entirely unknown'.[4]

She asked the Fund for £80 (about £4,500 in today's values) to help towards travelling expenses, hire of labour for digging, packing of specimens and for 'personal maintenance'. Her application was supported by her old friend and a Trustee of the Fund, Dr Henry Woodward, who in April 1911 wrote a long and detailed testimonial. He described each of her explorations in the Wye valley, in Cyprus, in Crete, and her previous visits to the Balearics, and the importance of the finds she had made. Woodward noted that 'Miss Bate has already been assisted in her researches by grants from the Royal Society and her anatomical knowledge has justified her employment (temporarily) in the Geology Department of the British Museum.' He revealed that she had also taken special lessons in comparative anatomy from the osteologist William Pycraft (who was in charge of the Museum's mammal skeletons and was well known for his popular weekly articles on science in the *Illustrated London News*) and from the Keeper of Geology, Dr Smith Woodward.[5]* Within a month she heard that the grant had been approved, and she hoped to leave in June, but by the third week of July the cheque had still not arrived. In a curious echo of the delay she had been caused in waiting for Henry Woodward to notify her of her Royal Society grant in Cyprus, the Fund had actually sent her cheque some weeks earlier to Woodward as her sponsor, but he did absolutely nothing about forwarding it to Dorothea until prompted by his fellow Trustees. She received it at last on 1 August 1911, but it was not until 23 September, much later in the year than she had planned, that she was able to set off on the three-day journey to the Balearics. Once again, it would be an expedition that tested all her physical and mental resilience.

* Dorothea turned her notebook for these lessons upside down and the unused pages became her diary for her third visit to the Balearics.

In Palma she was met by Padre Caldentey and his son Miguel who organized a man for her as a 'general factotum'. Also called Miguel, he was a horse-dealer by trade and was to become as indispensable to her as the faithful Nikola had been on Crete: as interpreter, bag carrier, companion wherever she went to explore, finder of places to stay and of extra workers to dig. Her plan was first to explore Ibiza, but not before she had spent a day in pursuit of black fossils at the lignite mines near Lloseta. It proved as frustrating an activity as the previous year. At every mine she was greeted with the same shake of the head and shrug of the shoulders. She caught the last train back to Palma, feeling 'V[ery] tired and I have a cold'.[6]

Ibiza, she hoped, would prove more fruitful. It was a journey of six and a half hours by steamer to the island, which the French traveller Gaston Vuillier described as 'distant and rarely visited'.[7] It was hilly and fertile, wooded with pines, figs, olives, and almonds, while its vast salt flats, or *salines*, had been an essential part of its economy since earliest times. Tiny villages were perched on green conical hills, a cluster of whitewashed houses that dazzled in the autumn sunlight with 'strings and strings of red peppers', Dorothea noted, 'hung out to dry sometimes nearly covering the wall and giving a very bright effect'. The island had a reputation in Majorca for being a dirty place, but it was the poverty that most struck her. 'All the people about this country seem rather poor and wretched,' she wrote in her diary; 'one wonders why when it seems a fertile spot. At first sight they do not seem so friendly or polite – but I believe really mean well.' She was told later that the law of primogeniture was responsible for many of the difficulties of the islanders, as younger sons tended to emigrate to America. There were, she discovered, fewer fishermen than on Majorca and even they preferred to tend their crops, keeping their boats as a sideline. The implications of this for her were serious, as it 'probably means that they do not know every stone of the coast as is often the case in Mallorca'.[8]

For ten days, with Miguel to help her, she climbed, sailed, walked, rode and scrambled her way round the island, never allowing herself rest while it was daylight, but investigating with unremitting zeal every likely cliff and cave. She purchased a pair of the locally made string-soled boots, only to experience 'rather a nasty climb as rocks wet and greasy and my string sole boots ditto'.[9] Occasionally she slept in 'quite a decent inn', but more often found herself in a flea-ridden room, only 'technically clean', with a 'v. bad supper = bread, bad! eggs and our own cheese and plenty of fruit and hard peaches'.[10] But of fossil bones there was no sign.

Pages of her diary are filled with detailed descriptions of the island and its people. She visited the great Carthaginian necropolis that was being excavated at Puig des Molins near the town of Ibiza. The 'man from the museum' showed her round the site, which she thought 'a marvellous place – the whole hillside thickly covered with tombs close together'. With a generosity unexpected (and possibly unwise) in a curator, he allowed her to take away some fragments of lamps and pots.[11]* She visited the fish market and noted that many of the women were wearing native dress: 'a very full print skirt sometimes so full it looks as if there was something to keep it out – short waisted little bodice with shawl or big handkerchief over it – a very small and often highly coloured apron – a handkerchief over head and above that a white round straw with low crown and black ribbon that streams behind'. The hair, she observed, was worn in a long plait, often tied with a bright red ribbon, and she adds, somewhat anatomically, 'they often bind the proximal end of plait round and round with tape as sometimes done to horses tail'. The men were only a little more soberly dressed in short full jackets, often dark purple, a bright handkerchief round their neck, tight trousers, which 'widened out round foot, something like costers', and a black hat a little like the women's but in felt.[12]

At the end of ten days, Dorothea had acquired considerable information on the culture, geology and geography of Ibiza, and of the Carthaginian civilization that had once flourished there, but in the countless caves and fissures that she had investigated, she had found nothing of *Myotragus*, not one patch of bone breccia or even a solitary bone in a rock. In fact, as she later wrote, there was 'no trace at all of any ossiferous deposit'.[13] This was the first island she had extensively explored that had produced absolutely nothing and to her disappointment was added considerable concern. She had been awarded a grant for this trip, based on her experience and reputation, yet in spite of her efforts she had failed to find any indication of an extinct fauna. That of course did not prove that one did

* Later excavations on the site have revealed more than four thousand tombs, of Carthaginian and later civilizations, and a wealth of Punic artefacts now in museums in Ibiza town.

not exist, only that she had not found it. No one else did either for another seventy years, when a most unexpected vertebrate fossil deposit was discovered on Ibiza in 1981 by a team of Majorcan palaeontologists.[14] Dorothea would have found it of immense interest. The earliest fauna, from the Upper Miocene (23 to 5 million years ago), included mammals and reptiles, but by the Upper Pleistocene there was not one single flightless mammal, only bats, lizards, sea fishes, and a great quantity of birds, of which fifty-two species have been identified so far. This suggests that a great natural disaster wiped out almost all the endemic animals of the island,[15] which, the palaeontologists note, makes the fossil fauna in the Pityusic islands unique in the Mediterranean.[16]

Reluctant to give up, Dorothea nonetheless decided that it would be 'wiser to turn my attention to Menorca',[17] and she left Ibiza for Palma. There she repacked her belongings and, pausing only to see the indispensable Padre Caldentey for introductions on the island of Menorca, headed by train and cart to the port of Alcudia in the north of Majorca, to catch the ferry. It was a five-hour crossing in the early hours of the morning. As they sailed round the Menorcan coast to the capital, Mahon, Dorothea stood on deck, her binoculars trained on the limestone cliffs she had come here to explore. 'Personally am much disappointed in my first view of island which,' she wrote that night, 'looks more or less like a flat mass of stones with a couple of small hills in the N[orth] E[ast].' Although the low limestone cliffs they passed were honeycombed with caves, they dropped sheer into the water. There were very few bays and, unlike in Majorca, 'no fishing boats to be found round every corner – this will probably add to my difficulties'.[18]

The approach to Mahon was spectacular, a three-mile-long deep-water harbour whose strategic possibilities had made the island such a prize in the wars of the eighteenth century. It was captured (and lost) three times by the British who extensively rebuilt Mahon. There were few places to stay, although Dorothea thought her hotel, the Central, was clean with quite good food.* After a mosquito-plagued night, she set off in a mule-drawn carriage for the town of Alayor about eight miles away along the splendid road whose construction had been ordered by Richard Kane, the first British Governor of Menorca. Padre Caldentey had given her a

* As late as 1950 there were only around two hundred hotel beds on the island.

letter of introduction to the local schoolmaster who, she hoped, had 'much information'. He could not be found. She was, however, told of a notary 'who understood fossils – still a bit early so had to wait while he dressed!'[19]

His interests were mainly archaeological, particularly of the prehistoric peoples of Majorca and Menorca whose culture is known as Talayotic, after the eponymous massive watchtowers scattered throughout the islands. Menorca is even richer in this culture than Majorca, with literally thousands of sites. In addition to the great *talayots* there are *taulas*, huge monoliths topped by a capstone, which may have been used for animal sacrifice, and *navetas*, stone structures like upturned boats, which are thought to have been used for both burials and dwellings. Pre-Talayotic man was a cave dweller, and a number of cliffs are honeycombed with caves, natural and man-made, which, like the *navetas*, were used for both dwellings and burials. The notary had excavated *talayots* and caves at the great troglodytic settlement at Calas Covas on the south coast. He showed Dorothea some of the artefacts he had found and also 'a block of stalagmitic stuff from the floor of one of the caves at Calas Covas,' she wrote. 'It seemed to contain human remains and a mass of skulls and bones of small rodents.'[20] Little was known of these cave dwellers. Some of the caves seemed to date back thousands of years; others had carvings that indicated they had been occupied in Roman times or later. Although Dorothea did not think the 'stuff' looked very old, she noted that it was 'certainly worth going to see'. In 1911 nothing was known of when man might have arrived in the islands.

The weather over the next two days was wet and blowy, and her only excursion in a boat was brief and uncomfortable. The waves made it impossible to land and her sufferings 'reached the acute stage!' She explored instead some artificial caves accessible from the shore a little way down the coast from Calas Covas. In one she found 'a great quantity of human bones in the few inches of sandy soil on the floor – grubbed about a little superficially found 3 lower jaws of different ages – the smallest quite tiny'.[21] As she reflected on the contrast between these bones lying loose in the soil and the human and rodent remains she had seen at the notary's house, an idea of such potentially astonishing significance occurred to her that she moved at once from the hotel in Mahon to lodgings in San Clementi to be as close as possible to Calas Covas.

'Hoping great things Mousterian men etc,' she wrote in her diary on 14 October, 'from that stalagmitic lump I saw at the Alayor Notary's.'[22] Little wonder she was excited. By Mousterian man she meant Neanderthal man and no human remains of such antiquity had ever been found in the Mediterranean islands. These people of the Ice Age were considered then by much of the scientific community to represent, as the eminent anatomist Sir Arthur Keith put it, 'the stock from which all modern races have arisen'.[23] Neanderthal skeletons, so called from their site of discovery in the Neander valley in Germany in 1856, were rare, although their flint tools were widely recognized at sites throughout Europe. These were distinctive, small, well-shaped flake tools, made by a well-aimed blow that shaved off a flake of flint from a pre-prepared core, giving the implement its characteristic shape and surface. This tool industry had first been identified in the 1860s in southwest France, at a rock shelter in the Dordogne called Le Moustier. As interest in such discoveries grew, further skeletal finds were made, in Belgium, Croatia, Gibraltar and, at the beginning of the twentieth century, in France. In 1908 and 1909, Neanderthal remains were found in three sites in the Dordogne, including Le Moustier itself. To the vexation of the British scientific community, so far no human remains of such antiquity had been found in Great Britain.

These discoveries in France and elsewhere of human remains were thought then to be hundreds of thousands of years old. With no scientific means of dating bone, absolute dating was impossible. The best that could be managed was relative dating (that, for example, the Late Palaeolithic was followed by the Neolithic) and estimates. In 1911, Arthur Keith could only write that the Early Palaeolithic period, which lay within the Ice Age, was 'one of very great but uncertain length . . . Its duration is variously estimated from 200,000 to 400,000 years.' Neanderthals were believed to be the dominant race at the beginning of that period, 'modern' humans at the end.[24] Current estimates now date Neanderthals from about 150,000 to 30,000 years ago. These discoveries had sharpened scientific and popular interest in the evolution of mankind and the possible discovery of a 'missing link' between man and the apes. If Dorothea could substantiate her idea that the bones in the notary's stalagmitic lump really were Mousterian, it would be an outstanding achievement for her and a splendid coup for the Natural History Museum to have in its collections such prized specimens, discovered furthermore by a British scientist.

With great anticipation, early on the morning of 14 October, Dorothea rode down the narrow track to the strange, deep cove of Calas Covas. The task she faced was enormous. The creamy-gold limestone cliffs on either side were pierced all over with caves, most of them artificial, like great black eyes. She thought there must be more than a hundred of them. She began a systematic search, working her way along the cliffs, entering five, ten, twenty caves. Each one mocked her with its emptiness. Many of the caves were inaccessible; as the writer Frederick Chamberlin noted, they could 'only have been reached across many yards of perpendicular cliff in any direction – with no surface to which any man who has ever lived could have clung'.[25] When she did find remains in a cave near the top of the cliff, 'the floor was covered with earth and a regular charnel house full of human bones – saw a number of others with bones too,'* but all these were burial sites dating from the Talayot people and later. A boy she had hired as a guide brought her a piece of bronze, claiming that it was part of a bell and that he had found it by a skull in a burial chamber. When she went there, she found nothing. He took her to caves with carvings and inscriptions, none remotely of Mousterian age, and her despair deepened. Finally the boy took her to the cave where the stalagmitic lump was said to have come from. Not only was there no sign of stalagmitic material, but 'almost doubt that piece came from here', she wrote miserably, 'though I don't know why the boy should lie about it'. She had explored every cave that it was humanly possible to reach and she was physically and emotionally exhausted. On a note of uncharacteristic bitterness she wrote, 'so my expected treasure ends in the usual way.'[26]

It is hard to imagine how difficult this diary entry must have been to write. Over four pages Dorothea detailed the failures of the day, yet still recorded with obsessive honesty how she began 'hoping great things'. She must have felt wretched. To date, only around four hundred Neanderthal individuals have been found and none on any Mediterranean island. The

* A similar find was made by J. E. Crawford Flitch, who travelled in the Balearics a year or so before Dorothea. Near San Cristobal he was taken by the local chemist to some low caves cut in the face of a ravine. In the floor of one the chemist revealed a 'charnel-houseful of human bones'. Crawford Flitch then found himself helping the chemist to reconstruct some of the skeletons. When they had finished, the Englishman wrapped some of the 'prehistoric skulls' in a copy of the *Daily Telegraph* and they left. (J. E. Crawford Flitch, *Mediterranean Moods: Footnotes of Travel in the Islands of Mallorca, Menorca, Ibiza and Sardinia* (London: Grant Richards, 1911).)

nearest of these discoveries to the Balearics is in southeast Spain. The stalagmitic material encasing the human remains found by the notary did not necessarily imply great antiquity, only that water had relentlessly dripped through limestone rock onto the bones during the centuries that they had been interred in a cave. In just a few years, in the right conditions, a small object can be encased in a hard, calcite coat. Dorothea would have known this, but the notary's staligmitic lump was the only one of its kind she had seen in the Balearics, and even though she had initially thought it 'did not look v. old', its contrast with the other human remains she had found made her quest unavoidable.

She had first considered the question of when man might have arrived in the Balearics the previous year. While excavating at the Cuevas de los Colombs on Majorca, she had come across something that intrigued her and made her consider the wider implications of her work. 'In two pockets,' she wrote, 'I noticed a great many bl[ac]k marks in the earth which to my imagination looked like traces of charcoal or some burnt material – but then how account for them except by human agency? Did early Mallorquinos feast on *Myotragus*?'[27] In the margin next to this entry, Dorothea has, at a later date written 'No', possibly as a result of this futile search for Mousterian man, but the coexistence of man and *Myotragus* is still a highly contentious issue. Over the years a great body of data has accumulated, subject to interpretation and reinterpretation as more palaeontological and archaeological excavations are made and methods of dating improve. It is now known that the animal survived for over five million years, adapting to climate and vegetation changes. The evidence available to Dorothea and her colleagues indicated that the species had died out before the arrival of man, but there is agreement today that *Myotragus* survived until approximately four thousand years ago, about the time that the first human settlers are now thought to have arrived in the islands. This suggests that, as with island extinctions elsewhere, there is a causal connection between the arrival of man and the extinction of *Myotragus*, either through hunting (so-called overkill), habitat change, or disease introduced by man. It also shows that, far from being inhabited by Neanderthal man, as Dorothea had thought might be the case, Majorca and Menorca may well have been the last islands in the Mediterranean to be settled by humans.[28]

For Dorothea, the pioneer of palaeontological excavation in the islands, failure to find anything even remotely resembling the stalagmitic lump was yet another setback in what she felt was a disastrous trip. After three weeks in the Balearics, she had absolutely nothing to show for it except acute disappointment and her fast-diminishing grant from the Percy Sladen Fund. She left Calas Covas with Miguel, passing a landscape so curiously full of stones from Talayotic ruins that in spite of herself she began to take note. There were tiny fields surrounded by enormous walls, piles of stones scattered everywhere, 'and how ploughing is done under the circumstances seems rather a marvel!' Miguel found a little land tortoise and she noted a drift of autumn flowers, tiny white jonquils, and the purple of autumn croci. She asked anyone they passed whether they knew of any bones in rocks and noted inconsequentially that 'unlike Mallorca many of the men here wear moustaches – foreign influences I suppose', but in between the flowers and stones, she writes sadly, 'No-one I meet about seems ever to have seen anything like my pattern bone.'[29]

The next three days brought no relief. 'Another disappointment again today – the fates seem against me – fisherman came up to say still too rough – and I don't think weather looks like improving.'[30] She drove to Fornells, a little fishing village in the north of the island. With some difficulty she persuaded a fisherman to take her out in a boat, discovering as soon as they were out of the harbour that 'it was quite rough – a heavy swell and a good deal of surf against cliffs which would have made it impossible to land – added to which I felt decidedly bad – though not actually seasick – lay on the deck on a wet sail.'[31] Thus supine, she observed that the limestone cliffs fell sheer to the sea, that there did not seem to be very many caves and, in any event, the surging waves barred any approach.

It was another three days before the weather improved. As she noted wryly, 'Not feeling very fit today so of course the verdict came good weather! Felt inclined to stay in but did not like to risk a fine day.' She went north round the Cap de Bajoli, the jagged coastline 'pierced by the mouths of barrancas'. These were gorges or ravines that ran down to the sea, although they also occurred inland in the limestone of the island. About midday they landed in a little bay some way beyond Cap de Bajoli for lunch, 'after which I climbed up the cliffs and strolled along rather aimlessly – at the next bit of barranco I was looking around when I suddenly saw a piece of bone in a piece of red rock – could hardly believe it'. She examined the ground carefully and found a sort of trail down the hillside of bits of bone in hard pale red earth. She found a few 'goaty teeth, I expect Myotragus', but nothing else seemed obvious. It was, she wrote, 'a great thing to have found one deposit though I fear not a good one – however'.[32] With Miguel she walked back to Ciudedela, feeling for the first time in nearly a month that her trip might be worth while.

By morning she was full of optimism. 'Went off in pony cart this morning with Miguel, a quarry man lots of tools and full of hope of all sorts of treasures!' They left the pony at a farm called Sestrucarias and walked on to 'my barranca', which she describes as being the third little bay northeast of the semaphore station. It had once been an enormous bone deposit, extending for over sixty yards, but all that appeared to remain now was 'but the merest remnant',[33] a long narrow strip of perhaps twenty to thirty yards, 'looking much like a mule track'. In the first section she looked at, there were unmistakable remains of *Myotragus* and, intriguingly, there were signs of some much larger bones, 'certainly ever so much larger than those of *Myotragus*'. They were also of a curious texture. 'What strikes me greatly about them is that they seem to be so very cellular and spongy-looking – if I had not actually seen them with remains of M. would almost wonder if they could be those of a cetacean [whale].'[34]

She hunted on, prising away at the deposit, which at most was barely a foot thick and in many places too narrow to contain more than a thin section of bone. Late in the afternoon, just below the top of the cliff she saw 'some red stuff' in a crevice in the cliff face and found that it contained 'a mass of rodents etc', but nothing more of what she calls the 'big beast'. Over the next few days she found three more deposits along this coast, mainly of *Myotragus*, although one of them also contained more of the big beast, but all the deposits were patches of scattered bone or just faint traces, with no sign of good limb bones or teeth. It was, she wrote, 'dreadfully tantalizing'.[35] Her frustration was all the greater as the cliffs had evidently once 'been v. bony',[36] but, having investigated every possible place, she found nothing more. Miguel managed to raise a smile from her by pointing out that the legally grown tobacco fields in this remote part of the island were actually of great interest to smugglers, but he could not lessen her disappointment. More in keeping with her mood was the 'horrid looking piece of coast' near by, where a cross stood in bleak memory of the 150 people who had drowned in the wreck of the French ship, the *General Chanzy* the previous year.

Packing her box of specimens 'such as I have', with Miguel driving the cart, she set off to explore the south coast, steeling herself against weather that was now atrocious and the unbelievable difficulty of finding anything in this island. In a 'sort of broken down cave', she found a tiny scrap of bone barely a quarter of an inch long,[37] which was encouragement of a sort, but then nothing more during two more days of endless searching. On 1 November, when she was exploring further west on top of the cliffs, they came to the remains of barracks built during the English occupation. Climbing down the rocks in a chill and driving rain, they found two caves. One contained nothing but a few land shells; the other, which was open to the sea and only a little above it, contained remains, 'evidently of *Myotragus*', and at last they were good ones. It was a large, high-ceilinged cave with great blocks of stones lying heaped up at the entrance. At the back, which was cold and damp, the floor was sandy, slanting up to within eight feet of the roof. A section of it had broken away, and in the top two feet she dug out land shells and splendid *Myotragus* specimens.[38] Just to get out these bones, however, had meant removing enormous quantities of sand and to dig further was beyond her physical and financial means.

Keeping to the same area, she walked for over a mile along the cliffs, then commandeered a passing fishing boat (and its crew), 'as it was impossible to examine this piece of coast properly from land – cliffs pretty high and land much broken up and covered with trees and thick scrub'.[39] There was no shortage of caves in the cliffs, but most were at sea level. The weather was quite good and she risked landing at a few caves, but there was 'a bit of a swell against rocks and not very easy landing – some rocks so sharp made my hands bleed!' She explored that part of the coast for four days, sustained by the thought that the next cave she found might reveal more of *Myotragus* or the big beast, until there was nowhere left to go. On 4 November she packed up her fossils and left for Mahon. She would allow herself just a few more days cave-hunting in the southeast of Menorca, before returning to Majorca.

Dorothea's determination was now tested even more. The wind-lashed sea was far too fierce for a boat and, hunched against the storms, she trudged the cliff-tops south of Mahon over another four days, negotiating 'such rough rocks' and not even finding many caves. Her thirty-third birthday on 8 November brought 'a great budget of letters', but no respite from the unrelenting weather. By 9 November she had found just one small deposit along this stretch of coast. It was a narrow strip, perhaps eight to ten yards long near Cala Binidali, and similar to the dreadfully hard deposit where she had found the big beast. Here, however, 'could only see many remains of little rodents and what believe to be *Myotragus* – not a sign of the big beast'. At its greatest depth it went down for two feet. 'Of course tried a bit to see if anything further underneath,' she wrote wearily, 'but with usual negative result.'[40]

On 10 November, exhausted with her efforts of the past days, Dorothea caught the overnight ferry for Majorca. Pausing in Palma just long enough 'for meal and clean up', she and Miguel travelled by train, then diligence, to Capdepera for a last flurry of cave-hunting. Before she left the Balearics, she wanted to re-examine the Cuevas de los Colombs, which had proved such a fruitful deposit the previous year.

The weather was now so dreadfully uncertain that she thought it hopeless to try to reach the caves by boat and dispatched Miguel to enquire about a suitable route overland. Miguel had co-opted his brother, 'with the latters mule and cart (springless)', and they set off across that inhospitable peninsula on a route that turned out to be little improvement on the previous year. 'The latter part of drive was rough in the extreme,' she wrote, 'and one would think must put the strength of the cart to the test even when empty.' After three hours of this the cart could go no farther and they had to walk for another hour and a half before finally reaching the caves. Drily she comments as she surveyed her work of the previous year, 'Find that, anyway in this instance, I am not a very satisfactory digger to follow! And there is not a great deal left.'[41] She extracted what she could, but the short winter day gave her few hours to work.

With the sun setting, it was too far to return to Capdepera that night, but Gabriel knew of a shepherd's house about half an hour's walk away. They arrived to find the stable 'occupied by 37 squealing little black pigs and the rest of the house locked up! Which did not,' she wrote, 'seem promising.' With some initiative and a log of wood, the men broke open the door and found an empty four-roomed house. A brushwood fire and a supper of bread, cheese, and 'a boiled egg apiece', which they had brought with them, improved things a little. 'Miguel made a bed for me near the fire – one of the grass mats off the cart and a sheepskin underneath and my burberry [*sic*] and another sheep skin on top.'[42] The men slept in another room on the shepherd's straw bed. The wind howled dismally throughout the night and they were all up at dawn. Once more they walked to the cliffs, but the waves were so huge they flooded the lower caves. They went into the higher bone cave to have breakfast, 'when suddenly a larger wave than usual dashed up drenching us all and as the wash fell back I saw our only loaf, my goatskin bag etc being washed down the rocks'. One of the men managed to retrieve the dripping objects, but it was clearly impossible to work there. After all the reverses of this trip, she is past anger and frustration. The best she can manage is: 'rather disappointing after all the trouble and discomfort of getting here'. And yet again, standing in that spray-soaked cave, her wry humour never quite vanishes: 'Waves were magnificent to watch but I was hardly able to appreciate them properly!' They lit a grass and brushwood fire 'and more or less dried ourselves – a comfort as it was dull work being wet'.[43] Behind the humour, however, was the knowledge that yet more days had been lost and her money was running out.

On 16 November Dorothea set off by herself for Lloseta in the west of the island, arranging to meet Miguel in a few days. 'Already miss Miguel very much,' she wrote, 'as find I have left behind my camera stand and my map!'[44] She was again in pursuit of the elusive black fossils, but at every mine it was the same story: nothing had been found. By the time she left Majorca, she had visited every single lignite mine and, in spite of offering a reward, had not seen one useful fossil. Feeling 'rather bad', with a headache and a chill, she drove up through spectacular mountain scenery to Lluc to examine a nearby plateau. It reminded her of Katharo, the high mountain plateau in Crete where she had found remains of *Hippopotamus pentlandi*, but although she searched the streams, looking for cuttings or sections that might reveal a bone deposit, there was nothing. She lunched with farmers on the local delicacy of 'rice and bits of blackbird, followed by roast blackbird or thrush on toast – v. good'. This had, she noted, become quite an industry (as indeed it still is). She caught the train to La Puebla where she had arranged to meet Miguel and woke in the morning feeling 'anything but well today – bad head, aches all over and spots on my face!'[45] This may well have been the songbirds' revenge.

Dorothea's last five days in Majorca brought her no new finds, despite her tireless searching. Near Formentor she braved the seas to examine the high and perpendicular cliffs, but 'there was nothing inviting in the cave line – and few of any sort.' It was too late to return to port and they landed at a lighthouse, manned by two friends of Miguel's. 'Had to trudge up 270 steps cut in the rock to begin with,' she wrote, 'then a steep path up to the lighthouse which is perched upon a wonderful eyrie.' The men were 'delighted to put us up for the night . . . went for a walk – saw over the lighthouse, supped, I answered innumerable questions! And went to bed!' But not before they had stood and marvelled at the spectacular view: ahead across the indigo sea towards Menorca, behind them the mountains of the Serra de Tramuntana with a 'splendid fiery sunset'.[46]

The sea was too rough to explore the cliffs further the next morning. She made a last attempt to explore by land with Miguel, but could neither see nor hear of any likely caves. She returned to Puerta de Pollensa, packed, and by carriage and train went south to Palma. It was time to go home. On 24 November she packed her last box of fossils, then found she was missing some specimens, which only arrived from Capdepera 'when M. was seeing me off on board the boat for Alicante – so left all boxes in his charge to forward to Cooks Barcelona'. Once again she observes of her companion of the last few months, 'I feel very lost without Miguel.'[47]

Before she went home, she arranged to have one last attempt at cave-hunting, this time on the mainland. North of Alicante are two promontories, Cabo de la Nao and Cabo Antonio, the points from which, millions of years ago, the great finger of land had extended from the Iberian peninsula into the Mediterranean basin. Their cliffs, she had been told, were cave-filled. She found a room in a fishing village, the place was too small even for an inn, and 'lived v. much en famille! Too many children and fleas but people all most kind and food quite good – fish fried, fish and rice, fish cold, fish small, fish large and so on!'[48] The fates were still against her. On the one calm day, her attempts to explore by boat were thwarted by a *carabinero* who, against all her arguments, forbade her exploring by boat for reasons she never discovered. Her writing fairly screams across the page as she writes of this. She packed and moved on by donkey and baggage mule to an even more remote place and with difficulty found a boat to take her out. The effort was not worth while.

They could reach only two caves; at the first it was too rough to land, while at the second they could row into the cave but the surf was so great they were beaten back. What made it all the more galling was that she was told there were promising caves round each of the promontories.

Dorothea had, finally, to recognize what had been apparent for much of the past two months: 'It seems hopeless trying cave hunting at this season as it must be done from a boat.'[49] As she wrote this she must have thought how very different her exploration might have been, had Henry Woodward remembered to send her the grant in May.

Out of the hundreds of caves and cliffs in the Balearics that she had climbed and scrambled to, crawled and might have drowned in during her three short visits, only six sites in Majorca and seven in Menorca had produced anything at all. In Menorca alone she had minutely examined a stretch of about *sixty* miles. Her diaries are filled with descriptions of the islands and observations of their flora and fauna. Now she was returning home with four more boxes of fossils from the Balearics. The difficulties and reverses that she had encountered in her exploration of the islands concealed from her alone something that was apparent to everyone else. Her visits had been a triumph. Against all the odds, she had discovered not only a remarkable species, but had obtained specimens that formed what Charles Andrews was later to describe as a 'large and complete' collection of *Myotragus.*[50] Furthermore, in Menorca she had excavated sufficient material to show that *Myotragus* on that island was appreciably larger than on Majorca, and differed sufficiently for Charles Andrews to recognize it as a new variety, naming it *Myotragus balearicus* var. *major.* As for the big beast that she discovered at Bajoli, that too was a new species. It was an extinct gigantic land tortoise, which she named *Testudo gymnesicus* Bate, while on both Majorca and Menorca she discovered unique species of giant dormice about the size of squirrels, and a shrew.*

* *Hypnomys mahonensis*, sp.n., *Hypnomys morpheus*, sp.n., and *Nesiotites hidalgo*, sp.n, respectively. Although Dorothea did not publish this, in the Cueva de la Barxa, the type site of *Myotragus*, there were, in the same deposit, a few frog-like bones. 'Huge quantities', according to the Majorcan palaeontologist Dr Josep Antoni Alcover, have been revealed in subsequent excavations. In 1979, these fossil bones were identified as a species of midwife toad, thought to be extinct on the island. However, in 1980 it was discovered to be very much alive, living by remote torrents in the Majorcan mountains. Dorothea, Dr Alcover wrote, was thus the first scientist to '"touch" the bones [of] this species'. The Majorcan midwife toad, or ferreret, is still extremely rare. (Dr Josep Antoni Alcover, personal communication.)

Dorothea's achievements are celebrated in the Balearics. The first tribute to her was published in 1920 in the journal *Revista de Menorca*.[51] Since then, scarcely an academic report relating to *Myotragus* has appeared without acknowledging her contribution, while a major palaeontological work on the islands, *Les Quimeres del Passat*, published in 1981, is dedicated to her.[52] In 2001, to honour the fiftieth anniversary of her death, a tribute to her was published in the journal of the Societat d'Història Natural de les Balears.[53]

Dorothea left the windswept promontory for Barcelona. At Cooks she found her four boxes of fossils had arrived safely, dispatched by the indispensable Miguel, and she caught the train for home. She arrived in London at 7.30 p.m. on 3 December 1911. At the end of this page in her diary she has written firmly, 'Finis', and that meant more than she knew. Although she was just thirty-three, she would never embark on such an expedition again. She would travel, excavate, and find new species, but those days of such brave and solitary pioneering exploits belonged to a passing age. The world would soon be at war and so would her family, not only against a terrible enemy, but against itself. Her return home may have marked the end of one phase of her life, but she was about to embark on another, an approach to palaeontology that she would make uniquely her own. The pioneering explorer was about to become the pioneering scientist.

CHAPTER 11

War and Other Reasons

Dorothea's diary for her third visit to the Balearics is the last of its kind that she ever wrote. After 1911, no journal or notebook is known to exist in which she recorded her activities in such detail for months at a time. In their place, to attempt to understand and reconstruct the rest of her life, I had her scientific reports and manuscript notes, a few brief and cryptic little notebooks, and files of her work-related correspondence, in some of which were tantalizing personal references. I felt at times as if I were trying to complete a vast jigsaw with unknown numbers of missing pieces, or even excavating a site as initially unpromising as anything Dorothea confronted. What emerged was a picture of a woman carving out for herself a unique and fascinating expertise of international renown, but who, of necessity, is defined by what she did rather than who she was.

Shortly after her return from the Balearic islands, Dorothea wrote to Smith Woodward to tell him that her four boxes of fossils were on their way from Barcelona, 'though I expect it must be some considerable time before they turn up'. Then she added, 'Have you thought of a nice and neighbouring island for me next year?'[1] Short periods of travel seem to have been acceptable to her parents; extended work at the Museum was not. Despite the boxes of fossils from her previous expeditions, which still demanded her attention, when Dorothea returned from the Balearics she went straight home to Berwickshire and did not return to London for two months. One of the consequences of these infrequent visits was the delay in determining her finds. It was not until 1912, for example, that she published a report on an extremely large mouse from Crete, part of the haul that she had brought home eight years previously.*

It was mid-February before she appeared again at the Museum, to find her fossils had arrived before her. She wrote at once to the Percy Sladen Fund with a brief report of her visit to the Balearics. She gave the Fund a short résumé of her finds and told them that she hoped to prepare descriptions of them shortly. Her report is professionally typed, and she was fortunate to have had access to a typewriter. There was a woeful shortage of such machines at the Museum but the Treasury had seized on the Museum's request to purchase a few more as an opportunity to decrease the levels of scientific staff. Perhaps it would be possible, the Treasury suggested, 'to reduce the number of assistants without injury to the scientific work of the Museum, by assigning purely clerical and copying work to assistant clerks and typists'. Such a change might then 'obviate the need for more typewriting machines'.[2] The Museum went without either the machines or a reduction in the number of scientific staff.

From the Museum's records, I know that Dorothea was at work on her collections on Friday, 1 March 1912. What I do not know as fact is her reaction to the news that several hundred suffragettes were storming through the West End and the City of London, smashing windows with hammers concealed in handbags or wrapped in stockings. Given her own activities and conduct, I am sure she wholeheartedly supported their aims and equally strongly abhorred their behaviour. Not that she wasn't herself capable of questionable actions when she believed they were justified, such as her disregard for the Cretan excavation laws, but it is impossible to imagine her condoning violence of any kind. Her strategy was to exercise persuasive charm, not overt aggression. On the advice of the Home Secretary, the Natural History Museum closed its doors to the public from the following Monday, when the suffragettes renewed their attack. This time they were very close to the Museum, smashing shop

* The Curator of Mammals, Michael Oldfield Thomas, had been examining the remains and drew her attention to one specimen, which appeared to be of special interest. From part of the lower jaw and a few teeth, two of which were in a perfect state of preservation, she determined that the creature 'must have been a mouse of very large proportions, one which, so far as I am aware, has no equal in size in the Mediterranean region'. She called this unique species *Mus catreus* Bate and presented these remains and an innominate (hip) bone to the Museum. (Bate, 'On a new species of mouse and other rodent remains from Crete', *GM*, 5: 9 (1912), 4–6.)

windows in Kensington and Knightsbridge – including those of Harrods. The Museum remained closed to the public for three weeks; only staff and accredited students were admitted. When it did reopen, the public were asked to leave muffs, parcels, and umbrellas with the cloakroom attendants. At one stage, unique and very rare specimens were withdrawn from display while some galleries were temporarily closed to increase the number of commissionaires on duty in those galleries which remained open.[3] As the aim of some suffragettes, *The Times* reported, was to make London 'absolutely unbearable to the average citizen',[4] the threat remained until they suspended direct action at the outbreak of war in 1914.

Dorothea, meanwhile, was beginning to plan another trip to the Balearics rather than to a new island,[5] but at the end of this unsettled March she was summoned home again. Her brother Thomas, who was serving with the Royal Artillery in India, had broken his leg badly, not in a skirmish, but while schooling a polo pony. Polo was a game de rigueur for officers. To augment his small allowance from Henry and his service pay, Thomas trained polo ponies for sale, but he still had to borrow from a 'Shroff', an Indian moneylender attached to the battalion.[6] Postponing all her plans, Dorothea remained in Scotland as a comfort to her parents. Thomas was too injured to be moved. Because of inadequate medical facilities in India, the leg had been badly set and would not mend. By the summer it was clear that he was still not well enough to return home and the family received a welcome invitation. It was from her uncle, Colonel Thomas Elwood Lindsay Bate (Henry's brother), to stay at the house he rented in Norway and frequently used for the spectacular salmon fishing.

Dorothea went and adored it, as she wrote in July to William Ogilvie-Grant at the Museum:

> I am always out fishing or roaming the country which is lovely and wild here – all wooded hills, rivers and lochs besides the tidal fjord that runs close to the house.
>
> From a fishing point of view the weather could not have been worse, persistently fine and hot so we breakfast about mid day and spend half the night out – on one occasion I did not return until 8am! – Not much water left in the river but they have managed to get 8 salmon and nearly 50 grilse and lots of trout and altogether we put in a very good time – and I love this sort of life.

But it was Thomas who was still her overriding concern; as she told Ogilvie-Grant, he had just had a second operation on his leg and it would be October before he could return home.[7] Even when back in England, he faced more months in hospital while his leg was broken again and reset with a metal plate to strengthen the join. He then spent more months convalescing, at Osborne House on the Isle of Wight.* Dorothea and her parents visited as often as the distance from Berwickshire allowed, spending hours in the exquisite formal gardens, first with Thomas irritably confined to a wheelchair, then painfully walking with crutches.[8]

With all this going on, by the autumn of 1912 Dorothea had still not been able to do anything about further travel, nor had she published anything on the Balearics. When Henry Woodward, in his capacity as a Trustee of the Percy Sladen Fund, wrote to her at Bassendean in October for a progress report on her collection, she could only tell him that her infrequent visits to London had delayed her work and she hoped to publish the following year.[9]

Dorothea was able to do some work at Bassendean. The naturalist and explorer Aubyn Trevor-Battye was writing a book on his recent travels in Crete and asked Dorothea, 'whose scientific work in the caves of Crete is so well-known', to contribute chapters on the caves and mammals of the island. This she did, taking the opportunity to pen, implicitly, her dislike of the exploitation of the Majorcan show caves: 'Although the caves of Crete do not equal in extent and grandeur the famous caverns of some other countries . . . [they] are all the more impressive from the absence of a guide armed with an acetylene or electric lamp, who glibly reels off the names given to the different chambers from some fancied resemblance to organ-pipes or the pillars of a cathedral.'[10] That concerts and *son et lumière* shows now take place in those caves would have appalled her even more.

It was almost another month before she could leave Scotland. She returned to the Museum on 9 November 1912, the day after her thirty-fourth birthday, and found the place in a state of intense excitement. Remains of great interest had recently been brought to the Museum by Charles Dawson, a Sussex solicitor and a regular visitor to the Geology

* One wing of Queen Victoria's much-loved 'place of her own' had been given to the armed forces by Edward VII as a convalescent home for officers.

Department. Dawson was also an amateur archaeologist and palaeontologist, and a friend of Smith Woodward's. Workmen digging for gravel near Piltdown Common, not far from the town of Lewes, had found fragments of what appeared to be a human skull, some fossilized elephant and hippo teeth, and a few crude flint tools in the gravel of an ancient river deposit which indicated that they might date from before the last Ice Age. Although his expertise was in fossil fishes, Smith Woodward had been impressed enough by these finds to agree to excavate the site further with Dawson in the early summer of 1912. Further remains were discovered, including more pieces of the skull and part of an apelike jaw with two teeth. Here, it seemed, was evidence of early man, earlier than Neanderthals, together with his tools and the remains of extinct mammals. Although the finds were extremely fragmentary, the indications were that it was all around half a million years old.

Smith Woodward concluded that the skull and jaw, despite the apelike appearance of the latter, belonged to the same individual and called it, to honour his friend, *Eoanthropus dawsoni*. Smith Woodward had the backing of a formidable array of scientists, whose opinion was that human ancestors would have had pronounced apelike characteristics as well as human. Piltdown Man fitted this theory quite neatly. It appeared that at last the scientific community in England could now compete with continental Europe in the race to find the missing link of the origins of man. On 18 December, at a meeting in Burlington House of the Geological Society of London, the details of the Piltdown discovery were announced to a capacity audience.

Almost immediately, however, doubts were expressed as to whether the human skull and the chimpanzee-like jaw were from the same individual – or even the same species. A sizeable body of scientific opinion thought that the early human skull and an ape's jawbone, although of very great age, were just coincidentally together in that ancient gravel bed. As more human fossils were discovered in Java, China and South Africa – in every case with the reverse characteristics of a more apelike skull and much less apelike jaw – Piltdown Man appeared increasingly as an oddity, though the great age of the specimens was not questioned. It was not until the development of more refined methods of dating after the Second World War that the fraud was finally exposed through the work of the Oxford anthropologist Joseph Weiner, and the geologist and palaeontologist

Kenneth Oakley, who had worked at the Natural History Museum since 1935. These two, with the anatomist Wilfred le Gros Clark, revealed not only the fraud, but also that the whole affair had been a fantastic hoax. All the remains had been infiltrated into the gravel beds at Piltdown, all had been stained to give the appearance of great age, and none could have found their way naturally to this area of Sussex. Indeed, the mammal remains had come from a number of very different sites, including Tunisia and possibly Malta. As for Piltdown Man, although the skull was reasonably old, the apelike jaw was actually modern. Kenneth Oakley and his colleagues determined that 'we are reasonably sure that the Piltdown specimen is actually that of an orang-utan – probably a female not yet fully grown.'[11] Arthur Smith Woodward – and all those other eminent scientists who had believed in Piltdown Man – had been brilliantly and appallingly duped.

At the end of 1953, Martin Hinton, who had progressed from unofficial scientific worker in the Geology Department in 1905 to the heights of Keeper of Zoology from which he had recently retired, wrote to *The Times* to explain how the fraud had escaped detection for so many years. He told the paper that he had seen the Piltdown material for the first time only at the great meeting of the Geological Society. 'As soon as I saw the jaw, and later the canine tooth, I knew that had they come into my hands for description they would have been referred without hesitation to the chimpanzee which was already known to occur in some of the Pleistocene deposits of Europe. Later I found that my future chief, Oldfield Thomas, then in charge of the Mammal Room, was of the same opinion.' The fraud had only been finally proved after chemical analysis, which had meant removing tiny samples of bone. Had that been suggested early on, Hinton wrote, 'such a proposal would have been rewarded with instant expulsion from the museum, and possibly death!'[12]

There are a number of candidates for authorship of the fraud. Whoever it was had extensive knowledge of palaeontology, archaeology, anthropology and anatomy, as well as access to fossil and recent remains. The list includes the distinguished anatomist Sir Arthur Keith; the Jesuit priest, philosopher and archaeologist Teilhard de Chardin, who knew Dawson well; the Australian anatomist Sir Grafton Elliot Smith; also Martin Hinton himself, driven, it is said, by his antagonism to Smith Woodward who refused to promote him. There were those in the Museum who thought his letter to *The Times* in 1953 was evidence of a guilty conscience. Dawson himself was suspected and there are others, including the writer and amateur archaeologist Sir Arthur Conan Doyle. The most recent research, by Professor Chris Stringer and Andy Currant of the Natural History Museum, returns to the most obvious man: Charles Dawson.[13] He had a record of dubious finds; indeed, a junior clerk in Dawson's practice remembers how in 1912, 'On occasion Mr Dawson boiled specimens in the office kettle. On these days I had to delay making the office tea.'[14] His motive may well have been the glory of discovery and the honour of having this 'earliest Englishman'* named for him.

In the absence of any written confession or incontrovertible evidence, however, the jury is still out. If Dawson was the guilty man, then the hoax element of the affair is much diminished; it becomes a case of fraud. Yet it has long been assumed that there was a malicious intent to Piltdown and that the intended (and in the event, actual) victim was the Keeper of Geology, Dr Arthur Smith Woodward. It may be that Martin Hinton, and perhaps even others in the Geology Department, were party to the fraud or exacerbated it, the motive being a bitter dislike of Smith Woodward. He was reputed to be an aloof and arrogant man with many enemies, whose department was viewed as being unhappy and fractured.

And where in all this is Dorothea? With her own recent search for ancient man in Menorca, she would have been fascinated by the Piltdown discoveries and the apparently associated fauna, but she also knew Hinton and Oldfield Thomas well and must have been aware of the scepticism in the Geology and Zoology Departments. Nothing exists of her own views on the matter, but I am left with an extraordinarily strong image of Dorothea: a still, calm figure in a seeming whirlpool of discontent, getting on famously with this apparently detested Keeper. He could undoubtedly be waspish about his staff; his departmental reports to the Trustees show a man impatient with administrative duties that interrupted his science and he evidently expected more of his staff than he thought they delivered. While he admired Charles Andrews' abilities, for example, Smith Woodward complained to the Trustees that he 'lacks neatness and methodical habits', was 'restless and travels' (although Smith Woodward acknow-

* This description was to be used by Smith Woodward as the title of his book on Piltdown Man.

ledges the 'great service' to the Museum of Andrews' fossil mammal discoveries) and he adds that the Keeper had to stop his own work to take on Andrews' responsibilities when he was away.[15] But these remarks are little different to those made by any of the Keepers, struggling to run departments with too few staff and a vast workload.

From Smith Woodward's correspondence with Dorothea, a somewhat different personality emerges. He appears as never less than supportive and encouraging; an influential, generous and kind friend. He seems to have given her every facility to work, ensured she was paid reasonable sums for her finds, and even offered to advance her money from his own pocket. There is nothing in Dorothea's surviving correspondence with him that even hints at a character so unpleasant that his colleagues desired his humiliation. Dorothea's friendship with Arthur Smith Woodward lasted until his death as, indeed, did her friendship with Martin Hinton.

In December 1912, however, Piltdown Man was viewed as one of the most momentous discoveries ever associated with the Museum. The exhibits were put on temporary display in the Central Hall, and a special guide was prepared, at a cost, the Trustees stipulated, 'estimated not to exceed £50 for 2000 copies'.[16] The publicity and universal interest was such that even Dorothea's mother came to the Museum, announcing herself at the gate as Mrs Fraser Bate.

Dorothea worked at the Museum to the end of that historic December and then throughout January, but Thomas's poor health continued to occupy the family. Again she had to go home and there she remained for most of the year. It was not until October 1913 that she was able to return. Charles Andrews had been making a detailed study of *Myotragus*,[17] the strange goat-antelope with ratlike teeth, and they worked on her finds together. At last she had the detailed measurements and references to enable her to write her much-delayed reports, one a survey of the bone deposits she had discovered in her three trips to the Balearics, the other on the 'big beast' from Menorca, *Testudo*, the giant tortoise. Both were published in the *Geological Magazine* in 1914. With their publication, her scientific career was slowly getting back on track, but her family was about to deal her a blow that had incalculable consequences.

Thomas's leg had healed but he was not fit enough to rejoin his battalion in India. Instead he was posted to the War Office in London and there, in a society drawing room, he met Mary Ulrica Alicia

Fitzwilliams, the girl whose plaits he had pulled as a child in South Wales. Now she was a young and attractive widow; family portraits show a sensuous beauty with wide calm eyes and a confident gaze. At an early age Ulrica, as she was known, had married a friend and business associate of her father's, Archibald Swan, a man much older than herself. After just a few years of marriage, in 1911 Swan died, leaving Ulrica with a daughter, a flat in London near Buckingham Palace, and a life interest in a trust he had prudently created. After a few months, Thomas and Ulrica became engaged – despite opposition from both families. The Fitzwilliams believed that Ulrica was throwing herself away on a penniless young officer with a 'game leg' who, they considered, had no prospects in the Army or anywhere else. This was not an unreasonable assumption in peacetime – Thomas could well have found himself retired from the Army as unfit with no qualifications for earning his living in civilian life. For Henry and Elizabeth Bate, the problem was very different. Ulrica had learned much from her businessman husband and, far from bringing a dowry to the marriage, insisted on a marriage settlement *from* Thomas. For the Bates this was painful and unacceptable. Unlike the Fitzwilliams family, the Bates had no wealth; all they had was a modest bequest left to Henry, and that they had intended to leave to Dorothea. Leila had married so well she needed no financial help from her parents, while Thomas, they thought, would have a reasonable standard of living as his career in the Army progressed. Dorothea had nothing except the small amounts she earned at the Museum, certainly no pension and, as they had always assumed, no prospects of marriage.

It was an impossible situation; they could ensure the happiness of their son only by destroying the future security of their middle daughter. Perhaps the decision was taken in the hope that Dorothea's circumstances might somehow improve, or perhaps, with the outbreak of war in August 1914, it was even made at Dorothea's insistence: that in the end the money should go to Thomas and Ulrica. The marriage took place on 28 October 1914 in the austere beauty of St Margaret's Church, Westminster and neither family attended. Henry never spoke to Thomas and Ulrica again.[18] With the outbreak of war, Thomas's career was rejuvenated. The Army could not afford to dispense with experienced officers and Thomas, promoted to major, threw away his crutches and stick, ignored his now permanent limp, and went off to fight.

The war also meant that, with the departure to military service of the more able-bodied scientists, the Natural History Museum had greater need of Dorothea. By the end of 1914, more than twenty of the staff, scientific and non-scientific, had gone to war. For Dorothea, the timing was perfect. Her parents had no choice but to recognize that, for Dorothea, her science was now an economic necessity as well as a passion. Almost immediately after that devastating decision to disinherit her, she wrote to Smith Woodward asking for work. From 5 October 1914 she was employed as an unofficial scientific worker in the Department of Geology, a 'temporary' position that was to last for the next thirty-seven years.

From that autumn Dorothea worked on preparing and arranging a great diversity of fossils: mammals, sea creatures, vertebrates and invertebrates, some Pleistocene, others millions of years old, from elephants and rodents to starfish, birds, and fishes. She was paid at a rate of about £1 (about £45 in today's values) per one hundred fossils. Depending on how many she prepared, she took home every six weeks or so anything from £10 to £25.[19] Scientific assistants, second class, who were members of staff, were paid between £12 and £24 per month. For a short time before her trip to Crete she had done similar work; this, however, was a full-time and exacting job and, as far as the Museum was concerned, her contribution to the war effort. From it Dorothea built on her knowledge of anatomy and developed skills that were to lead to her original and pioneering work on the relationship between fauna, climate and the environment. Throughout the war, day after day, she carried the main burden of the preparation work of the department. She also had the satisfaction of seeing a second of her unique Pleistocene creatures put on display. Her reconstruction of the tiny hippo from Cyprus had been a focal point in the Geology Galleries for some years. It was now to be joined by the reconstruction of a skeleton of *Myotragus*, created from bones of different individuals. It went on display in the Fossil Mammal Galleries together with the actual type skull and jaw mounted on a separate stand.

Where she lived at this time is not at all clear. She gives as her permanent address her parents' home in Scotland, and apart from a mention of an address in Pimlico where she seems to have stayed for a few days, and another in Earls Court, which also seems temporary, there is no record in these years of a London address. It was not until 1928 that she rented on a long-term basis a flat a mile and a half or so from the Museum on the West Cromwell Road in West Kensington. It was a neighbourhood where accommodation was inexpensive. Dorothea's flat was in the basement of a converted house, demolished some years ago to make way for the multi-laned A4.

The Museum needed Dorothea for more than the preparation of fossils. Collections sent to the Museum still had to be examined and described; with so few staff, that work began to come to Dorothea. She had of course reported on her own material; for the first time now she was being asked to report for publication on the work of others. The first collection that Smith Woodward asked her to deal with drew on her knowledge of the extinct fauna of the Mediterranean islands. It was a small assortment of mammal and bird remains, which she was told came from the Ghar Dalam cave (the Cave of Darkness) on Malta. It was the first of a stream of collections from Malta that would continue throughout her life. The phenomenon of dwarfism and gigantism on the island was very marked, as Dorothea well knew from her reading and conversations with Forsyth Major. Among the species of fauna already identified were three species of elephant ('large, smaller and pygmy', Dorothea noted[20]), two species of dwarf hippo, gigantic swans and dormice, and small deer.

The collection had been sent by Guiseppe Despott, the Curator of the Malta University Museum of Natural History, who had begun excavating in various sites in the limestone of Malta in 1912. Much of his work had been in the Ghar Dalam cave, a vast cavern near the sea, about four hundred feet long, quite narrow, with tunnels and chambers branching off. Through the middle of the cave a river had once flowed, which probably collected faunal remains from the ground above on its journey and deposited them in a deep central channel in the heart of the cave, or washed them into fissures and crevices. Archaeologists and geologists had excavated Ghar Dalam at various intervals from 1865 and had begun to reveal an extraordinary volume of animal and some human remains.

This first collection of Despott's, given to Dorothea to identify, ranged from a dwarf elephant to the gigantic dormouse, *Leithia melitensis*. Most numerous were the bird remains. She identified from a wing bone (its right metacarpus) a previously unknown species of small swan* and a number of birds not previously found in fossil form. She trawled through

* *Cygnus equitum* sp.n.

the extensive, if scattered, literature on Malta, and to her report on Despott's finds she added an exhaustive list of all the vertebrate remains found in the Pleistocene of Malta.[21]

Dorothea must have longed to excavate in Malta herself; all the more so when she received a devastating letter from Despott. He thanked her for sending him a copy of the report, asked for two more for the university, and then added: 'I note however that there is a mistake about the locality where the majority of these bones have been found. They were not from "Ghar Dalam" but from a fissure known as "Tal Herba".' This site is a few miles to the northwest of Ghar Dalam, and well away from the sea. There had been nothing with the collection to indicate this. Furthermore, he told Dorothea, 'a short report by me [on Tal Herba] was published in the Proceedings of the British Association of 1915,' and he offered to send her a copy.[22] Dorothea's reaction to this letter can only be imagined. Not only had the specimens come from more than one site – and most of them not the one she had been writing about – but she had no means of knowing what had come from where. It is perhaps an indication of Despott's less than structured method of working that not only had he omitted to send a copy of his 1915 report with the bones, but he had also not thought it necessary to attach to them any identifying labels.

Unaccountably, Despott also failed to mention to the Museum that, during his excavations at Tal Herba, he had made a discovery that was potentially of enormous interest. According to his 1915 report, in the bottom layer of this fissure, among remains that included some fossil deer teeth, he found other teeth, 'Two human ones, belonging to the upper jaw'. With evident excitement he continued, 'I am not aware that in Malta human remains have been previously found associated with the above-mentioned animal remains. The discovery is thus one of great importance.'[23] The teeth were taurodont (bull-toothed): that is, the roots were fused – or almost fused – together, a characteristic of Neanderthal teeth. If indeed he had found human teeth among Pleistocene remains, it would not only be unique to Malta, but it would also be the first indication that humans and these extinct creatures had coexisted on any Mediterranean island. It would be the evidence that Dorothea had so desperately tried to find on Menorca.

Yet there is no later reference to these two human teeth from Tal Herba. Curiously, however, in 1917, when excavating in Ghar Dalam, Despott made a remarkably similar find. In a trench about one hundred feet from the mouth of the cave, he found a first upper molar and then a human milk molar in the next layer down, both 'conspicuously' larger than modern ones.[24] Unlike his earlier report of 1915, this one would be noticed. Despott had received a grant of £10 (about £450 in today's values) for his excavations from the Anthropology section of the British Association for the Advancement of Science, whose president was Sir Arthur Keith, the conservator of the Hunterian Museum at the Royal College of Surgeons. One of the foremost experts on human evolution of the day, he had been involved in the determination of the Piltdown material. Keith had no hesitation in declaring that these were indeed the teeth of Neanderthal man and that Despott's find was of truly great importance. He must have assumed that further excavations would reveal evidence of an associated tool industry of which there was, as yet, no sign.[25]

There was a limit to how much time Dorothea could spend on the Malta problem. The pressure of work on everyone in the Museum was enormous. By now conscription had emptied the Museum of all men of military age, with the exception of those unfit or whose jobs were described as 'indispensable'. At the beginning of 1916, the government declared that all museums were to be closed to the public on the grounds of national war economy. There was of course a great outcry, with protests in Parliament and in the newspapers. As *The Star* rumbled, 'The museums and galleries are the universities of the people, and it is false economy to rob the people of their moral and intellectual food . . . the staff required for a closed museum is not materially smaller than the staff required for an open museum. If the museums are to be closed in order to remove and protect their contents against the danger of destruction by aircraft, that is another matter.'[26] But that was not the issue. It was to save money, about £50,000 a year, which would, according to the *Evening News*, 'pay for the war for *very nearly a quarter of an hour!*'[27]

Every newspaper pilloried this policy. Under such a concerted onslaught it was moderated and sections of a few museums were allowed to remain open, including the more popular galleries of the Natural History Museum. There were, of course, security measures, which now sound all too familiar, to prevent bombs 'being conveyed' into the building; the most valuable objects were removed from the galleries and sent out of London. It was also suggested that nine inches of sand should be put on the spirit building roof as a protection against aircraft attack, although it was then calculated that not only might this be inadequate protection, but, worse, that the weight of the sand might do even more damage.[28] The Museum closed to the public at 4 p.m., while Dorothea and her colleagues were allowed to work on in their studies, using the dimmest possible light because of fear of aircraft attack. In 1917 there was some minor damage to the Museum, but Dorothea was appalled to hear of Henry Woodward's experience. His house on the far side of Hyde Park from the Museum had all its windows shattered by a bomb, which left, as Woodward wrote, a 'private HOLE about the size of a lime kiln' in his garden.[29]

In June 1915, Dorothea had taken a rare weekend off and gone home to Bassendean. There she composed another application for a grant to the Percy Sladen Memorial Fund. Unable to travel abroad, she had read of a bone cave much nearer to home that she wanted to excavate.[30] It was in the Assynt mountains in the far northwest of Scotland. Digging had taken place there in 1889, but Dorothea believed the excavations were only superficial. She asked the Fund for a grant of £30 to help towards her living and excavating costs. Within a month of her application, not only had her grant been approved, but she also received the money.

She wrote immediately to Henry Woodward who was still a Trustee of the Fund. The tone of her note must have been blissfully elated for it elicited a humorous if fatherly response from her old friend. He addresses her as 'Lady Dorothea' and writes, 'I was glad to receive your note and entirely echo your prayer for the success of the Scottish exploration.' He then cautions her not to raise her expectations too high: 'But we cannot hope for another *Myotragus (scotius)*. The Trustees are very well satisfied in having supported your past explorations which have yielded such splendid results!' And then in a delightful display of whimsy, 'Only imagine the sensation you would make if you could walk down Piccadilly leading by a string your Pigmy Elephants, *Hippopotami*, *Myotragus*, Tortoises etc etc all in one long queue, the little elephant blowing his trumpet, and the Hippopotamus wagging its tail.'[31] He might to this impressive procession have added ranks of all the unique squirrel-sized mice, voles, and shrews she had discovered.

In spite of all her excitement and Woodward's enthusiasm, she did not go, not then nor at any other time. In all her papers, there is just one reference to Assynt; it is mentioned as a brief aside in another application

Dorothea made to the Percy Sladen Fund nearly twenty years later in 1934. She wrote then that the grant for the Assynt cave had been made early in 1914 (*sic*) but that 'owing to the outbreak of War and other reasons the work could not be carried out that year . . . My £30 was put on deposit and is now £43.18.0.'[32] It was in fact nearly a year after war was declared that Dorothea applied for the grant; the outbreak of war had nothing to do with it. Her memory may have confused the actual date of application, but it was clearly the 'other reasons' that in the end prevented her going.* There is no record of what these reasons might have been and it is very curious. Although the Assynt district was remote, transport was simplicity itself compared to travel in the Mediterranean islands. She knew the precise location of the bone cave and it was also early autumn so the weather was not an issue. What cannot be ruled out is some devastating personal crisis.

Although it is possible to reconstruct what Dorothea was doing from her reports and the Museum's records, there is nothing to illuminate how she coped with the awfulness of the war as well as her fears for her brother and so many friends and colleagues away fighting. In her earlier years there is no indication of any spiritual or Christian dimension to her life; the only mention of religion is when she attended church almost as part of the social round in Cyprus, or as a source of great irritation when her workmen chose to attend church on a Sunday or feast day rather than continue excavating with her.

Yet in 1915, Dorothea became a Christian Scientist, a religion whose whole emphasis for believers is on the healing of both body and mind 'through the power of Truth',[33] and it does seem very possible that whatever made her adopt this faith must surely be linked to her abandoning her excavation of the Assynt caves. Her nephew (Thomas's son) believes that it was the continuing shock and pain of being disinherited by her parents in favour of her brother, but in the absence of Dorothea's own thoughts, it has to be conjecture.[34]† Nowhere in her papers is there any sense of what it meant to her, nor has any subsequently emerged.

* In 1926 another team of palaeontologists excavated the caves, but they published only preliminary notes and today work on the caves is still unfinished. See T. J. Lawson (ed.), *The Quaternary of Assynt and Coigach: Field Guide* (Cambridge: Quaternary Research Association, 1995).

† Christian Science had been founded in the nineteenth century by the American Mary Baker Eddy, a woman whose life had been crippled by physical and emotional pain. Widowed shortly before the birth of her only child, when she remarried, her new husband would not tolerate

Dorothea was not the only scientist at the Museum to be attracted to this new faith. Bruce Cummings, a young second-class scientific assistant in the Entomology Department, also considered joining the Church, although it is not clear how far he took his interest.[35] A brilliant and self-taught zoologist, he had joined the Museum in 1912 with hopes of revolutionizing the study of systematic zoology. Instead he found himself, 'God save the mark – in the insect room!'[36] He was exempted from military service because of chronic ill health and was one of the scientists whose work in the war was vital for more than the continued functioning of the Museum.* Advice was given to the armed forces on matters ranging from how to deal with a leech up a nose to the type of wood suitable for airship construction and the identification of seeds responsible for fodder poisoning of horses.[37] The Museum also suggested to the armed services that they might like to look at the general principle of protective coloration of animals, an exhibition of which was on display in the Central Hall. By early 1917 groups of soldiers were regularly attending at the Museum to receive instruction and demonstrations; the life-saving camouflage of animals was to be imitated for the first time by the British military.[38]

Practical advice was also given to civilians, particularly on matters of health and hygiene. Visitors to the Museum were confronted in the Central Hall by exhibits graphically illustrating the life history of the housefly and warning of its dangers to man. A trayful of appetizing-looking food, modelled realistically in wax, 'is made to look repellent', wrote a reporter

---

the child in the house and he was put into foster care, although her continuing ill health was also a contributory factor. Then without warning, the foster family with her son moved hundreds of miles away and she had no contact with him again. She founded the First Church of Christ, Scientist, in 1879 and within ten years more than one hundred branches had been formed in America. The movement spread to Britain in the 1890s, when the first branch was established in London. Close to the Museum there was a church and a reading room, a place for study and quiet contemplation. Although medical consultations are not forbidden, believers who are sick are treated by Christian Science practitioners, or healers, those who have devoted their lives to healing others through prayer. By the 1930s there were more than three hundred practitioners in London alone.

* Cummings apparently suffered from multiple sclerosis, which was partly responsible for thwarting his ambitions as a zoologist. But his disappointment was also with the Museum, which, he believed, because he had no academic training, refused to consider him for posts that he felt himself more than capable of filling. Even such a basic tool as a microscope had been denied him. He poured out his anger in a diary, the brilliant and deeply moving *Journal of a Disappointed Man*, which he published under the pseudonym W. N. P. Barbellion.

for the *Daily Telegraph*, 'by reason of the house flies which swarm upon it', while a mouldy heap of kitchen waste (also, happily, a model) was accompanied by labels listing the fatal diseases that the flies may carry, 'having settled on unburied filth'.[39] Cummings' contribution, between attacks of increasing ill health, was to study the habits of houseflies and mites and he wrote a penny pamphlet on lice, which were a serious problem in the trenches. In it he included a variety of preventive and remedial measures, among them 'a horrible mixture of hog's blood with wine and essences of roses', although the best preventive, he wrote, was 'strict personal cleanliness and the careful avoidance of those on whom the insects are likely to be found'.[40] It is, however, a serious work with practical and useful advice, and it was well received in the press.

In the reduced community of the Museum it is inconceivable that Dorothea and Cummings did not know each other, particularly given their interest in Christian Science. For Dorothea, this new faith appears to have become so necessary that on 7 November 1919, the day before her forty-first birthday, she joined the Mother Church, an act only undertaken by those totally committed to the movement. Two weeks prior to that, Bruce Cummings, who for the last few weeks of his life had been so weakened he was scarcely able to move, died, leaving a widow and child. He was just thirty years old.

CHAPTER 12

Man and Mammals

The pressure of work on Dorothea actually increased after the war. It took years for normality to return to the Museum. Those who had been on war service returned during 1919 to be confronted by great accumulations of specimens that had arrived in their absence. New staff had to be taken on. In November 1919, another woman joined the Geology Department as an unofficial scientific worker, Helen Muir-Wood, known to her family and friends by her middle name, Marguerite. She was seventeen years younger than Dorothea, but in the predominantly male environment of the Museum the two became friends. Born in Hampstead in north London, Muir-Wood had a first-class degree in Geology from Bedford College, London, and was working on a doctorate. Like Dorothea, she spent her days preparing thousands of fossils, although her speciality was brachiopods, marine invertebrates. In spite (or possibly because) of her academic qualifications, Muir-Wood found it hard to be accepted; even when she was awarded her doctorate, there were those of her male colleagues who would refer to her only as plain 'Miss' Muir-Wood and profess not to know a doctor of that name.[1]

In 1919 it was out of the question for Dorothea and Helen Muir-Wood as women to have applied for permanent posts, although the Museum was more than happy to exploit their talents on a temporary basis, yet there was, at the time, a critical shortage of staff scientists. The salaries were so low, however, £150–£300 (about £3,500–£7,000 in today's values) for a second-class assistant, £300–£500 for a first-class assistant, that it was impossible to attract educated men of the right calibre unless they had a private income. *The Times*, which was particularly outspoken about the Museum's recruitment policy, blamed a system of promotion that was 'slow and capricious'.[2]* In 1921, Helen Muir-Wood was made a part-time assistant and given full charge of the brachiopod collection, but was still barred from the permanent staff. It was not until 1928, the year in which voting equality for men and women was at last achieved, that women were permitted to apply for assistantships. The advertisement read: 'Candidates may be of either sex, but women will note that their scale of salary, notably as regards annual increments, is less than that of men.'[3] Men, it was assumed, had families to keep. Women, apparently, did not. Helen Muir-Wood applied for an assistantship, but was not appointed for another eight years.† Dorothea, who by then was almost fifty, did not apply. Some institutions had long taken a broader view of the merit of women scientists; in 1919, Dorothea, sponsored by Dr Arthur Smith Woodward and his deputy, Dr Francis Bather, had been elected a Fellow of the Zoological Society of London. It was recognition of her abilities although it did nothing to enhance her income.

Between September 1921 and April 1922, Dorothea endured the loss of three people dear to her. On 6 September, her old and valued friend Henry Woodward, who teasingly called her 'Lady Dorothea', and who had been so vital a support in her early career, died at the age of eighty-eight. Three months later Henry Bate, who suffered from bronchitis, became very ill. On 9 December, Dorothea travelled up to Bassendean and was there when he died on 17 December. He was eighty-three. Thomas and Leila were there too. Although Dorothea regularly went to stay with her parents and with Thomas and Leila, this was the first time that the

* The fees paid for the most scholarly work were also much criticized. Charles Sherborn, on whom Dorothea, and everyone else, relied for information about the most abstruse paper published in the zoological field, completed the second section of his *Index Animalium* in August 1922 to universal acclaim. The *Museums Journal* noted that the 'only unsatisfactory feature in the book' was that, in the last twenty-nine years, Sherborn had received from the Museum just over £100 a year (about £3,000 in today's values). 'But then, of course, bibliographers have no trade union. The zoological world ought to feel ashamed at accepting such work for such a paltry recompense.'

† In 1955 she became Deputy Keeper of the Department of Palaeontology (as the Geology Department was renamed), the first woman to reach such a senior post.

family had all been together since Thomas's marriage, seven years earlier. Her brother had emerged with honour and decorations from the war and had retired as a brigadier-general in 1920. Leila had also been decorated, receiving an OBE for her war work in Cambridgeshire, chairing committees that helped prisoners of war and war pensioners, and was a benefactor of the women's Land Army. In 1917 her husband Tansley had bought a sizeable mansion, Waltons Park at Ashdon near Saffron Walden in Essex. At weekends there were house parties with shooting in season, at which Dorothea was a regular visitor. Just four months after Henry's death, at Easter 1922, Tansley died. It was utterly unexpected. Dorothea came immediately and stayed with her sister for three weeks. From then on Dorothea became Leila's '"prop" in life',[4] dropping everything to be with her elder sister in times of need. Leila was left a wealthy widow and travelled widely, to India and Afghanistan as well as Europe.

That was not an option available to Dorothea. Early in 1922 she received a letter from Hugo Obermaier, the Professor of Prehistoric Archaeology at Madrid University, and a great admirer of her work in the Balearic islands. He told her of a cave in Majorca that promised much and suggested that she should explore it, but her workload made that possibility very remote.[5]

Guiseppe Despott was still busily excavating in Malta and sending material to the Museum,* but others were at work there too. In the early 1920s, the island attracted a British team of archaeologists led by Margaret Murray, a lecturer in Egyptology and archaeology at University College London. She put Gertrude Caton Thompson, then an archaeology student at Newnham College, Cambridge, in charge of the excavations at Ghar Dalam.† To learn about faunal remains, Caton Thompson made an appointment at the Natural History Museum and was introduced to

* He also sent material to the Hunterian Museum at the Royal College of Surgeons for determination. As she was an acknowledged authority on Mediterranean fossil fauna, even some of that ended up on Dorothea's desk. One collection in particular baffled the Physiological Curator at the Hunterian Museum, Richard Higgins Burne, and he wrote in desperation to Arthur Smith Woodward. 'I have worried thro' them and have made them out so far as I can. The Canid bones are as like a Fox as two peas, but so small. Is there a little fox in those parts? The Birds are beyond me except so far as they are Birds.' (Letter from R. H. Burne, Royal College of Surgeons, to Dr Arthur Smith Woodward, 26 July 1922. Malta file, P MSS BAT.) The collection was passed to Dorothea who identified the specimens for Burne, including the little fox, in less than a week.

† Caton Thompson was later to become renowned for her work in Egypt and Zimbabwe.

Dorothea. Both Margaret Murray and Dorothea, Caton Thompson wrote later, 'seeing I was in earnest, gave richly of their time and patience'.[6]

What interested Gertrude particularly in Ghar Dalam were the two taurodont teeth that Guiseppe Despott claimed to have found in his excavation of 1917. She found it inexplicable, however, that no trace 'of the characteristic and prolific Mousterian industry has survived on the island',[7] and it was that which took her to Malta. After a season digging in Ghar Dalam, there was still no evidence. She found rich faunal remains, which she sent to Dorothea for identification, but far from Mousterian finds tens of thousands of years old, the human material dated mostly from the Bronze Age, some three thousand years BC, and consisted mainly of pottery fragments. A second season brought similar results, a splendid variety of fossil mammal remains, but absolutely no sign of Neanderthal man. Nor did any of the bones that Dorothea examined show any indication that they had provided food for ancient humans. All Gertrude Caton Thompson could say about the two teeth was that they had been discovered by Despott 'in circumstances incapable of satisfactory interpretation'.[8] A few years later the great French palaeontologist, Professor Raymond Vaufrey, who corresponded regularly with Dorothea on palaeontological matters, found no evidence of early human occupation in Sicilian caves either, which, as Dorothea wrote to him, 'seems to make those ["Neanderthal"] teeth from Malta, still more of a puzzle.'[9] It is now known that taurodontism was not confined to the Neanderthals, as Sir Arthur Keith and others of the time believed, but is still found today in some modern populations.[10] In 1953, two years after Dorothea's death, Kenneth Oakley, using the techniques that had revealed Piltdown as a forgery, subjected the teeth to fluorine analysis and concluded that they were no older than Neolithic.[11] Although there are still those who argue that Neanderthal man did inhabit Malta,[12] the most likely explanation for the teeth apparently being found with much older fossil mammal remains is Despott's poor excavating techniques. However, given Despott's mysterious initial claim to have discovered two human teeth with Pleistocene remains in Tal Herba in 1915, there may be a less benign scenario. No incontrovertible evidence has ever emerged of Neanderthal habitation anywhere on Malta.

In the spring of 1924, Dr Arthur Smith Woodward retired and was awarded a knighthood for his thirty years of service to the Museum. He was world renowned for his expertise on fossil fishes and his scientific contribution had been immense. The tea parties that he and his wife Maud gave were famous; after cucumber sandwiches and little cakes, the eminent scientists being entertained would be asked to sign Lady Smith Woodward's celebrated tablecloth (reminiscent of Lady Gregory's wonderful signature tree at Coole Park in the west of Ireland) and she then embroidered the signature in coloured silks. There are more than three hundred of these distinguished names. Dorothea's is in the bottom left-hand corner, just underneath Smith Woodward's own.* He retired to a house in Sussex close to the gravel pit where the Piltdown skull had been discovered and, although nothing more had been found since Charles Dawson's death in 1916, devoted his time to further excavations there. If others were involved in the hoax, it would be nice to think that any satisfaction or amusement they obtained from the futile labours of this outstanding scientist was tinged with guilt, as he wasted his retirement in pursuit of a chimera.

Smith Woodward's successor, Dr Francis Bather, inherited a department that was overworked and understaffed, an endemic problem that suddenly became critical. Dorothea's great friend and mentor, Charles Andrews, was to have been promoted to Deputy Keeper on Smith Woodward's retirement, but in the early spring of 1924 he became seriously ill. On 20 May he wrote to Smith Woodward, with whom he had worked for more than thirty years, regretting the Keeper's retirement. 'I think it is bad for the department and no good to you,' Andrews wrote. 'I am very sorry to be away and to have no idea when I shall be back. It will be the greatest break in the continuity and traditions of a department that has ever happened I should think . . . I would give anything to be about again even for a year.'[13]

Five days later he was dead. Bather made Dorothea the new Curator of Aves (fossil birds) and of Pleistocene mammals, although in her case the post was not official and she continued to be paid according to the number of fossils she prepared. Two new assistants were appointed to take on the rest of Andrews' work. The department also received temporary and distinguished assistance from Professor D. M. S. Watson, later a

* The tablecloth now hangs outside the office of the Keeper of Palaeontology in the Natural History Museum.

Trustee of the Museum, and Clive Forster Cooper, who was to become its Director. The vertebrate palaeontologist Guy Pilgrim, a senior official with the Geological Survey of India and friend of Dorothea's, was also drafted in on a temporary basis. On leave from India, he was greeted with relieved familiarity by Bather. 'Welcome home,' Bather wrote. 'I have just been making arrangements to clean up some of the exceedingly filthy specimens on which you are to start work.'[14] Nevertheless, these temporary measures could only partly relieve the pressures on the department, which continued to be overwhelmed with new collections.

Bather wanted to reorganize the whole department and was forthright in his condemnation to the Trustees of the existing situation. He complained of everything from 'Small pay and prospects' to overcrowded galleries with 'many specimens of no public interest'. Their removal from show would make them more accessible to researchers and students and allow for the cases to be rearranged. There was plenty of space in the basement storerooms, Bather argued, but it was not well utilized. The wall-cases were hopelessly unsuited for many displays; better lighting was needed and better ventilation. 'At present they [the storerooms] are like bakers' ovens,' Bather complained, 'except for the excessive dirt that accumulates in them.' Added to this, there was nowhere to unpack the collections that were constantly arriving from all over the world. For Dorothea and her colleagues, change could not come soon enough. It was a waste of their time – not to mention that of researchers, national and international, who came to study the collections – to have to hunt for specimens in storerooms stacked high with 'old bones' for which there was inadequate storage, or in boxes covered in more than forty years of dust, some still unopened since arriving from the British Museum in Bloomsbury in 1881.[15]

It was to take Bather the whole of his four years as Keeper just to set in motion the changes needed. He had first to acquire the support of the Trustees, and then push his plans through against all the obstructions endemic in a large, underfunded institution. 'Every plan in turn,' he raged in 1925 after a year in office, 'has in turn been brought to a standstill by (presumably) lack of funds to employ artisans. If this continues, the collections of the Department will, for a large part, be as useful as books in a furniture depository.'[16] Eventually the need for change was acknowledged and the reorganization slowly took effect, although that of course had its own problems. Writing from her brother's house in Herefordshire at Christmas four years later to Dr Lang, Bather's successor as Keeper, Dorothea ruefully remarked, 'I hope the Department is more peaceful and that I shall no longer have to share my room with a pneumatic drill or even the British workman!'[17]

Dorothea's work was commended by Bather in his first Keeper's report to the Trustees. She was, he wrote, 'one who was willing and able to report on the unceasing flow of mammalian and avian remains from the Pleistocene deposits of this and other countries, and to attend to the numerous callers who bring such remains'. He went further in acknowledging how very hard she worked. 'A very great deal of time is taken up with the determination of bones from cultural sites and relatively recent deposits. Twickenham, Torquay, Jersey, Malta, Syria and the Sudan are among the localities that have furnished such specimens.' It was a field of enormous potential: 'This work occasionally brings in desirable gifts and may at any moment lead to another discovery of such supreme importance as the Piltdown skull. Miss Bate performs it more than competently.' This was much-deserved praise and not dispensed lightly; Bather, new to the job and less waspish than his predecessor, assesses Dorothea's colleagues in just a few concise phrases. So Dr William Lang, his deputy, 'an excellent investigator and curator, is learning the administrative duties'; Wilfred Edwards, a palaeobotanist who was to become Keeper of Geology, 'is competent and business-like', while the Curator of Molluscs, Leslie Cox, 'a modest hard-worker, has learnt his job rapidly'. Each of his scientists receives similar succinct approval.[18]*

From Dorothea's studies of the almost countless collections sent to the Museum, she acquired an astonishing expertise in fossil and recent mammals and a reputation that was becoming international. In the summer of 1924, the Professor of Geology at the American University in Beirut, Alfred Ely Day, who had been sending specimens for identification

* Of the administrative and clerical staff, Bather is kind about all but two, one who was not likely to become an adequate secretarial clerk because of a 'defective education . . . a pity, because he has brains'. The other, E. A. R. Bush, 'has everything except brains: polite – even polished – faithful and willing, with a remarkable phraseology and some capacity for simple clerical work, he perpetually gets hold of the wrong end of the stick. Probably, when his mind has caught up with the over-rapid growth of his stature, his amusing but aggravating defects will be remedied.' As he was still in the department many years later, presumably they were. (Keeper's report to the Sub-Committee on Geology etc., 1924, DF 104/55.)

to the Geology Department since 1909, sent a quantity of bones, flints and teeth from various sites in what was then Syria, with a request for their identification. Dorothea was asked to work out a considerable amount of material from a site called Ksâr 'Âkhil. This was a rock shelter under an overhanging cliff near the village of Antilyâs a few miles north of Beirut and about a mile from the sea. 'It must have been an agreeable spot in the bleak winter,' wrote Professor Day, empathizing enthusiastically with the early inhabitants of the site, 'though rather warm in summer.' The rock shelter had been discovered by two men from Beirut who were following a local rumour that treasure was buried there. They were convinced that the bone breccia near the entrance to the rock shelter was in fact the remains of a concrete wall built to seal up the shelter, and would not listen to Day's explanation of what it really was. 'I told them repeatedly,' Day wrote happily to Bather, 'that under this pre-historic mass they could hope to find no treasure buried a few centuries ago, but they did not give much credit to what I said, and we finally went into partnership in the digging on the agreement that they should have all the treasure and I all the flints and bones.'[19]

The collection looked of great interest, but Dorothea was hampered in her work by the lack of recent material from the region for comparison, and wrote to Day accordingly. He promised to 'try to get you the skeletons you desire',[20] but before he could do so, Dorothea was presented with the most wonderful opportunity. At the end of 1924, she had heard from an old friend from her Cyprus days, Sir Thomas Haycraft, who had been President of the district court of Larnaca. Now he was the Chief Justice in Palestine (as it then was) and he and his wife invited Dorothea to stay with them in Jerusalem. The Museum agreed to spare her from her duties for a few weeks and the visit was arranged for August 1925. The timing, as it transpired, could not have been better.

The systematic exploration of certain cave sites in Palestine, where flint tools had been identified just lying on the ground, had been initiated by the British School of Archaeology in Jerusalem and its founder, Professor John Garstang. The site for the first excavation was on the northwestern shores of the Sea of Galilee, about six miles from Tiberias. The work was entrusted to a young archaeologist from Oxford, Francis Turville-Petre, who began to dig there in April 1925. His finds were of such significance that they were reported that summer in *The Times*. The first caves he excavated, a group of three small rock shelters known as the Mugharet el-Emireh, revealed important flints and fossil fauna, but no human remains. Close by, high in a limestone cliff, was another cave, the Mugharet el-Zuttiyeh or Cave of the Robbers. As he worked down through the floor of that cave, Turville-Petre found plentiful evidence of Bronze and Early Iron Age habitation through to the present day. Below these levels and almost at the bottom of the Palaeolithic stratum, Turville-Petre made an outstanding discovery: four fragments of a human skull, found at the same level as Mousterian flints.[21] The British School were convinced that the skull was Neanderthal and of major importance, but to their consternation the find seemed to attract little academic attention.

Philip Guy, the Inspector of Antiquities for the Palestine government, on 24 August fired off a passionate broadside to *The Times*, 'in the hope that English palaeontologists and prehistoric field-archaeologists as well as geologists may be induced to pay greater attention than hitherto to an area for whose antiquities England is, under the mandate, responsible'. There were nearly three thousand Palaeolithic sites in Palestine, and his department had too small a staff and budget to undertake research and excavation work themselves. 'Discovery must be left to independent bodies,' Guy declared, 'but though Palestine under the English mandate offers admirable opportunities, English societies have been notably backward in taking advantage of them.'[22]

If only for the *amour propre* of the Natural History Museum, Dorothea's journey could not have been more timely. Quite shamelessly, the Museum seized on the fortuitous coincidence of her pre-planned holiday and the day after Guy's letter was published, this report appeared in *The Times* in answer to his tirade:

> Mr PLO Guy, in his article on Prehistoric Man in Palestine in *The Times* yesterday, stated that English societies had been notably backward in taking advantage of recent discoveries in Palestine. The Natural History section of the British Museum has, however, been greatly interested in the recent discoveries, and Miss DMA Bate, of the Geological Department, left for Palestine about ten days ago to investigate them and to help in determining the remains of the fossil animals found.
>
> Miss Bate, who has done similar work in Cyprus, Crete, and

> other Mediterranean islands, is engaged at the British Museum in connexion with the determination of mammals of the pleistocene period. The British Museum has little money to spend upon investigation work, but in view of the importance of the work in Palestine it has been arranged that Miss Bate shall be away for two months. [The report then added:] Mr Turville-Petre brought home some boxes of bones which are being sent to the British Museum, and they will probably have to await the return of Miss Bate.[23]

This unaccustomed publicity came as a complete surprise to Dorothea, as indeed was the welcome news that she was to work on the animal bones, as she wrote to Dr Bather from Jerusalem.

> My sister sent me a little cutting from the 'Times' in which I think I recognise your pen!!
>
> I am very glad to see that the non human remains from the Galilee caves are coming to the Museum and I want to thank you for arranging to keep them for me to work out on my return – and I shall certainly be able to do so with added zeal after being out in this country.
>
> In working out the collection from Syria [from Professor Day] I was much hampered by lack of recent material for comparison and it is impossible to make a really satisfactory job of it without this – so I am trying to get hold of recent skeletons of some of the mammals here that I think will be most useful.[24]

Bather's response to this speaks volumes about the tensions between departments within the Museum. 'It is a very good thing,' he replied to her, 'that you should get hold of some recent skeletons, although I am a little uncertain how far we ought to accumulate recent material in this Department. I presume we have to do it in self defence if other departments do not do it for us.'[25]

From Jerusalem, where she met Philip Guy and Professor Garstang, Dorothea covered hundreds of miles, observing, noting, and collecting live and extinct fauna, and absorbing as much as she could of the geology and history of the Holy Land. In Beirut she visited Professor Day at the American University and examined faunal material in his collections. She visited Turville-Petre's excavations near Tabgah on the Sea of Galilee, observing both the stratigraphy of the caves and sixteen species of birds, several of which were unknown to her. A week or so later, travelling from Amman to Zarqa in Jordan, she jotted down another seventeen, including kingfishers, kestrels, goldfinches, and an eagle. She also spent a pleasant few hours by the fish-filled Lake Huleh observing ospreys, yellow wagtails, pied kingfishers, and cormorants by the hundred.[26]

On her return to the Museum, Dorothea expected to start work immediately on the animal remains sent by Turville-Petre, only to discover that not only had they failed to appear, but that the Museum had no idea where they were. The human skull had been examined by Sir Arthur Keith at the Royal College of Surgeons, and it was to him that Dr Bather wrote in some irritation. 'Miss Bate returned expecting to find the boxes here, but apparently they have never arrived . . . I shall be glad if you can give me any information as to the present position of affairs. Miss Bate has visited the cave and has made herself familiar with the situation in Palestine and will be rather disappointed if the original proposal has fallen through.'[27] It had not. A week later, on 6 November, Keith belatedly sent the specimens, together with a quite unreasonable demand. 'I do hope Miss Bate will be able to contribute her share of the report soon for I do think we should try to get it out before Xmas,'[28] which gave her little more than a month to prepare, sort and identify the bones, and write her report. For any assemblage of fossil fauna it would have been a hard deadline to meet, but with one as complex as the Palestine caves, with little comparative material to help, it was impossible. Dorothea was not prepared to compromise her work and she thought it far more sensible to delay her report until she had the fossil bones from Turville-Petre's final season at Zuttiyeh.

This collection, found beneath the Mousterian layer ('Zuttiyeh,' Turville-Petre told Dorothea, 'is now quite cleared out'[29]), arrived at the Museum in October 1926. As far as possible she put aside other collections until her report was completed. It had been work of the most satisfying sort, as she wrote to Professor Garstang: 'What a wonderfully rich cave that has proved to be and the results of the excavations in it and the Emireh caves should provide a very helpful basis for comparison in further cave work in Palestine and Syria.'[30] As for the region itself, she was utterly fascinated by its unique 'religious, historical and geological interest. Its fauna is no less interesting, and reflects the wonderful variety of climate, vegetation and altitude in this comparatively small area.' The fossil fauna of the caves she found to be extraordinarily rich and diverse. She identified about forty species of mammals, birds, and reptiles. As these could hardly have represented the entire fauna of the country, they indicated, she wrote, 'the existence of faunas even more abundant than that of the present day'.[31] Some of the bones and antlers had cross-cut incisions or longitudinal markings, and while some could be attributed to the gnawings of carnivores or rodents, others could not. Dorothea showed this material to the eminent French archaeologist and prehistorian, Abbé Henri Breuil, who confirmed that these were indeed examples of the early 'handiwork' of Palaeolithic man.[32]

Dorothea's enthusiasm for Turville-Petre's discoveries is unbounded. It was her first opportunity to work on faunal remains that had been found unquestionably with early man. She made copious notes and aides-mémoire on the advance of man and his developing tool industries and art from the Palaeolithic to the Neolithic, recording as well from her own work the fauna and climate that seemed to be associated with the successive stages. Visiting sites in Palestine and seeing for herself the context in which Palaeolithic remains were found had been a seminal experience; she was in her element and it was to be the area that would produce her greatest work.

Each month Dorothea had to write a report on her work for the Keeper, and part of her handwritten draft for July 1928 has survived. In that one month she registered Pleistocene mammalian remains from Manchuria and Crete; examined and wrote reports on collections of Pleistocene and Neolithic mammal or bird remains from southern India, Teddington, Yorkshire, Wiltshire, Crete, Glamorganshire, and the Thames valley; and looked after visitors from Eastern Europe, Germany, and France who wished to study in the department. In between these activities, she visited excavations taking place in Islington in north London to examine mammalian remains, and this is just a small part of one month's work.[33]

As I read through this list, Dorothea seemed to be centred in a great web of place and time, drawing in evidence to illuminate the changing nature of the world she inhabited. Her own excavations in the late 1920s were limited. Between 1925 and 1929, she returned to the bone cave in the Wye valley that she had first excavated over a quarter of a century previously,[34] and made brief visits to Sardinia and Corsica, but she did not write anything about the islands until the mid-1930s, such was her volume of work.*

Some of the collections she examined were from ongoing projects, such as Malta and two major sites in Britain. One of these was the Meare Lake Village, a Celtic Iron Age site that dated from about 300 BC to 50 BC near Glastonbury in Somerset. For several years the Somersetshire Archaeological and Natural History Society sent her small, largely fragmentary collections of bird bones, from which she identified about thirty different species, mainly waterbirds, game birds, and birds of prey, an indication of at least part of the diet and activities of these wetland people. Over the years, Dorothea sent the Society more than a dozen short reports, more often than not accompanied by a note apologizing for a delay as she had had to deal with more urgent work first.[35]

The same problems occurred with material from a very different excavation in Somerset. This was Gough's Cave in Cheddar Gorge, owned by the Marquis of Bath.† In 1903, 'Cheddar Man' was discovered there, the skeleton of a young male adult buried about 9,000 years ago. At the end of 1927, the Marquis's agent, Richard Frederick Parry, a knowledgeable amateur archaeologist, began excavations at the cave entrance. Dorothea, who had identified a small collection of fossil fauna from the cave in 1924,[36] was attracted to Parry's work because of his attention to stratigraphy (the cave produced finds extending over thousands of years, from the Romano-British to the Pleistocene), but she was concerned at the size of the collections he sent her. A note to Parry in April 1928, for example,

* Charles Forsyth Major, the palaeontologist who had inspired her all those years ago, had excavated in the islands and there were a number of unanswered questions concerning the fossil mammals he had collected, not least of which was the identity of an antelope he had discovered in Sardinia. He had found neither skull nor jaw, essential for a definite determination, but in her notes Dorothea has queried whether this might be *Myotragus*, the unique species with rodent-like teeth that she had found in the Balearics. Dorothea visited a number of the sites he had excavated, doing some digging of her own. She did find more remains of the antelope, and although the specimens were far from perfect, it was apparent that they were substantially different from the peculiar Balearic creature. A note in her hand on these remains in the collections of the Museum states simply, 'This is not Myotragus.'

† The cave had been leased in the 1890s by Richard Cox Gough who saw its money-spinning potential. He turned the spectacular cavern into a show cave with electric lighting and charged a shilling entrance fee. When Gough's lease on the cave expired, management of the cave reverted to the Marquis of Bath. (D. J. Irwin, 'The exploration of Gough's Cave and its development as a show cave', *PBSS*, 17: 2 (1985).)

graphically illustrates the pressures she was under. 'I am extremely sorry that I have not yet been able to identify all your specimens from Gough's Cave. I am enclosing you a little preliminary and tentative note which perhaps may be of use . . . I have had great pressure of work lately and am at the moment working out collections which came in last Spring.'[37] Four months later she has still not been able to complete the report. 'I had hoped to have it ready by the end of last month but we have had so many foreigners working here lately and so much other work on hand that it was not possible.'[38]* The Gough's Cave correspondence of more than fifty letters together with Dorothea's notes on the fauna, represents the only surviving detailed record of Parry's four years of excavation at this very important site as his original site notes have been lost or destroyed.[39] It is a correspondence characterized by Dorothea's unceasing apologies for delay in examining the specimens, and Parry's unfailingly polite entreaties for her to hurry up.

Quite the oddest specimen to reach Dorothea came from an experienced archaeologist, Sylvia Benton, who was excavating a cave deposit at Covesea near Elgin in Scotland.[40] Sculptor's Cave, so-called because of apparently Pictish markings on its walls, had been inhabited by humans from the Bronze Age to the Roman occupation. Benton had discovered human and animal bones; many artefacts including Bronze Age jewellery, coins and pottery; and about two dozen bone needle-type instruments, which Dorothea identified for her, even though, she told Benton, the

* Once Dorothea had identified the remains, there was the question of what then should happen to them. 'Perhaps you will kindly let me know what you would like done with the specimens?' she asked Parry in September 1928. 'Most of the scraps do not appear to be worth preserving.' (Letter from Dorothea Bate to R. F. Parry, 11 September 1928, Gough's Cave correspondence, Accession no. 2004/87.) Parry told her she was welcome to keep any that were of use to her, 'and if you would return just a few distinct bones, I should be glad'. (Letter from R. F. Parry to Dorothea Bate, 14 September 1928, Gough's Cave correspondence, Accession no. 2004/87.) She put together a small selection for him, and sent them off. 'I shall be glad to know if they are suitable for your purposes,' she wrote, 'and if so I would then keep a few of the better specimens for our collection as you kindly suggest – and I suppose the remaining fragments need not be kept.' (Letter from Dorothea Bate to R. F. Parry, 19 September 1928, Gough's Cave correspondence, Accession no. 2004/87.) Parry agreed and, to the regret of modern palaeontologists, they were not. 'Their agreement sealed their fate,' wrote Andrew Currant, a successor to Dorothea as Curator of Quaternary Mammals at the Natural History Museum, 'of what would now have been a very valuable taphonomic collection had it survived intact.' (A. P. Currant, 'The late glacial mammal fauna of Gough's Cave, Cheddar, Somerset', *PBSS*, 17: 3 (1986), 286–304.)

material was 'more recent than I generally deal with'.[41] Among the items was a small, pebble-like object 'with purple teeth on it', which seemed to have fossil-like features.[42] Dorothea suggested that Benton show it to someone in Scotland, and she duly took it to J. Graham Callander, the Director of the National Museum in Edinburgh. The correspondence at this point takes a surreal turn. 'The Director,' Sylvia Benton wrote to Dorothea, 'is confident that the alleged pebble you saw is the top of a hot water bottle. There must be someone in the world,' she continued reasonably enough, 'who can analyse the purple stuff and say whether it contains modern glaze or not. Can you help me to find him?' And then, less reasonably, 'I must know before I die.' And finally, 'So sorry to bother you again but I feel you are one of the few truthful men [*sic*] in a lying world.' Then to reinforce the point she adds a postscript, 'I don't believe the h.w.b. theory and I'd love to disprove it.'[43]

It was a ridiculous situation, but one that was also mortifying for Benton, and it became worse. 'The pebble is really serious,' she wrote to Dorothea a few days later. 'That ass of a Director disputes my stratification [of Sculptor's Cave] on the ground that the top of a hot water bottle was found in an alleged bronze age layer.' Dorothea probably thought the director had a point, but the truth, as so often, was simply tiresome. It was not Benton's archaeology that was at fault, but her trust in her workmen. 'I have since found,' she told Dorothea, 'that my foreman has mislead [*sic*] me and the exact find spot of the pebble is uncertain.'[44] As Benton was still reluctant to accept that the pebble was not a fossil, Dorothea offered to get it analysed. She was thus a little bemused to receive from Benton just the label identifying it. 'Would you also send pebble again,' Dorothea wrote. 'I can't do much with label only!'[45]

It was nearly another year before the pebble was finally determined. The verdict, Dorothea told Benton wickedly, 'May not please you!'[46] But Benton was by now resigned to humiliation. 'I am sorry Callander is right,' she wrote to Dorothea, 'but I admire his acumen. I expect it really is a hot water bottle top. My foreman finally confessed that it did not come off the barrow. What he was doing with it when he was supposed to be sifting barrow earth I can't think. It is all very annoying. I was so positive that I had seen all the earth dug out and exactly where.' Her last words on the subject were 'Please give the "pebble" to the dustman with my love.'[47]

Dorothea's interest in material from archaeological sites (hot water-bottle tops aside) was something quite new for palaeontologists, and Dr William Lang, Francis Bather's successor as Keeper of Geology, drew the Trustees' attention to this in his first report in 1928: 'Miss Bate deals with the increasing number of enquiries accompanied by specimens of Pleistocene and post-Pleistocene mammal and bird remains,' he wrote, 'and writes reports on these for the various scientific and archaeological bodies which send the specimens.' At least one hundred such reports survive among her papers, mostly unpublished. 'She thus attracts much new material to the Museum, and forms a link with the Archaeological world, which otherwise might pass over much material of palaeontological interest. Miss Bate often gets into personal touch,' Lang added, 'with her enquirers and thus establishes very pleasant relations between the Museum and the Public.'[48]

The great variety of collections from amateurs and professionals on which Dorothea worked strikingly illustrates the fascination that archaeology and palaeontology held for so many people in the 1920s. For the public, interest had undoubtedly been aroused by such outstanding discoveries as the tomb of Tutankhamun in Egypt in 1922, and it was a subject to which the press and the new medium of the radio gave considerable coverage. In 1929, Dorothea was sent for identification mammalian and bird remains from the Roman fortress and amphitheatre at Caerleon in Wales. Those excavations had initially been financed by the *Daily Mail* 'to the tune ultimately', according to the archaeologist Sir Mortimer Wheeler, 'of some thousands of pounds in return for exclusive news'. While Philip Guy and John Garstang in Palestine were bewailing the lack of trained archaeologists, Mortimer Wheeler's mind 'had been turning increasingly to the need for systematic training in a discipline which was now emerging from the chrysalis stage and was incidentally now in the public eye'.[49] In ten years, his great scheme of an Institute of Archaeology was established in London.

One of the driving forces behind Dorothea's work, and clearly the reason she took on so many reports on faunal material from archaeological sites, was her conviction that excavators and researchers should take seriously this source of scientific evidence. As she examined those numerous collections, it became apparent to her just how much faunal remains could reveal about the life and environment of a past society.

Over the years, her detailed examination of hundreds of collections, both Pleistocene and more recent, began to hone a unique expertise. She could tell archaeologists and prehistorians the animals with whom earlier people may have coexisted; which creatures they hunted and ate and whose skins they wore; which animals may have been ejected from cave shelters with the arrival of man and which had become domesticated. Dorothea was a pioneer of this work, which her contemporaries called comparative zoology but much later was to become known as archaeozoology, the study of animal remains from archaeological sites.

It became part of the Dorothea Bate legend that if archaeologists and prehistorians sent her sufficient quantities of bony fragments, she would give them a complete account not only of the fauna but of the climate and environment as well. It was essential to her to have remains of modern mammals and birds for comparative purposes, to identify the extinct forms and compare their evolution. From marks on bones she could tell whether animals had been butchered by man or beasts, or whether the marks were decorative. The great boxes of bones that were sent to her from all over the world became, with her extraordinary knowledge, a window into the past; a means of understanding the changing climate and environment of the time and areas of the earth she studied, and how man and animals either adapted or became extinct. Her greatest work in this field was in association with a brilliant archaeologist whom she met in the early 1920s.

On 9 January 1923, a small, dark-haired young woman arrived in South Kensington to study the Museum's collections from certain British caves for a thesis on the Upper Palaeolithic in Britain.[50] She was Dorothy Annie Elizabeth Garrod, a newly qualified archaeologist, anthropologist and prehistorian. Dorothea showed Garrod the collections in what was to be the beginning of a professional collaboration and friendship that lasted until Dorothea's death. Dorothy Garrod was thirteen years younger than Dorothea and, like her, had been largely educated at home. With the support of an academic family she had been able to go to Newnham College, Cambridge, where she read history and then after the war achieved a distinction in the anthropology diploma at Oxford University, where Francis Turville-Petre was a fellow student.[51]

Garrod had just returned from a year of working with the Abbé Breuil and the great palaeontologist and anthropologist Marcellin Boule in the caves of southwest France. It was here that Garrod first encountered the Mousterian culture of Neanderthals, which was to become pivotal to so much of her work. In 1925, Abbé Breuil suggested that she excavate a Palaeolithic site in Gibraltar that he had discovered in 1917, a rock shelter opposite a ruin known as Devil's Tower. Garrod excavated there in 1925 and 1926, discovering, to great acclaim and publicity, the skull of a five-year-old Neanderthal child and the associated fauna and artefacts. She sent the faunal remains to Dorothea for determination, apart from the fossil voles, which went to Martin Hinton who had just completed a much-praised study on the species.[52]

Dorothea found the collection a model of archaeological method, with all the specimens clearly labelled with the layers from which they came. It was a welcome contrast to the presentation of much of the material she dealt with, particularly Despott's collections from Malta. Pleistocene fauna from Gibraltar was already well known, but Garrod's was the first excavation in which the animal remains were found associated with Palaeolithic human remains and artefacts. 'It is this definitely established association,' Dorothea wrote, 'that gives the present material its chief value.'[53] She examined Abbé Breuil's material as well as Garrod's, and identified twenty-five mammals altogether, while from the 'thousands of bird bones' she distinguished thirty-three species. Much of the bird material, however, had still to be cleaned and examined. All the birds still lived in Europe, except one, and that was of particular interest. She identified it from 'a single and quite unmistakable fragment' as a great auk, the large flightless bird that had become extinct in the mid-nineteenth century.* Dorothea's report, together with another by the palaeo-ornithologist Anne Eastham in 1968, provided the most complete

* The great auk was commonly known in historic times only in the far North, although, as Dorothea notes, it was not an arctic species. It died out through a combination of being hunted for its meat and feathers, while good specimens could be sold to museums and private collectors for hundreds of pounds. It was not altogether unexpected to find its fossil remains so far south, as it had recently been identified from two other sites in southern Europe; in America its bones had been recorded in Florida. Dorothea suggests that its retreat from the Mediterranean may have been through human agency, rather than climate change; as a flightless bird it was easy prey for early man, particularly during the nesting season, although it would have been much more difficult to catch on the water. Ever practical, she notes that 'the fact that Cormorants are still eaten in the Mediterranean region is proof that a fishy flavour is no deterrent.' (Bate, 'The animal remains', in Garrod *et al.*, 'Excavation of a Mousterian rock-shelter at Devil's Tower, Gibraltar', *JRAI*, 58 (1928), 108.)

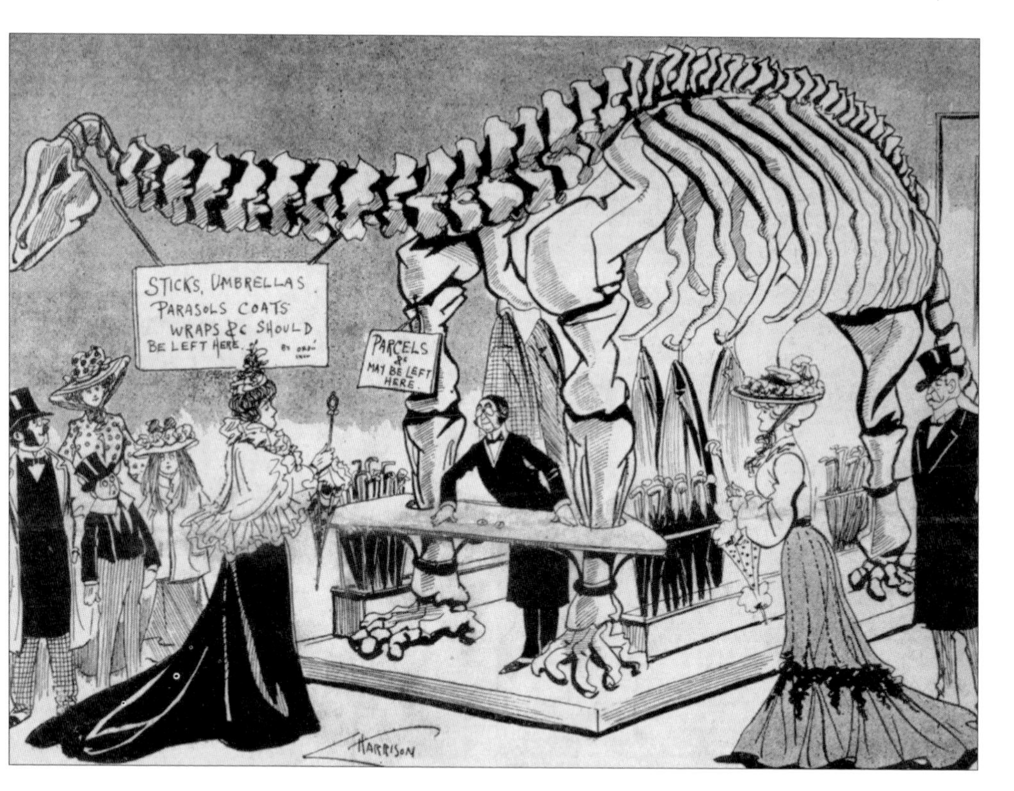

A practical suggestion for how the gigantic cast of *Diplodocus carnegii* might be used.

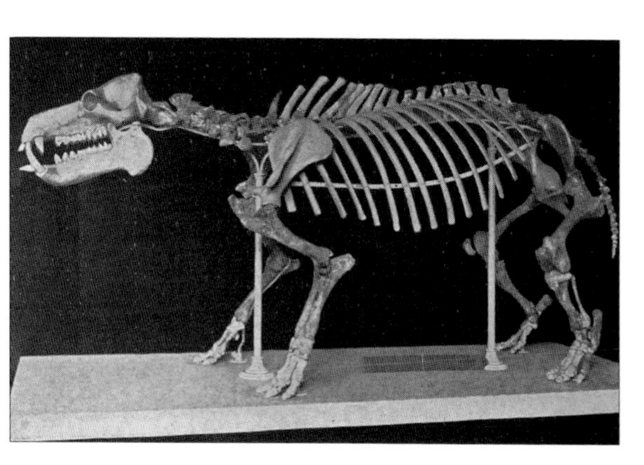

The Museum had galleries full of casts and actual skeletons of fossil and recent mammals, but the arrival of *Diplodocus carnegii* in 1905 may well have inspired Dorothea to reconstruct from bone and plaster the little hippo from Cyprus, *Hippopotamus minutus*. Completed in 1906, it was displayed in the Geological galleries for many years.

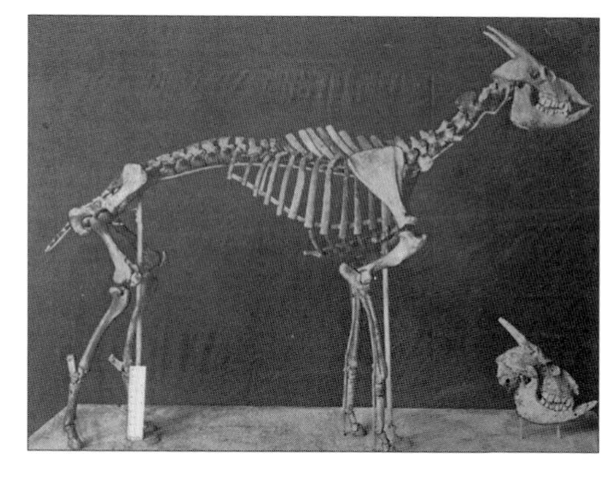

The reconstruction of *Myotragus balearicus* Bate went on show in 1915 and was displayed with the type specimen skull and jaw 'of an old individual' that Dorothea excavated on her first brief visit to Majorca in 1909. This extraordinary extinct goat-like antelope she discovered in Majorca and then Menorca is still the subject of vigorous research and debate.

The lower caves of the Cuevas de los Colombs in north-east Majorca, one of the cave-sites where Dorothea discovered fossil remains of *Myotragus*. To reach the caves could take two hours by (calm) sea, or an exhausting six hours over land.

Myotragus is so specialised there are many different opinions as to what it might have looked like in the flesh. These are just some of them.

(segons A. Azzaroli)

(segons A. Bonner)

(segons J. Clutton-Brock)

(segons R. Attenborough)

(segons M. Köhler)

(segons V. Sastre)

Searching for Piltdown Man. View of the site with inset portraits of Charles Dawson (left) and Arthur Smith Woodward (right). Smith Woodward retired in 1924 and spent his time fruitlessly digging for further remains, despite the fact that nothing had been discovered since Dawson's death in 1916.

Dorothea with her brother Thomas (who was convalescing after his polo accident) at Osborne on the Isle of Wight, 1913.

The shooting party. This photograph, undated, shows Dorothea in a jauntily feathered hat, standing behind Leila holding a gun. Although the picture is indistinct, the man seated on the left is Tansley, Leila's husband.

It was not until after the First World War that the Luddingtons moved to Waltons Park. The house survived the Second World War unscathed, only to be destroyed in a catastrophic fire in February 1954, while Leila was away. Almost all the family's personal effects were destroyed. She never returned to Ashdon.

It was only with great reluctance that Dorothea agreed to sit for this photograph in 1930 when she was fifty-one. It was used for a Hungarian journal article to celebrate the work of women scientists. Its title was 'Women and the Study of Fossils'.

The archaeologist Gertrude Caton Thompson.

Dorothy Garrod (centre) with Theodore McCown (left) and Francis Turville-Petre (right), at her camp in the Wadi el-Mughara, 1931.

The Bethlehem pit dug in the Hasbun brothers' fruit garden on the highest point in Bethlehem. The scale of this excavation is astonishing. The find spot of the giant tortoise carapace is left of centre in the photograph. Note the increasingly difficult climb for the basket carriers on the right.

Dorothea and a workman clearing two fossil elephant teeth. One of the ridged teeth can clearly be seen to the right of the workman.

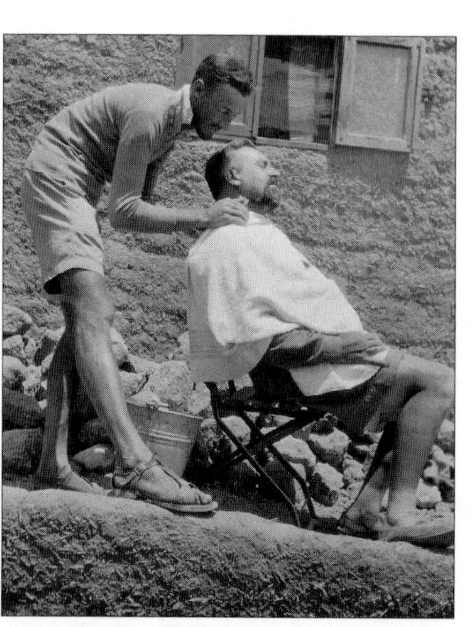

James Starkey, who was murdered in Palestine by Arab insurgents in 1938, being shaved by Gerald Lankester Harding, the Director of Antiquities for Transjordan, in 1932. There is a tragic irony here: Starkey subsequently grew a luxuriant beard which contemporary newspaper accounts suggest may have given him a Jewish appearance and so led to his death.

Dorothea, on the far right, at lunch in the Museum in 1935 with (from left) the palaeontologist Guy Pilgrim, philosopher, archaeologist (and Piltdown suspect) Father Teilhard de Chardin, a Mrs Sahni and the back view of Wilfred Edwards, Dorothea's colleague who was to become Keeper of Geology.

The Geology Department in 1938 just before the retirement of the Keeper, Dr William Lang (centre front row). Dorothea is front row right, Helen Muir-Wood is front row left. Wilfred Edwards is fifth from left. Kenneth Oakley is second row, fifth from right.

The contrast between this picture (enlarged from the Departmental photo) of Dorothea in 1938 after her illness and the robust woman of earlier photographs, is stark.

Dorothea in Africa in 1947 when she was sixty-eight, in photographs taken by her colleague, Kenneth Oakley.

Dorothea being carried ashore by a (clothed) 'native', the most practical, if inelegant, means of reaching dry land.

Relaxing after a day's collecting. (From left) Mary and Louis Leakey, Dorothea, and John Waechter.

descriptions of Gibraltar's avifaunal assemblages until the twenty-first century.[54]

The Devil's Tower excavation established Dorothy Garrod's reputation. As a direct result, she was invited to become a student at the British School of Archaeology in Jerusalem (BSAJ) and to undertake the excavation of a cave near the village of Shukbah, northwest of Jerusalem in the Judaean hills. Dorothea's formidable expertise made her the obvious palaeontologist to determine the fossil vertebrate remains. Although it was the BSAJ which had asked Garrod to excavate, their financial position was such that they faced the possibility of closure and Garrod had to rely on the generosity of benefactors. Just a few weeks previously in November 1927, Dorothea had attended the Annual General Meeting of the BSAJ in London and had sat listening to the anger of successive speakers that such a situation should have been allowed to occur. Field Marshal Lord Allenby, without whose military skills there was unlikely to have been a British Mandate in Palestine at all, thundered that it was 'nothing short of disgrace if we of this country who were entrusted with the Mandate for Palestine, who took such a large share in freeing Palestine from Turkish dominion, did not accept the duty incumbent upon us to explore the antiquities of that country . . . All that is wanted,' he concluded on a note of sudden optimism, 'is enthusiasm. If we have enthusiasm, money will come.'[55] But it came only slowly and not from the government; industry and individuals would have to support the excavations of the British School, not the public purse.

Dorothy Garrod, with two American colleagues who had their own funding, excavated at Shukbah from April until the middle of June 1928.[56] The cave was more than a thousand feet above sea level and was sited above a broad wadi, known as the Wadi-en-Natuf. In her ten weeks of excavation, Garrod discovered two archaeological levels. The lower layer was Mousterian, 'which resembled', she reported in the autumn to the British Association meeting in Glasgow, 'that already known from the Galilee Caves', while the upper layer contained a microlithic industry of a type never before discovered in Palestine.[57] The flints were of a curious crescent shape, which Garrod calls 'lunates'. There were also parallel-sided blades, of which a number, Garrod wrote, 'have on their edges the peculiar polish produced by cutting corn or grass', and that might imply that the hunter-gatherers who had once lived in these caves were developing the practice of agriculture. Eleven human skeletons were discovered buried in this layer, including an adult male skeleton 'found in a sitting position under a large fallen rock, with the remains of two children on his knees'.[58] It was a remarkable site, which Garrod left in June with every intention of returning the following year. The faunal remains were shipped to Dorothea for identification.

Just before the Shukbah collection arrived, Dorothea received an enquiry from John Crowfoot, the Director of the British School, concerning the whereabouts of the Emireh and Zuttiyeh material. 'The collections are here,' she told Crowfoot with some surprise,

> and I shall be very glad to show them to you if you have time to come here when you are in London next. There are not a great many specimens and we should be most grateful to have them for the National Collection. Up till now, we have had no Pleistocene vertebrate remains from Palestine, [so] this collection would not only be a nice acquisition in itself, but will also be of the greatest use when working out any future collections from that part of the world. Of course, if you would like something more official Dr Lang, Keeper of the Dept. will write to you.[59]

However, it was not that simple. To Dorothea's horror, Crowfoot seems to have been told that the Museum had been asked to return the material and he wrote again, asking her to check the Museum's official correspondence files for what had been agreed. By this time the Shukbah material had arrived and the Zuttiyeh–Emireh specimens were crucial for comparative purposes. Dorothea could find nothing conclusive in the files and resorted to persuading Crowfoot of the importance of the collection to the Museum. 'Apart from merely possessing these remains,' she wrote, 'it would seem very desirable to have at least the described and figured specimens here. They are of considerable palaeontological interest and will be most useful for purposes of comparison in all future work on the Pleistocene fauna of Palestine.' As a compromise she proposed making a small representative collection for exhibition in the Jerusalem Museum. 'I might perhaps also remind you,' she added in brisk defence of the requirements of her own institution, 'that a great deal of time and labour has been given here to the study and preparation of these collections.'[60]

The compromise was accepted and the representative collection was sent to Palestine. Crowfoot had also asked Dorothea to make a similar collection for the Jerusalem Museum of the bones from Shukbah, 'when the examination of them is complete'.[61] In the margin of this letter she has written tartly, 'This will not be begun until further material is sent,' which she expected after Garrod's second season at Shukbah. At the beginning of 1929, however, the Department of Antiquities in Jerusalem asked the British School to undertake immediately the excavation of a group of caves threatened with destruction by quarrying in the Wadi el-Mughara on Mount Carmel.[62] Rock was needed to build a new harbour at the port of Haifa, which lay at the northern point of Mount Carmel, and the limestone bluff containing the caves was one of two possible quarry sites under consideration. The whole area was known to contain antiquities. When preliminary excavations at the Wadi el-Mughara caves revealed human artefacts and remains, their importance was such that Garrod and the funding for Shukbah were transferred there immediately and Dorothea was asked to determine the faunal remains. In collaboration with the American School of Prehistoric Research, work was begun at the Mugharet el-Wad, the Cave of the Valley. It was to be the first of seven seasons for Garrod of outstanding excavations on Mount Carmel and all plans to quarry at these caves were abandoned.* Garrod did not return to Shukbah at all and neither did anyone else for another seventy-two years, when an Anglo-American archaeological team began new fieldwork there in 2000.[63]†

In February 1929, Dorothea's mother died. Elizabeth was seventy-six and appears to have been living in London since her husband's death, but not with Dorothea. As it was Dorothea who registered the death, it seems unlikely that either Thomas or Leila was present. Elizabeth died intestate, leaving just under £285 (about £8,000 in today's values). Nothing is known of her later life, but Dorothy Garrod's message of condolence in June indicates that it had been a difficult time for Dorothea: 'I am so sorry to hear that you have had trouble and loss in your family and send much sympathy,' Garrod wrote, but the main purpose of her letter must

* For a full account of this fascinating saga, see Jane Callander and Ofer Bar-Yosef, 'Saving Mount Carmel Caves', *PEQ*, 132 (2000), 94–115. Wadi is the modern spelling; in the 1930s wady was the norm.

† After their preliminary survey, however, the political situation in the region prevented further work.

have considerably cheered Dorothea's spirits. 'We have had a very good season here, and one of the results has been rather a harvest for you; I don't know whether you will be glad or sorry! The general result of the dig has been to get a far more complete succession than has so far been found in the Near East.'[64] She listed the layers, beginning with the most recent, 'Historical' and ending with 'Mousterian', a time-span of 100,000 years.*

In August 1930, Dorothea received a second collection of animal and bird remains from the Mugharet el-Wad and produced a handwritten report in just eleven days. The most common species were gazelles, wild goats, and deer.

> Most of the animals must have been introduced into the cave to be used for food, and it is owing to this that practically all the bones were found to be broken . . . Remains of large oxen, horse and wild boar are scarcer, possibly because though valuable as food they would be more difficult to obtain. The comparative scarcity of such small animals as hares which might be useful both as food and for their skins is noticeable.[65]

Garrod's excavations revealed that the Mesolithic humans who had lived in the cave had produced a microlithic industry which so resembled that of Shukbah that it became 'abundantly clear', Garrod wrote in 1932, that it 'would not fit exactly into any of the pigeon holes already existing'. She therefore gave it a label of its own, 'Natufian', from the Wadi en-Natuf at Shukbah.[66] It was a culture that had advanced beyond a simple tool industry. Two carved objects were found; one, discovered in 1928 in the preliminary excavation, was the statuette of a young deer, sculpted on the end of a piece of long bone. The other, which Garrod had described to Dorothea in 1929 as 'our most important individual find',[67] was a carving of a small human head in calcite. Jewellery was also found on human skeletons: headdresses of dentalium shells, a necklace of 'curiously shaped bone pendants strung in pairs', while the skeleton of a young child 'had the remains of a cap of bone pendants'.[68]

* In between these, Layer 2, Garrod noted in her letter, was 'Mesolithic of Shukba type, but more developed'. Layer 3 was 'Upper Palaeolithic? type. (New and rather strange).' Layers 4 and 5 were 'Middle Aurignacian α and β Western European type' and layer 6 was 'Lower or early-middle Aurignacian (unfamiliar Type)'.

At the end of 1930, Garrod discovered evidence of the Natufian culture in a third cave, which she had noticed in the cliffs to the east of the railway line as she travelled from Haifa to Jerusalem. As she could not work both sites simultaneously, the British and American Schools in 1931 asked Francis Turville-Petre and Charlotte Baynes (who had worked with him in Galilee) to excavate this cave, the Mugharet el-Kebara. It too was a rich deposit, yielding hundreds of lunates, sickle blades and exquisitely worked ornaments, bone tools, and decorated sickle hafts.* Turville-Petre also found quantities of faunal material, which, with the latest specimens from the Mugharet el-Wad, were sent to Dorothea. She received these collections in September 1931 and ten weeks later had written a preliminary report, which was published the following year.[69] It was a short report but dramatically different from anything she had produced before, and it was written with a new and energetic confidence. In it she introduced radical notions of the successive changes in the climate and ecology of the region over a period of 100,000 years, from the Mousterian period associated with the Neanderthals to the present, demonstrating just how far her interpretation of material had advanced.

Dorothea identified twenty-five mammals, which she divided into two categories: those that are now extremely rare or have disappeared within historic times, such as the leopard, red deer, Persian fallow deer and gazelles; and those that are now unknown anywhere in the region. Among these are a very large ox and the spotted hyena, which is now found only in Africa south of the Sahara. She constructed her arguments on climate on the preponderance in different layers of the two most plentiful animals, deer and gazelles. 'In view of the fact that Deer are typical inhabitants of forest and jungle country,' she wrote, 'while Gazelles are equally typical desert dwellers, it seems legitimate to suggest that the transition from Deer to Gazelles as the dominant species, indicates a change from moist conditions in Mousterian times to a dry climate in the Mesolithic, with a consequent alteration from a wooded to a more open country.' And it

* Turville-Petre had been due to return to Kebara the following year to continue excavating, but did not. Ill health was given as the reason. In fact his behaviour was such that Garrod could no longer work with him. There is no record of what actually happened, but there is the suggestion of a clue in the empty whisky bottles discovered some years later in his archaeological dumps outside the cave of Kebara. See Ofer Bar-Yosef and Jane Callander, 'A forgotten archaeologist: the life of Francis Turville-Petre', *PEQ*, 129 (1997), 2–18.

was a gradual change, she discovered; the overwhelming predominance of deer in the earliest levels of the caves, the Mousterian, was followed by the appearance of gazelles and deer in equal number in the succeeding period, known as the Aurignacian* (which was associated with Cro-Magnon man), followed in turn by a predominance of gazelles in the Mesolithic layers, about 8,500 years ago.[70] This was the first time Dorothea had attempted such an analysis, and she would develop this argument on climate change further and famously in her monograph on the Mount Carmel caves. Crucially, what her faunal analysis would also suggest was the chronological position for the human finds.

In 1931, Garrod began excavations at two other caves in the Wadi el-Mughara, the Mugharet et-Tabun (Cave of the Oven) and the Mugharet es-Skhul (Cave of the Kids). She put her American colleague, Theodore McCown, in charge of Skhul where over two seasons he uncovered first the skeleton of a child associated with Mousterian artefacts and then nine more skeletons including two infants and seven adults. The breccia was too hard to extract the skeletons on site. Instead, McCown and his colleagues used 'light hammers and fine chisels' to expose the fragile bones and then, to ensure their safety, 'quarried' them out, still supported in great blocks of limestone. To protect them, the bones were greased or painted with a layer of shellac, then wrapped in silver foil and thickly encased in plaster of Paris. For transport to London, each great slab was put in a box, and the narrow space between the limestone block and the box was filled with concrete. The largest of these weighed over two and a half tons.[71] By lorry and steamer they were conveyed to the Royal College of Surgeons in London. There McCown worked on the bones with Sir Arthur Keith. It was to take two and a half years to extricate the skeletons from their concrete and limestone casings and another year and a half before every bone and fragment was measured, studied, and photographed and the report written.[72]

Their discoveries were dramatic. After their initial study of the infant's skeleton in London in 1932, McCown and Keith concluded that it appeared

* Archaeological terms range from the general and comprehensible Stone Age (Old, Middle, and New or Palaeolithic, Mesolithic, and Neolithic) to the more specific and somewhat opaque (for the uninitiated) Acheulian, Mousterian, Aurignacian, and so on, derived from the names of sites of different ages that had revealed the first artefacts of the particular cultures. Aurignac is in southwest France.

to show a nearer approach to modern humanity than European Neanderthals, and so did the other skeletons. They therefore proposed to designate them *Palaeoanthropus palestinus*, together with the Zuttiyeh skull discovered by Turville-Petre in 1925.[73] Keith, with his customary enthusiasm, called the skeletons from Skhul 'the most imposing specimens of fossil humanity I have ever seen'.[74]

The cave of Tabun, which had originally appeared quite small, turned out to be the largest, deepest, and most complex cave of all. Garrod had found Mousterian tool industries associated with abundant fauna and the skeleton of a woman of 'a type closely related to the Neandertal', Garrod wrote, 'but with certain well-marked peculiarities'.[75]* This and all the skeletons with their unique characteristics were to overturn accepted notions of human evolution in the region and are fundamental to the evolutionary debate today.

Dorothea found the species of fauna in the layer underlying the woman to be varied, and included both rhinoceros and hippopotamus. As she suggested to Garrod, 'there must have been permanent rivers of some size, with the low country consisting of open grassy plains and bordered by wooded hills; a great contrast with the Palestinian landscape of today.' Garrod proposed in her next season to dig in the area 'close to the rock where the bone is usually best preserved', and this, she hoped, would 'add considerably to the fauna of the deep layers'.[76] With the prospect of a rich haul, Garrod needed expertise on site, and she and Dorothea began to plot how Miss Bate could join the final season of excavating in the spring of 1934.

* Garrod at first assumed this skeleton was the same species as the Skhul skeletons, *Palaeanthropus palestinus*, but then discovered differences that seemed to suggest astonishingly that it might be another human type from approximately the same archaeological level. Arthur Keith, however, believed all the skeletons were the same race: *Palaeanthropus palestinus*. These and other questions surrounding 'Tabun woman' (not least its stratigraphic position in the cave) have been explored by Ofer Bar-Yosef and Jane Callander in 'The woman from Tabun: Garrod's doubts in historical perspective', *JHE*, 37 (1999), 879–85.

CHAPTER 13

The Bone Beasts of Bethlehem

The finds from Mount Carmel were of such importance that the British Museum in Bloomsbury mounted a temporary exhibition on the prehistory of Palestine in February 1934. Dorothea spent much of the early weeks of the new year with Dorothy Garrod and Theodore McCown working on this, while Garrod and McCown doubtless also took the opportunity to persuade Dr Lang, the Keeper of Geology, of Dorothea's importance to their forthcoming season in Palestine.

The argument that may finally have persuaded the Museum to give her leave was Dorothea's suggestion that she should visit Malta on her way to Palestine as she had been associated with the island's faunal past for so many years. The Director of the Malta Museum, Sir Themistocle Zammit, was delighted at this possibility and by February matters were looking decidedly hopeful. 'It does seem as if I might perhaps be able to come about the end of March,' Dorothea wrote to Zammit. 'I hope ever so much to be able to carry out this plan. It might make it easier,' she continued persuasively, 'if I could feel sure of being able to do some collecting of Pleistocene fossils to bring home for our collections, which, with the duplicates you are kindly giving us, would add considerably to our material from Malta.' Her approach to Zammit was all flattery. 'Your deposits are so rich that I expect this would not be a difficult matter if Dr Baldacchino is kind enough to help me – and I of course, would come with the hope of doing anything I can to help your Museum.'[1]

That was sincerely meant. In January 1933, Guiseppe Despott had taken sick leave and left the Malta Museum six months later. His place had been taken by Dr Joseph Baldacchino, an enthusiastic natural historian and archaeologist who was unsparing in his condemnation of the disasters that confronted him on taking up his new post. His annual report on the Malta Museum reveals a natural history section crowded and ill organized, with an almost entire absence of labels. Exhibits were arranged without meaning 'and consequently useless for purposes of instruction'.[2] Even worse was the state of Ghar Dalam material, something Dorothea may not have been entirely surprised to learn. 'An enormous quantity of fossil bones,' Baldacchino wrote, 'mostly in a fragmentary state, was heaped up in utter disorder in one of the rooms above the cave.'[3] Despott's retirement may not have been entirely due to ill health.

By March, Dorothea's leave had been granted and the Trustees of the Museum agreed to pay her salary for one month. She had also applied for a grant from the Percy Sladen Fund, which was already contributing to the Mount Carmel excavations. After her researches in Malta, she told the Fund, she proposed to proceed to Palestine, 'to assist Miss Garrod in collecting Pleistocene vertebrate cave remains . . . These remains are expected to throw valuable light on the early forms and climatic conditions in Palestine. Miss Garrod,' she added, 'has offered me hospitality during my stay in her camp.'[4] Garrod, now a research fellow at Newnham College, Cambridge, wrote in support of Dorothea's application: 'As I have not been able to include a trained palaeontologist in my party it would be very valuable if Miss Bate could spend a few weeks on the site, supervising the removal of bones and selecting those which are worth further study. In this way,' Garrod argued, 'a good deal of preliminary work could be done in the field and it would not be necessary to transport to England such a large mass of material as usual', which would, of course, cost much more than a modest grant for Dorothea.[5] Within a week the Fund had made its decision and awarded her a grant that totalled £80 (about £3,000 in today's values).[6*]

In the last week in March, Dorothea left England for Malta. In 1933 a new site to the southwest of Valletta had been discovered, a fissure known

* £43. 18*s*. 0*d*. of that £80 was the grant plus interest she had received from the fund in 1915 for her proposed excavations at Assynt in northern Scotland, which she had never carried out.

as Tal Gnien, which she was particularly keen to see. Baldacchino had found there the remains of dwarf elephants, giant tortoises, extinct birds, and a carnivore, 'a large Mustelid hitherto unknown from Malta'. Mustelids, which have long bodies and short legs, include weasels, pole-cats, and badgers and, in this case, an otter-like creature, which had adapted to living in water. It was, Dorothea noted, highly specialized for burrowing and aquatic habits, indicating a climate considerably wetter than Malta is now. She called this extinct creature *Nesolutra euxena*.[7]* Over the course of a week Dorothea found at Tal Gnien remains of a few tiny shrews associated with a large tortoise, a swan, and *Leithia*, the giant dormouse about the size of a squirrel. She also found a second previously unknown extinct creature, a vole, which she named *Pitymys melitensis* Bate. At Ghar Dalam she examined a site that Baldacchino had excavated just before her arrival. The collecting had been carried out with such meticulous care that she hoped Baldacchino had at last brought modern methods to excavations on Malta.

At the beginning of May, after three intensive weeks' work, she sailed from Valletta to Port Said and then travelled by train to Jerusalem, to find that, before she could join Dorothy Garrod at Mount Carmel, the Department of Antiquities had urgent need of her expertise. Some fossilized bones had been brought to them by a doctor from Bethlehem who had found them protruding from a dump of earth, the material excavated from a well being dug by the town council. She examined the bone fragments and recognized them as the unmistakable remains of a fossil elephant. She asked to see where they had been found, and was driven the five miles to Bethlehem. The site was at the highest point of the town, nearly eight hundred metres above sea level, in a large, walled garden full of splendid fruit trees. Against the north wall was a pit some fifty feet

* Charles Forsyth Major had discovered a similar creature in Sardinia in 1901, which he had called *Enhydrictis galictoides*. He believed the creature to be aquatic, but subsequent work had thrown doubt on this. Forsyth Major's papers relating to the creature had been found at the Natural History Museum in Basle and also some limb bones, and these were sent to Dorothea as the acknowledged expert on the subject. She worked on this material through 1934, publishing substantial reports on both the Malta and Sardinian creatures. (Bate, 'The habits of *Enhydrictis galictoides* with description of some limb-bones of this Mustelid from the Pleistocene of Malta', *PZSL*, 2 (1935), 241–5; and Bate, 'Two new mammals from the Pleistocene of Malta', *PZSL*, 2 (1935), 247–64.) She concluded, from her detailed examination of *E. galictoides*, that Forsyth Major had been wrong 'and that this animal was not specialized for an aquatic mode of life'. The pupil had now outstripped the teacher.

deep and ten feet square with a great pile of earth heaped against the wall alongside it.

Dorothea first examined the dump and, like the doctor, found fragments of bone protruding from it. She then examined the pit. Its sides were of sticky, damp clay and, as she noted ruefully, 'My skirt and shirt had to be washed after my descent!'[8] But the prize was there: at a depth of twelve feet was a piece of elephant tusk sticking in the wet clay with flints and gravel.[9] Dorothea was asked by Ernest Richmond, the Director of the Department of Antiquities in Jerusalem, to deal with the remains and 'see what I could arrange about further work'.[10] As it was the department that issued licences for excavations, Dorothea believed that this site, which seemed so wonderfully full of potential, would be her responsibility. Ernest Richmond agreed to loan the elephant tusk and other bones she had found to the Natural History Museum, subject to a formal agreement. She was committed to joining Dorothy Garrod at Mount Carmel immediately and, apart from arranging to have the fossils sent to South Kensington to await her return, there was nothing more she could do until she returned to London.

From Jerusalem, Dorothea took the train north to Haifa. For the last part of the journey the track ran near the Mediterranean. To the east stretched the green flanks of Mount Carmel, rising up from the plain to over eighteen hundred feet at its highest point. To the west, on its rocky promontory, was the ruin of the Crusader castle at Athlit. Just three miles to the southeast of the castle was the Wadi el-Mughara at the foot of the great limestone bluff that commanded the coastal plain, the caves clearly visible. The archaeologists' camp, known as Tibn (the Arabic for straw) Towers, stood among the few trees scattered below the cave. Straw-roofed and built of mud bricks, it was vulnerable to the seasons. In winter, according to Garrod, it could be 'vile! We were alternately flooded out and blown to bits, with thunderstorms keeping up a more or less perpetual accompaniment.'[11] Early summer brought the onset of heat and humidity. Fleas were a problem and when Dorothea arrived, improvements had only just been made to the very basic outdoor lavatory.[12]

The Mugharet et-Tabun, when Dorothea clambered up the hundred feet or so of cliff, presented her with a truly wondrous sight. The entrance to the cave, once small and hidden from the ground by a cluster of carob trees, was now, as Garrod described it, 'a vast irregular opening filled with a totally unsuspected thickness of archaeological deposit'.[13] Over five seasons, Garrod had dug down through seventy feet of stratified Lower and Middle Palaeolithic deposits, a unique and monumental sequence of continuous human habitation, which Dorothea estimated at 100,000 years.[14] Today that figure is put at 600,000 years or more.[15] By the end of this final season in August, Garrod had discovered well over 55,000 flint implements in Tabun, 25,000 in el-Wad while Skhul had revealed another 10,000, a staggering total of more than 90,000 flints for Garrod to record, catalogue, examine and identify.[16]

The fauna excavated that season more than justified Dorothea's presence. She identified more rhinoceros and hippo, confirming the warm, damp period she had noted from the fauna sent to her the previous year. Extraordinarily, a piece of elephant tusk was also unearthed, just a few weeks after Dorothea had identified the first elephant to be found in Palestine at the highest point in Bethlehem. Conditions in Tabun were uncomfortable; 'unbearably hot during the time that the sun entered it', Garrod wrote, 'and unbearably draughty when there was a wind',[17] although that was no more than Dorothea was used to. Whenever she had a moment, she noted birds in the wadi, 'a v. handsome bunting', a vulture, buzzard, and a bird like a large long-tailed wren with a 'Short cheery little songlet'.[18] In the evenings at six o'clock out would come the sherry bottle and Garrod could sometimes be persuaded to play her flute,[19] though it was unlikely that she did so during Dorothea's three weeks in camp, as Garrod had broken her hand.[20]

On 26 May, Dorothea left Tibn Towers, returning home via Beirut to examine specimens at the American University. Ahead of her were two immense projects: the report with Garrod on the monumental discoveries from the Stone Age of Mount Carmel, and the hugely promising new site at Bethlehem, the excavation of which, after her conversations with Ernest Richmond, she was eager to set about organizing. She had also to work out the material she had collected in Malta. It had been a most fruitful trip.

Dorothea arrived home to be told of the death of Gertrude Caton Thompson's mother, whom she knew well from dinner invitations to their flat in Albert Hall Mansions, just up the road from the Natural History Museum. She hurried round at once to a devastated Gertrude who railed against the 'hideous suffering' her mother had been forced to endure from a cancer left too long undiagnosed.[21] Gertrude found the emptiness of the flat unbearable after living there for so many years with her mother, and asked Dorothea 'constantly' to dine with her or would go to supper at Dorothea's simple flat in the West Cromwell Road, anxious for company, particularly when it was as acerbic, interesting and informative as Dorothea's, until she could cope with her loss. Gertrude was to note later what a 'particularly good friend' Dorothea was to her during this painful time.[22]

In June, Dorothea's visit to Malta and Palestine was discussed at the Museum Trustees' regular meeting. She had from Malta, it was minuted, 'collected some valuable Pleistocene vertebrate remains. She had also selected from the specimens collected at the Wady Mughara caves a series very acceptable to this Museum.' And then, inevitably, came the subject of cost. 'Dr Lang stated that during four weeks of Miss Bate's absence she had been paid at her usual average rate but that owing to a request from the Dept of Antiquities that she should investigate a new discovery of Pleistocene mammals [the Bethlehem finds], she had been ten days longer in Palestine than anticipated, as a result of which a collection of new material was being sent over for her to study. In view of these circumstances the Trustees agreed that Miss Bate be allowed to count a further week for pay at the usual rate.'[23] Dorothea is silent on her loss of three days' pay.

A few weeks after her return, Dorothea received a letter from Dorothy Garrod. 'You will want, I know,' Garrod wrote, 'to hear all the news from Tibn Towers.' She told Dorothea of her latest faunal finds, that her hand had now mended, and regaled her with stories about the workers at the dig. It was a long, gossipy letter, which reflects the respect and friendship between the two and gives a lively insight into life at Tibn Towers. 'Our behaviour has been exemplary since you left; the strictest temperance, very little bad language, and Anne [Fuller] has given up smoking. It was nice having you here,' Garrod ended, 'I hope you enjoyed it as much as I did. I wish you could just look in on us from time to time, via a magic carpet!'[24]

With great speed, Dorothea published reports on two new species discovered in Tabun, *Trionyx* (a river turtle), and a crocodile.[25] She also published a paper on a fossil warthog from Palestine, which Theodore McCown had discovered in Skhul cave in the spring of 1932. It was the first time that the remains of the creature, either fossil or recent, had been discovered outside Africa and she described it as a new species. She called it *Phacochoerus garrodae* 'in honour of Miss Dorothy A. E. Garrod, whose splendid work has done so much to enrich our knowledge, not only of the Pleistocene fauna of Palestine, but of the early history of Man'.[26] Garrod must have been vastly amused to have immortality bestowed on her through association with a warthog. There are similar such honours in many of the names that Dorothea devised for the new species she identified from the caves. An extinct vole was named *Microtus mccowni* Bate in honour of Theodore McCown; an extinct dormouse was named *Philistomys roachi* Bate after Edward Keith-Roach, the District Commissioner for Northern Palestine; while it was inevitable that Dorothea's friend and colleague at the Museum, Martin Alister Campbell Hinton, who had written the acclaimed monograph on voles and lemmings, should lend his name to a primitive vole, *Microtus (Chonomys) machintoni* Bate. On a more allusory level, a large extinct pig was splendidly named *Sus gadarensis* Bate, while a shrew that Dorothea had originally called *Crocidura xanthippe* after Socrates' ill-tempered wife, she had later, and more charitably, to rename *Crocidura samaritana* when it was discovered *xanthippe* was already in use.

Dorothea's priority on her return from Palestine had been to set in motion plans for the excavation of the Bethlehem site. The specimens, however, did not arrive from Jerusalem until the end of July, and that is possibly why she did not announce the momentous discovery of the extinct elephant to the scientific press immediately on her return. But the discovery was far from secret. Word of the find had already spread rapidly around the now considerable and highly competitive archaeological community in Palestine.

By 1934, the angry complaints of Professor Garstang and Philip Guy on the lack of interest in archaeology in the region had been answered. Nine schools of archaeology had been established in Jerusalem, from Britain, France, Germany and the United States. Added to these, museums and universities from Europe and America mounted their own expeditions and in the early 1930s around a dozen sites were being excavated.

In 1932 excavations had begun at the Tell ed Duweir near Hebron, about forty minutes' drive from Bethlehem. This great mound, some seventy feet high and eighteen acres across, was the site of the biblical town of Lachish. The director of the dig was the archaeologist James Leslie Starkey, known to his friends as Starks, whose passion for the Near East extended to naming his house in Walton-on-Thames 'Badari', after the famous Egyptian site. Like so many archaeologists of the time, he had worked with the great Egyptologist William Flinders Petrie. By 1930, however, he was eager to direct his own excavation and persuaded the American pharmaceutical entrepreneur and philanthropist, Sir Henry Wellcome, to fund an expedition. It cost Sir Henry the very substantial sum of between £5,000 and £7,000 a year (about £180,000 to £250,000 in today's values) and was known as the Wellcome Archaeological Research Expedition to the Near East, or WARENE. It was archaeology on a grand scale with purpose-built headquarters at the site and even its own rather splendid headed notepaper. It was a far cry from Tibn Towers. In successive seasons, Starkey revealed a great fortress town of the kingdom of Judaea, dating from the sixth century BC, while subsequent excavations showed the site to have been occupied since the Bronze Age.*

Dorothea's fossil elephant finds at Bethlehem, so near to his own excavations, intrigued Starkey and he suggested to the Department of Antiquities that he would try to persuade Sir Henry Wellcome to fund an expedition. The possibility that such substantial resources as Sir Henry's might support this excavation was obviously attractive to the department, and without Dorothea being consulted, Starkey assumed control of the Bethlehem project. Unaware of this, when Starkey asked Dorothea for information on the elephant, she gave it freely. Only then does Starkey appear to have told her that the dig was now his. He wrote to her in terms that she found barely tolerable, underplaying totally her vital role in the whole enterprise: 'It was most fortunate,' he wrote, perhaps a little disingenuously, 'that at this moment I heard of your visit to Palestine.' He then informed her that 'the Expedition [WARENE] had been promised the concession by the Department.'[27] In rage she wrote on this letter in a large, angry hand: 'It must be remembered that before Mr

* Starkey also discovered eighteen ostraka – clay pottery fragments used as a writing surface. These Lachish Letters, as they became known, bore some ninety lines of clear and readable writing and were the first personal documents found from this era. They were first deciphered by Harry Torczyner, Professor of Hebrew at the University of Jerusalem. A few fragments of these ostraka and other antiquities from Lachish can be seen in the galleries of the British Museum.

Starkey approached me the Dept. of Antiquities had asked me to deal with the remains from Bethlehem and to see what I could arrange about further work.' But it was too late. She had been outmanoeuvred in a game she did not know she was playing, and to the end of her life felt she had effectively been forced to waive her claim to the Bethlehem deposit.

She did not surrender it, however, without first exacting two very specific guarantees from Starkey.[28] With no lessening of her anger she scrawled on Starkey's letter, 'Mr Starkey cabled the Palestine Dept of Antiquities after I had agreed that he could work it – with proviso that I work out the material and the Museum gets specimens.' What made Starkey's audacity worse in her eyes was that he had still to secure the funding from Sir Henry, for which her participation was essential. To that end, Starkey told her that he hoped 'we shall succeed in putting up a really good show with your close co-operation'.[29] The emphasis is Dorothea's. If she wanted any further involvement in the project there was nothing she could do except cooperate. She duly met Sir Henry and convinced him of the importance of the site, showing him specimens she had dug from the damp, sticky clay of the pit.[30]

Dorothea may have been forced to waive her claim to be in control of the dig, but it was her expertise that had revealed its significance and she would be the one to announce it. Accordingly in August she wrote to the editor of that most authoritative of scientific journals, *Nature*: 'I am enclosing a letter to announce a discovery of considerable importance in Palestine, and I shall be very glad if you would kindly have it included in your next number of *Nature* so that the first record may be in a scientific journal – and not in a newspaper.'[31] *Nature* complied. 'This is,' Dorothea wrote, 'the first time that such remains have been discovered [in Palestine], nor have they yet occurred in Syria, and it will be of interest,' she added, referring to Mount Carmel and Zuttiyeh, 'to see if this find has any connexion with the faunas of so-called African types which have been found in several Palestinian caves associated with a Mousterian culture and human remains.'[32]

Once Sir Henry had agreed to fund the project, Starkey immediately set about organizing it. He himself would remain in overall control of WARENE, overseeing both Lachish and Bethlehem. As geologist, he secured the services of Elinor Gardner, a research fellow from Lady Margaret Hall, Oxford, who had worked extensively in the 1920s with

Gertrude Caton Thompson in the Fayum in Egypt and in Southern Rhodesia. By the end of September, matters had advanced sufficiently for Starkey to suggest a meeting with Dorothea, who proceeded to give him 'an excellent lunch'. He then wrote her a formal letter, covering the points that he thought they had agreed – and which Dorothea promptly returned for alteration.[33] It was only after a further meeting at the Museum with Wilfred Edwards, the Deputy Keeper of Geology present, that satisfactory arrangements were resolved. These included giving Dorothea adequate travel and living expenses and, vitally, complete control over the palaeontological work. All her discoveries, however, Starkey wrote, 'should be treated as strictly confidential, and that such information should not be communicated to others without reference to the director of the Expedition, the Department of Antiquities or Miss Gardner'.[34] Everyone connected with WARENE was bound by the same strictures, but how she must have wished that similar discretion had been applied to her finds in the elephant pit in May. But this time she could agree to the arrangements and wrote an official reply to Starkey, although, as she makes a point of recording in her covering letter, the all-important question 'of the British Museum benefiting from the resulting collections will arise after you have seen Sir Henry'.[35]

Wellcome, not unreasonably, expected to keep and exhibit in his own Museum the finds for which his philanthropy had paid. Within a few days, even that was resolved, with Sir Henry writing to Dorothea, 'I am hoping the results will be such that I shall be able to offer the Trustees of the Natural History Museum some specimens as a momento [*sic*] of your assistance.'[36] 'Momentoes' were not quite what Dorothea had in mind, but the principle had been established. She was now, with Elinor Gardner, the 'geological section' of WARENE.

Wellcome required that full reports of the progress of excavations should be sent to him regularly and Starkey had now to write separate reports for the work at Tell ed Duweir and on administrative matters at Bethlehem. The field reports on the elephant pit were to be written by Elinor Gardner. In his first Bethlehem report, Starkey noted to Sir Henry Wellcome that, in addition to their work on the pit, Dorothea was also to be asked to identify animal bones found at Tell ed Duweir and Elinor Gardner to examine the geology of the Tell. Interest in geology in Palestine at the time was also for higher stakes than prehistory; as Starkey noted in his report, the Iraq Petroleum Geological Survey of the south of Palestine had been under way for two years. The company had built at great expense a pipeline from their oil fields at Karkuk to Haifa and Beyrouth, and, Starkey reported to Wellcome and everyone else who received copies of his 'strictly confidential' report, 'Their survey in S. Palestine has, I believe, located a very large structure, larger than any at Karkuk, which is of sufficient promise to warrant a bore.' He ended this first geological report to Sir Henry a trifle gnomically: 'Thus it is that the geology of Palestine is the talk of the moment and may your own cause be observed on the crest of the advancing waves.'[37]

With the prospect of digging at Bethlehem now a reality, Dorothea gave Starkey every help she could. The licence to dig, which he held from the Department of Antiquities, gave Starkey free access to examine the land, but the support of the owners of the fruit garden (the Hasbun brothers and their aged mother, whose house was right in its centre) was still essential to the smooth running of the operation. Dorothea, who had met the family, advised Starkey that not only were they very amenable folk, but would be willing to help in any way possible. They would of course be compensated for damage to or loss of any of the fruit trees, which included three large almonds, one fig, apples, plums, pistachios, vines, a young olive, and several apricot trees.[38] There was an unoccupied house near the garden, and the owners of this, wrote Starkey in his report to Sir Henry Wellcome, 'are at present living in South America, where it is the custom of many of the inhabitants of Bethlehem to make their pile'.[39]

In December 1934, Starkey wrote to Dorothea of his progress, and she replied with her customary charm and enthusiasm, while managing to hint that the enterprise might not be as straightforward as Starkey supposed. It was a note that was to characterize much of her correspondence with him.

> I was very interested to get some news regarding the activities of the 'Geological Section' – and glad to know that you had retrieved some further specimens from the well dump – especially the remains of the smaller species – I have evidence of at least two animals besides the elephant which indicates decided possibilities for the 'gamble'! – the house just beyond the pit should be delightful if it can remain

> uninvolved in the excavation – I hope you feel that the prospects are encouraging – I am so glad that you also found the owners of the site to be friendly ... I am looking forward to the Spring adventure.'[40]

Starkey had wanted control of the dig and Dorothea, with a few well-chosen phrases – 'gamble', 'adventure', 'I hope you feel the prospects are encouraging', as well as raising the possibility of their living quarters sliding into the pit – ensured that he knew his responsibilities too.

Whatever the potential drawbacks, Starkey proceeded to rent the house just beyond the pit. It had been newly whitewashed and included in the rental were six chairs and two iron bedsteads, although Starkey did arrange for new mattresses and bedding. Elinor Gardner arrived there on 5 March 1935, to find the place extremely draughty and so cold paraffin stoves were brought in to make the place habitable.[41] Using local men, 'ten picks and hoes and forty basket carriers',[42] she began work immediately on digging down through the garden, anxious to reach the bone bed before Dorothea joined her.

On 28 March Dorothea arrived to find a much-enlarged pit and the first bone deposits exposed at a depth of about eighteen to twenty-three feet. She kept two notebooks for this excavation. In the first she has noted down the exact position of the finds, the time they were found and snippets of essential information, such as 'muktaf = basket'; this notebook is small and practical with a blue plastic cover with a loop for a pencil, much the sort of thing for jotting down shopping lists, but it serves as a brief diary. The other is, literally, a school exercise book with space on the cover for name, address, subject and class. This she used as her 'Register of Specimens' for two of the three seasons she excavated in Bethlehem.[43]

Almost immediately she found an elephant tooth and tusk, and what she thought might be a rhinoceros tooth, but the quality of most of the bones and teeth was mushy or very fragmented. Some of the remains lay in bands, others were isolated specimens, and all were difficult to extract from the coarse gravel and clay. Every fragment found was given a number and Dorothea notes precisely where it lay and the type of matrix in which it was found. Elinor Gardiner drew large-scale plans of the pit, tracing the contours from a map she had had photographically enlarged, telling the photographer, 'contours most important'. These charts showed the distribution of the bones and the relationship between the different types of matrix.[44] Mixed into the gravels were flints, varying in size from great blocks to small slivers. Given the remarkable finds that Dorothy Garrod had made at Mount Carmel, the possibility that there might be evidence of early man in Bethlehem was ever present. Certainly Dorothea notes all flints that might be man-made as carefully as the bones.

By 4 April they needed to employ more men to cut new steps into the pit, which was now so deep that a derrick was rigged to haul up the baskets of matrix. It worked well most, if not all, of the time. More of a problem was securing the right quality steel pick, as the flint deposits were so unexpectedly hard. The picks had to be sharpened at the end of every day, 'and tools', Starkey noted, 'which were new six weeks ago, are mere stumps now and worthless'.[45] Four new picks and ten baskets had to be sent over from Tell ed Duweir.

For most of the time Dorothea worked at a depth of thirty feet in the pit, only occasionally travelling out into the wild countryside to note birds and to try to make sense of this extraordinary site. How the bone deposit came to be some two and a half thousand feet above the level of the Mediterranean on one of the highest points in Palestine was Elinor Gardner's task to determine. The previous year, Dr Leo Picard of the Geological Department of the Hebrew University had visited the site and told Starkey that he knew of nothing like it in Palestine.[46] Nor had Elinor 'in Palestine, Egypt or elsewhere ... seen anything exactly like these deposits and Mr Blake, the Government Geological Advisor, and Mr Daniels of the Iraq Petroleum Company are equally puzzled'. All Elinor could say at this stage was that whatever the conditions had once been to allow such a deposit to form, 'they were certainly very different from those now obtaining.'[47]

They continued working until 24 April, when Easter and the need to close down the site for the end of the season stopped further excavation. In the last few days, a group of elephant bones and teeth had been exposed. 'They are in a much better state of preservation than the first mushy bits of bone we have found,' Dorothea wrote with relief, 'perhaps preserved from surface drainage by more flinty and harder matrix.'[48]

These and the other more important finds had still to be plastered, a technique to protect and strengthen specimens before removal from the matrix. The bone had first to be hardened, and some long or large bones – and even teeth – had also to be supported with a stick or iron rod. Shellac could be used, but Dorothea preferred celluloid, or cellulose nitrate, which until the 1940s was widely used to protect and consolidate fragile fossil bones, generally being painted onto the specimen with a brush. A literally explosive mixture of acetone and amyl acetate, it was highly flammable, and also expensive.

Dorothea, with her usual resourcefulness, often made her own mixture, as she told a correspondent. 'I found celluloid dope most efficient; equal parts of acetone and amyl acetate. Cut celluloid (you can get old films from photographers and wash off film) into small pieces and add to above mixture – Keep a treacly thickness for stock and thin with mixture. When specimens are first doped use a very thin celluloid so that it will penetrate. Always have a jar of dope on the excavation. When brought to the surface specimens must never be left exposed in the sun.'[49] Once the 'dope' had hardened, she soaked tissue paper in water and laid that over the specimen, which was then wrapped in hessian bandages that had been soaked in a bucket of plaster. Only then, once the plaster had hardened, would she extricate the strengthened specimen from the matrix. Removing the plaster once the bones had reached their destination was a lengthy process, much to Starkey's irritation. The precise details of some 'good teeth of elephants and their kindred species', as Starkey vaguely described the major finds, had to await preparation and examination in London. 'This will be a slow business,' he wrote with some impatience, 'and it will be some months before they are in a condition for exhibition.'[50]

By the end of this first season they had dug a main pit some thirty feet deep and a western extension about seventeen feet deep, with three sets of stairs. Ditches were dug around the excavation to carry off the winter rains. With everything as secure as they could make it, they left, Elinor to Oxford and Dorothea to London. 'The Expedition has been fortunate in securing the services of these two experts and I hope,' wrote Starkey in the manner that so infuriated Dorothea, 'that we shall be able to induce them to continue their researches next season.'[51]

The bones arrived in London towards the end of June 1935. Most were so fragile that cleaning and preparing them was a difficult and gradual process. From her initial examination, Dorothea identified at least six species of mammal, including elephant, 'a very primitive one', hippopotamus (rather than rhino), and a large ruminant, but the quality of the material was poor. 'I feel sure,' she wrote to Starkey in July, 'that the fauna would be of great value and interest if only we could get good specimens. Miss Gardner and I talk of "when we reach the bone bed" – if that does happen I think that the work and expense would be well rewarded.'[52]

With this latest barb directed nicely at Starkey, Dorothea turned her attention to her other work, which included her involvement with two famous projects. In New York the former President of the American Museum of Natural History, the palaeontologist Henry Fairfield Osborn, was completing, after twenty-seven years, a lavishly and beautifully produced monograph on mastodons and elephants of the world. Dorothea had corresponded with him on Pleistocene mammals for more than twenty years. In 1931 she had paid a short and woefully undocumented visit to the United States. Apart from a couple of postcards she sent, there is absolutely no information about this trip. She went to 'Boston, New York, Washington etc – a delightful experience',[53] and she must have visited Osborn at the American Museum of Natural History, but what else she did is unrecorded. Osborn certainly came to London that autumn, visiting the Museum to examine the elephant collections. He died at the end of 1935, before completing the second volume of his monograph. His colleagues in New York at the AMNH undertook the task, and acknowledged their indebtedness to Dorothea for her work on the chapter on extinct dwarf elephants of the Mediterranean islands.[54]

The remarkable find by Alvan Theophilus Marston, a London dentist and amateur archaeologist, of part of a human skull of great antiquity in a gravel pit at Swanscombe in Kent interested her greatly. Associated with the skull were Acheulian* flint tools and faunal remains, and these he brought to the Natural History Museum where he met Dorothea. She examined the faunal remains and suggested that he show the skull fragment to the co-discoverer of the Piltdown skull, Sir Arthur Smith Woodward. Marston's house, as Smith Woodward described it to Dorothea, was chaotic, with 'flints and bones spread everywhere, as he is a great enthusiast and he is not likely to allow the gem of his collection to depart yet for any museum'.[55] This showed some prescience, for although Marston did give the skull pieces to the Natural History Museum, the terms on

* The Acheulian culture preceded Mousterian and lasted from about 1,500,000 to 150,000 years ago.

which he did so were the subject of years of contradictory and often angry correspondence.

Smith Woodward himself, meanwhile, continued his own sad search for further evidence of Piltdown Man. Dorothea was a regular visitor to his house in Sussex, and brought down with her on one occasion Gertrude Caton Thompson. 'The gravels at Piltdown,' Gertrude recalled, 'were pitted with hopeful and unsuccessful trial holes and in one of these Sir Arthur Smith Woodward was digging alone. It was a hot day, he looked on the elderly boil in shirt and braces and was glad to resume his coat and return with us to tea with his wife.'[56] His digging would continue until age and loss of sight brought such physical exertions to a halt.

In the early autumn of 1935, among the many visitors to the department was the Swedish palaeontologist, Professor Carl Wiman, an expert on fossil birds. Dorothea had come to know him while writing a paper on extinct birds from China and Mongolia with her colleague at the Museum, the ornithologist Percy Lowe.[57] Wiman had come to London for a fortnight to study at the Museum and, as Curator of fossil birds, Dorothea was responsible for guiding and advising him, but there are few curators who can subsequently have been called 'My dear Ange tutelaire', as Wiman henceforth referred to Dorothea. On his return to Sweden he wrote to her at least twice a month about his work, in letters full of charm, addressing one letter to 'The Ange Tutelaire of the frivolous professor emeritus',[58] while in another he proclaimed, 'I clasp you to my heart (chocking!)'[59] Dorothea made time to advise him when she could and offered to help him with his English although she was deeply immersed in determining the Wadi el-Mughara material and was also preparing for her next Bethlehem visit. She had received her contract in February 1936 from Starkey, engaging her for a further season at Bethlehem.

Elinor Gardner arrived back at the pit on 12 March 1936 to find beautiful weather and 'our garden was looking particularly lovely with apricots in perfect flower.'[60] The recent heavy rains, however, had wrought havoc with the pit. Walls had collapsed and the western staircase, which had been cut in clay, was damaged. Elinor immediately began remedial work, which quickly turned into further expansion of the pit.

Dorothea arrived a week later. She had sailed to Port Said in Egypt, then, as before, catching the excellent overnight train service to Jerusalem. The Palestine Railways advertisement in the *Handbook of Palestine and*

Trans-Jordan reads now as a poignant historical commentary. 'Comfortable and interesting travel in Palestine. The Trains of the Palestine Railways, equipped with Corridors, Restaurant and Sleeping cars travel over the historic Sinai Desert, and connect Egypt with Palestine, providing views and means of reaching many of the most interesting scenes and places in the Holy Land.'[61] As it grew light she watched the now familiar landscape appear; Israel's history, recent and ancient, flowed past the windows. There was the fourteenth-century Muslim Tower of the Forty Martyrs, nearly a hundred feet high, and from the top of which Napoleon was said to have directed the siege of Jaffa. That was succeeded by the British First World War cemetery at Ramleh, a blur of palest grey headstones marking the graves of more than three thousand men. Shortly after was the RAF airfield and in the middle distance a long hill, the site of the ancient city of Gezer. First occupied by cave dwellers 5,000 years ago, it then successively became the Canaanite city of the Bible, then Israelite, Roman, Christian and Arab cities in that order, 'twelve in all'.[62] Nine miles farther she passed the Vale of Sorek, 'scene of fierce fighting in November 1917'. In the distance was the Valley of Elah, 'where the armies of Saul and of the Philistines lay opposite to each other and David slew Goliath'. At Bittir they passed further sites of war, the Turkish trenches of 1917 and the setting of the 'terrible slaughter' of Jews by the Romans in the reign of Hadrian in AD 134.[63]

From there the train began its climb towards Jerusalem. Within a month of Dorothea's arrival, a number of factors combined to fracture the brittle stability of the country. Arabs felt a growing resentment at increased Jewish immigration as the threat from Nazi Germany intensified, and this, together with the sale of land to Jews and the desire for a national Palestinian government rather than British rule, led in April to an attack by Arabs on a Jewish bus and the beginning of the Palestinian revolt. The Balfour Declaration of 1917, which supported the establishment of a Jewish homeland in Palestine without prejudicing the rights of the existing non-Jewish population, seemed increasingly unworkable, and the aspirations of Jew and Arab disastrously irreconcilable. Absorbed in their work, few of the archaeologists seemed aware that they might be in danger.

So far had Elinor enlarged the pit that it now extended ten feet into the garden of the house they were staying in, much as Dorothea had forecast. 'This brings one corner within a few metres of our house,' wrote

Elinor, 'and we cannot therefore go further in this direction at present without sacrificing our very comfortable quarters, which thanks to Mr Starkey and his carpenter Hasan have many improvements this year.'[64] Broken windows had been mended and Starkey had bought floor mats, basket chairs, new bedding, and two new iron beds as the old ones were finally declared too dilapidated. Most of the workers from the previous year returned, including the Hasbun brothers who owned the garden, and the cook-housekeeper, a Bethlehem woman called Katherina.[65] This season fifteen 'pick and hoe' men and between fifty and sixty basket carriers were employed, the new men and boys being quickly taught 'our methods' by the old. 'Labour is much more plentiful this year,' Elinor observed, 'and each morning a row of applicants is waiting for work.'[66]

It was a glorious spring, with the fruit trees in blossom, and what was left of the garden was bright with blue iris and scarlet ranunculus. Dorothea collected bulbs and noted great tits and chaffinches in the orchard. There was a constant stream of visitors: James Starkey, Charles Inge and other archaeologists from Tell ed Duweir; officials from the administration and their families, while Mortimer Wheeler came for Saturday lunch.[67] Elinor Gardner, meanwhile, went into Jerusalem to get her licence for the small lorry lent to them by Starkey, a means of transport that was rapidly taking over from donkeys, camels, and horses.[68] It was a Ford pick-up, and on the Monday morning Dorothea experienced it with great enthusiasm: 'Miss G and I went into Jerusalem,' she wrote, 'in our Ford Pickup – 30 hp – 8 cylinder, a delightful vehicle.'[69] The road to Jerusalem was good, but the maximum speed permitted was 30 mph and just 15 mph in towns.

Work started in earnest on 24 March and every day brought fragments of elephant bone or teeth. Down in the pit the sun was intense. Dorothea spent hours there, delicately scraping away at the bones or painting them with her explosive and strong-smelling celluloid mixture. She did not wear a hat. After days of this she notes, 'Spent day mostly in bed ? [*sic*] sun touch,' and then in the only reference that has survived in all her papers to her faith in the power of prayer, she scrawls, 'turned to c s [which must be Christian Science] and by evening got up,' as she notes rather endearingly, 'for supper of cup of Marmite'.[70]

Still mystified by the position of the Bethlehem site, Dorothea and Elinor made 'shorter or longer excursions' to get a better understanding of the geology of the region. It gave them an excuse for a day out and it enabled Dorothea to examine faunal remains from other sites. They went to Megiddo, known more widely from its biblical name, Armageddon,* where they spent a day exploring the site and another with Dorothea examining a collection of animal remains. She did the same for Professor John Garstang, who since 1930 had been excavating ancient Jericho, long buried under the great mound of Tel al-Sultan.

On Sunday 19 April, they set off at dawn to visit the famous hot springs at Zarqa Ma'in, a gorge opening onto the northeast end of the Dead Sea. 'This expedition,' Dorothea noted with undimmed enthusiasm for their vehicle, 'was only possible owing to our having the light Ford Truck and it could hardly have been carried out with a less efficient car.' Taking two men with them, Elinor drove them through Jerusalem to Jericho and then to the Allenby Bridge on the Jordan river, 1200 feet below sea level. From there one of the men took over the driving and 'I sat in the back of the open truck,' wrote Dorothea happily, as they drove into Jordan. At the town of Madaba, they turned onto a one-track road, which descended to the hot springs at Zarqa Ma'in. It was so precipitous and narrow that, Dorothea noted, 'cars must start from Madaba before mid day – and may not start the return journey before 2pm.'[71] A police permit was also required. The road was moderately steep at first, but then became a direct descent of the mountainside by means of over thirty hairpin bends, so steep and close together that most cars had difficulty negotiating the turns. 'Our Ford,' noted Dorothea, 'splendidly driven by Gamil Jalouk, went down without any difficulty and made the upward journey in like manner.'[72]

The scenery was wild and grand, the gorge of red and mauve sandstone, while electric-blue lizards darted among the rocks. 'The whole scene,' Dorothea wrote, 'is dominated by a towering mass of chocolate coloured columnar basalt which rises straight above the entrance to the narrow defile below the Springs.' They crossed one of the many hot streams, 'too hot to plunge one's hand into', and lunched on a rocky platform. After 'a preliminary geological inspection', they returned to Madaba (it was after two o'clock) where they stopped to examine the famous sixth-

* The translation of Armageddon, prosaically, is simply the hill of Megiddo. The site had first been excavated by German archaeologists in the early part of the century; then in 1925 teams from Chicago University and the British School at Jerusalem under Philip Guy had begun work, revealing a city of continuous habitation, subsequently believed to date from 7000 BC to 450 BC.

century mosaic map of Palestine, which had been discovered beneath the floor of a Greek Orthodox church. From there they left the road and drove across country, continuing upwards until they had a splendid evening view of the north end of the Dead Sea. The Ford, of course, 'proved equal to all that was required of it'.[73] They reached the main Jericho–Amman road, just at 'lighting up and raced to Jericho where we stopped for coffee'. Dorothea's verdict on the day was that it had been 'a fine expedition'.[74]

Over the next two days the temperature steadily rose and the finds were poor and scanty. Elinor was confined to bed with a cold and the Ford pick-up had to be taken to a garage with a puncture. On 22 April, Dorothea noted, one of the workmen, Mahmoud, 'had found, and stupidly smashed' a good elephant molar, which she thought might be part of two. He redeemed himself a day or two later, however, when he struck a very large bone and then found more extending over six feet in a patch of clay at the base of the north wall of the pit. Dorothea thought it might include an elephant skull, but it was all 'frightfully wet owing to the clay, so may be impossible to extract'.

There was just a week left before they were due to leave Bethlehem and close down the pit for the season. Mahmoud was left to work on the elephant remains and he disclosed yet another slab of bone, then some pieces of rotted bone, 'disintegrated into a thousand pieces [but] still in position'. With enormous difficulty they managed to extract the large slab of bone, so heavy that it took four men to carry it on a sack out of the pit. It was the last full day of excavating. It was also 'pay day and many of our nice people were paid off', Dorothea wrote, 'as we only want a small number for early part of next week'. Their main task now was to prepare the pit for the winter rains. The newly discovered bones at the base of the north wall were carefully covered in sacking. By Wednesday 29 April they had completed the rainwater channels round the pit and surrounded the whole area with several strands of barbed wire.[75] They were going home not a moment too soon. In the previous week there had been strikes by Arabs, which, Dorothea noted, had closed all the shops in Bethlehem. Riots had caused the postponement of at least one archaeological dig in the north of Palestine and might soon have affected them.

Dorothea returned to the Museum from Bethlehem to find more material from Malta. This was becoming a great test of her diplomacy and a source of professional anguish. Collections would arrive accompanied by an enthusiastic letter from Dr Baldacchino on the careful methods he had used in his excavations, although from the evidence of the collections he sent that was not always apparent. She also received a most alarming letter from the palaeontologist Charles Taylor Trechmann, who was visiting Malta. 'Baldacchino,' he wrote with a frustration that overwhelmed his syntax, 'is well, too well for the deposits in the Ghar Dalam cave, which he is causing to be dug out with picks and then smash up is pitiful to behold.'[76] Asking Trechmann to try to 'bring your influence to bear on this matter',[77] Dorothea went to great pains to instruct Baldacchino in basic collecting techniques, suggesting to him, for example, that he should draw diagrammatic sections, marking the thickness of the different layers represented and showing the exact position of specimens of particular interest. In her examination of the Mount Carmel material, evidence was emerging of a sudden major change in the Pleistocene fauna. She called it a 'faunal break' and was anxious to discover whether anything similar was apparent from the Malta deposits, although that would only be possible, she told Baldacchino, if he used the utmost care in his excavations. 'With your help in this way,' she wrote, with perhaps no real expectation of improvement, 'I am beginning to hope that we may, in the not far distant future, get a real knowledge of the different faunas represented.'[78]

In the midst of all this came very serious news. On 25 July 1936, Sir Henry Wellcome died and with him, very nearly, the continuing excavations at Lachish and Bethlehem. WARENE had been entirely a personal concern of Sir Henry's that had nothing to do with the Wellcome Foundation. It came as something of a shock to his Trustees to discover that the 1935–6 season had cost about £7,000. The industrialist Sir Charles Marston, who listed biblical archaeology among his interests and was the sponsor of Garstang's work at Jericho, had contributed small sums to the expedition. What the Trustees dearly wanted was for him to undertake, if not all, then at least a good part of the costs. On behalf of the Trustees, G. Hudson Lyall wrote to Marston, expressing concern at the cost of the enterprise and reminding him that all agreements with the archaeologists expired at the end of August.[79] Marston was, according to Hudson Lyall, 'more than distressed' to learn that the excavations might be discontinued. Sir Henry had contributed six-sevenths of the costs and, Marston told Hudson Lyall, compared to Sir Henry he was a poor man. (Marston was the chairman of the Villiers Engineering company in Wolverhampton and his father had founded the famous Sunbeam car and motorcycle company.) However, if the expedition were to be called the Wellcome–Marston expedition, then, Marston told Hudson Lyall, he would be prepared to find one half of the costs. He added that Starkey was 'far and away' the best practical archaeologist living and 'was also extremely cheap at the figure we were paying him'.[80] Starkey was paid £550 a year (about £19,000 at today's values) and had been since 1933.

After deliberating for another month, the Trustees decided in September 1936 to go ahead on the basis of costs being limited to a total of £5,000, which meant that their contribution would be a maximum of £2,500. In order to keep within this figure, Starkey refused a pay rise, although his team all received modest increases. Just three weeks later, the Trustees nearly reversed their decision when news reached them of an Arab raid on Lachish. Starkey made light of it, telling Hudson Lyall it was no more than the theft of a torch and a broken scullery sink, although the gang had also tried to set the place on fire with paraffin.[81]

The future of Dorothea's Bethlehem excavations, however, was still not resolved. Sir Charles Marston was delighted to be associated with Lachish and all its attendant publicity – he was particularly happy to be the one to inform the press and the BBC of developments there, but the geological work was quite another matter and Starkey did not want him involved in that. Marston was, according to Starkey, 'a rather rabid fundamentalist' whose belief in a literal interpretation of the Bible was at odds with 'anything appertaining to anthropological research'.[82] Still no decision was made by the Wellcome Trustees. In February 1937, Starkey informed Hudson Lyall that Elinor Gardner was already in Cairo and was expecting to resume work in Bethlehem about the middle of March. He added that 'Miss DMA Bate will be leaving London at about the same time.'[83] Hudson Lyall, who supported the Bethlehem work, appealed to his fellow Trustees 'for a limit of £500 in respect of it'.[84] With less than a month before work was due to start, they finally agreed. Dorothea's and Elinor's travel and living expenses were assured, as were the wages of their workmen. They would also receive help for equipment from the industrialist Sir Robert Mond, who had been more than generous in earlier seasons. However, such luxuries as Dorothea's beloved Ford pick-up were a thing of the past.

CHAPTER 14

The Stone Age of Mount Carmel

Dorothea arrived in Haifa early on 18 March 1937. Delays in customs made her miss her train to Jerusalem, but the police advised her not to travel by road, but to wait for the next train at midday.[1] In spite of Starkey's protestations to the contrary, the security situation was worsening. There was an extraordinary atmosphere in Palestine for the visiting archaeologists: the intellectual excitement of great finds against a background of violence and unrest, and a constant police and military presence. A 7 p.m. curfew was now strictly in force and Dorothea only just reached Bethlehem in time.

At first light she hurried out to the pit and was horrified at the devastation from the heavy winter rains, particularly the collapse of the north side where the elephant remains were, although even this disaster could not quite subdue her sense of irony. 'Except for the disappearance of my cliff face with specimens the pit is now really a much better shape, no overhanging places, and no clay slides!' While the men cleared the debris, Elinor, or EWG as Dorothea now refers to her, surveyed the pit for a new plan, while 'I,' wrote Dorothea, 'assisted mildly. While out there a Bubbul flew from one of our trees to another, but did not stay long. Slacked in afternoon . . . and I to bed early as I was not feeling quite top notch.'[2] After the constant pressure of the last few years she was exhausted; despite her age (she was now fifty-eight), her passion for her work still drove her almost beyond endurance. That night she slept badly and, although she was still unwell in the morning, nonetheless went down into the pit. Nothing was found except scrappy fragments of bone. Even Gertrude Caton Thompson's arrival after tea on the 24th, having flown from Egypt, only went some way towards lightening her spirits.

It was another week before the remains began to emerge from the rubble; first scraps of bone, then one or two larger pieces, which Dorothea plastered. Interspersed with the bones were flints, some of which seemed to show distinct flake marks and appeared highly polished. Gertrude Caton Thompson (who was amazed at the 'vast hole' they had dug) examined them and thought, as Dorothea noted with engaging optimism, 'some *may* be artefacts – What Ho!'[3] However, much work needed to be done to confirm this. Flints that seemed to promise much in the previous seasons, 'suspicious Characters', Elinor called them, had turned out not to be man-made.

The large pieces of bone, which Dorothea at first thought might be part of the elephant, seemed to belong to something quite different as more was uncovered. Dorothea found herself in the hot sun on her hands and knees, her shiny straw hat protecting her head and neck, her sensible suit jacket loosely belted and her blouse neatly buttoned to the neck, 'clearing slabs of bone'. From the shape and sculpturing of the bones, it became clear to her that they were part of the carapace of a giant tortoise, a 'beast – or beasts' as Elinor called them, which 'must have been nearly two metres long by half as broad'.[4] Some of the carapace that was not too wet or embedded could be removed with relative ease and was placed in a drying shed to harden, but much was firmly encased in the stony matrix and needed skilled, gentle handling. Every piece of tortoise they found was numbered and its depth plotted on a chart to facilitate reconstruction. On 10 April there was almost disaster. After lunch it began to rain. The wind increased to gale force and by evening there was a violent thunderstorm. At 1.30 a.m. the storm woke them with its ferocity. Dorothea rushed out through the torrential rain to the drying shed. Buffeted by the wind, squelching in mud and drenched through, she desperately tried to rescue her best specimens, moving those she could lift and trying to protect the rest with whatever was to hand. At daylight she inspected the damage. There was, Dorothea wrote sadly, 'a great pool of water at bottom of Pit and all around the ground wet and greasy – impossible to go down – hope my beautiful bones have survived!'[5]

It would take the pit several days to dry out, and as digging was impossible she seized the opportunity to visit other important sites. Organizing a car and driver, Dorothea, Elinor and Gertrude set off to Tabgah on the Sea of Galilee, to the Zuttiyeh cave, which Dorothea had visited in 1925 and where Francis Turville-Petre had found the Galilee skull. They stayed the night at a hospice to the north of Galilee, with a spectacular view of Lake Huleh. The next morning they visited a site that Dorothea had been associated with for a couple of years and which she considered of very great importance. In 1935, Professor Leo Picard of the Hebrew University had sent her bones found near the bridge at Jisr Banat Ya'qub (known now as Gesher Benot Yaacov) on the Jordan river, about four kilometres from Lake Huleh,* and these Dorothea had identified as elephant bones and molars from the Pleistocene of Palestine. On a beautiful morning, the women crossed the Jordan at Jisr Banat Ya'qub and walked upriver on the Syrian side towards Lake Huleh. The Jordan was in full spate as the snows melted on Mount Hermon, the river banks a mass of pink, white and green with oleanders, papyrus and reeds. 'It is,' noted Gertrude Caton Thompson, 'the habitat of many rare birds and flowers which delighted my two companions who knew more about both than I did.'[6] At the site they were met by Picard's colleague, Dr Moshe Stekelis (a frequent visitor to the excavations at Bethlehem), and they examined the artefacts and faunal remains he had recently unearthed.

* The site, which contained flint tools as well as faunal remains, had been revealed in 1933 when drainage work had uncovered part of the river bed. Dorothy Garrod and Elinor Gardner had been given a permit to excavate there by the Department of Antiquities and found fossilized bones as well as flint tools just lying in debris on the river bank. Because of her Mount Carmel work, Garrod decided not to proceed and suggested that Picard should take it on. Fieldwork under Picard and the archaeologist Dr Moshe Stekelis, planned for 1936, had had to be abandoned because of the Arab revolt. Stekelis also feared that, unless something was done, further proposed drainage work in the region would damage the site. 'It would be very, very kind of you,' he wrote to Dorothea in May 1936, 'if you could write to the Dept. of Antiquities . . . There is no doubt of course that your letter will have more influence than anything else.' (Letter from Dr Moshe Stekelis to Dorothea Bate, 23 May 1936, folder 'Yst Benat Jacoob, finished', Box Misc. Palestine sites, P MSS BAT.) Within a few weeks, Dorothea had the matter in hand. 'I have recently had the opportunity of stating my interest and favourable opinion about the site in a helpful quarter,' she told Stekelis, and added, 'The Dept. of Antiquities already know that I consider the site one of great importance in the investigation of the Pleistocene of Palestine.' (Letter from Dorothea Bate to Dr Moshe Stekelis, 22 July 1936, folder 'Yst Benat Jacoob, finished', Box Misc. Palestine sites, P MSS BAT.) In 1937, Stekelis resumed his work.

They would, Dorothea hoped, prove to be a link between the cave faunas of Palestine and what she was now sure were the much older deposits at Bethlehem.

Excavations in the pit resumed as soon as the weather improved, and the men were soon finding more remains, particularly from the tortoise bed. Other animal remains emerged: a tooth Dorothea thought might belong to *Bos* (ox), a foot bone of an antelope, part of a tooth of a small carnivore and, in contrast to the great beast, the humerus of a 'tiny tortoise'.[7] This brought the total number of species found to ten. Some of these came from the main pit but Elinor had been supervising the digging of a new pit in the hope of finding more species to help them determine a reliable date for the deposit. With the western extension of the main pit, there were now three vast holes in the fruit garden, encroaching even more dangerously on their house.

The weather continued to plague them with fierce winds, heavy showers and thunderstorms, and they had barely a week left of the season. The largest piece of tortoise carapace, which measured some sixty centimetres square, had still to be removed. It had taken Dorothea many days to free it, working with a tool as fine as a needle to extract the matrix between the tightly packed bones. Once that was done, the removal of this large piece was, according to Elinor, 'comparatively simple'. The surface and edges were plastered and, when that had hardened, the whole block was removed, only to reveal, Dorothea noted, 'a v large limb bone lying underneath'.[8] It was so wide and solid that she thought it must be from an elephant rather than the tortoise. It was too damaged to make a good specimen, but she dared not move it with so few days left in case more bones were exposed. They covered the area in paper, baskets, and sacks, which they topped with a thick layer of powdered white chalk from the dump, a technique they had found to be good not only at preserving the bones, as it set hard, but also because it marked where the bones were. They covered everything with wet gravelly clay, 'and under all this', wrote Elinor, 'the bones should remain unharmed by rains and rockfalls'.[9] On the very last day of digging, the remains of a small carnivore were uncovered, but Dorothea decided to leave that too until the next season. The pits were made as safe as possible, with any overhangs cut back and drainage ditches dug around and inside the pits. On 24 April they paid off the workers and continued packing up the specimens, among them some eighty-two flints, which might possibly be artefacts. Four days later they left Bethlehem with the hope, Elinor wrote, that they would 'find the pit in good condition in 1938'.[10]

Before she left for Bethlehem, Dorothea had sent to the publishers the corrected proofs of her faunal report for *The Stone Age of Mount Carmel* and the final proofs were waiting for her on her return to London. Garrod was going through the same process with her Mount Carmel archaeological report, as she wrote to Dorothea. 'I am beginning to feel I shall never want to open the book or refer to it again in all my life! Do you feel like that by the time you have read your stuff about 20 times in manuscript, typescript and proof?'[11] Dorothea undoubtedly did, but she was about to be confronted, for the first time, by a damaging and deeply hurtful attack on her professional competence. It came from, of all people, Professor Leo Picard of the Hebrew University in Jerusalem, the geologist and palaeontologist with whom she was collaborating on Jisr Benat Ya'qub. In the journal of the *Proceedings of the Prehistoric Society* for January to June 1937, he published an article entitled 'Inference on the problem of the Pleistocene climate of Palestine and Syria drawn from flora, fauna and stratigraphy'. In this he drew heavily on Dorothea's published works, criticized her conclusions on climate – he based his own conclusions on climate on stratigraphy, not fauna – and damagingly misquoted her findings on the dating of some extinct fauna.

She was appalled at his attack and on 29 June wrote a refutation with a furious covering letter, addressed to the editor of the journal, the archaeologist Dr J. G. D. (Grahame) Clark.* 'Soon after copies of this number arrived here several of my colleagues came to me about this article – which certainly appears to contain a good deal of acrimonious criticism of my work on the Palestine Pleistocene vertebrate fauna. This naturally can be ignored, but the paper also contains a number of misstatements of facts which have been published by me.'[12] Picard had compiled a chart showing a list of fauna of Palestine, the locality in which they had been found, and the period of time that they had inhabited those localities. Of the 269 references, 93 were from Dorothea's published writings and, as she told Clark with unfettered anger at this misuse of

* Dr Clark was to succeed Dorothy Garrod as Disney Professor of Archaeology at Cambridge in 1952. He became Master of Peterhouse and was knighted in 1992.

her work, had 'only been described by myself, so these references stand as quoting from my writings. But the time duration of a number of species is given quite differently to what I have published and would obscure some of the interesting results which are appearing through the study of the fauna.' What Picard had done on the chart was extend the stratigraphic range of certain species, such as the hippopotamus, into ages long after they had in fact been extinct. This would therefore, among other difficulties, bring into question both her conclusions on climate change and the faunal break.

She continued:

> I thought that my account of the fauna of the Wady Mughara Caves, which is about to appear in book form with Dr Garrod's account of the Industries etc, might be a sufficient refutation of Professor Picard's 'mistakes', but it has been pointed out to me that the 'Proceedings' have a wide circulation and that the articles therein may be accepted by many readers, who may perhaps not see my account, which will be contained in an expensive volume.* I have been urged to send some protest regarding these misstatements of published facts, and in this the Keeper of my Department concurs. [Not surprisingly, as any criticism of Dorothea's abilities also reflected badly on the Museum.] I dislike and deprecate personal controversies [Dorothea continued] and have tried to make my note an impersonal one. I enclose this note and shall be very glad if you will kindly have it included in the next number of the proceedings. You will find [she added] that it gives some information, as well as criticism.

The information it gave was corrections to his dating and a brief summation of her main conclusions from *The Stone Age of Mount Carmel.* Her criticism of Picard is caustic. 'Misstatements of published facts,' she

* Dorothea had wanted her part of *The Stone Age of Mount Carmel* to be published as a separate volume to make it more easily available, but the cost would have been an extra £50 (about £1,700 in today's values), which the publishers would not meet. She approached Dr Lang, who thought this would be money well spent, but his department could not provide it and Dorothea herself could not afford such a sum. Garrod's and Bate's reports were published as one volume.

wrote, 'are not what we expect in published papers.' And of all people, Dorothea implies, Picard should know better as he 'is one of those to whom we look for aid in the solving of problems still outstanding, notably geological ones. For he has unrivalled opportunities living, as he does, in the midst of his own special field of labour.' She ended with the hope that her account of the fauna from the Wadi el-Mughara caves 'will persuade Professor Picard that a study of this fauna may be a helpful auxiliary to the investigation of the Pleistocene climate of Palestine'.[13]

Having vented her anger, refuted Picard's arguments and corrected his errors, Dorothea did not send either note or letter. Nor, interestingly, did she destroy them. Possibly she had no appetite for vengeance once her anger had cooled; perhaps she may have relied on her faith to heal the damage done by Picard's criticism. She had turned to Christian Science for support after her family difficulties; it may have made her rethink what her note might actually achieve. The following day, 30 June, she wrote to Dr Clark about another matter and simply stated at the end, 'I cannot help feeling sorry that Professor Picard's paper should have appeared in this form in the *Proceedings of the Prehistoric Society* . . . it does not give a correct view of our knowledge up to date of the faunal changes that have taken place in the Pleistocene of Palestine – However this will be dealt with in my account of the fossil fauna of the Wady Mughara Caves,'[14] and simply added the publication details.

In Dorothea's introduction to her section of *The Stone Age of Mount Carmel*, there is a short paragraph in which she acknowledges that much further work is needed in the investigation of climate, including 'results from detailed geological and soil investigations',[15] exactly Picard's field. Whether this was in her original manuscript, or was added at the last moment at proof stage, there is now no way of knowing. It had been colleagues who had urged her to protest at Picard's misstatements. Her initial reaction, and the one that in the end she followed, had been to allow her account of the fauna of the Wadi el-Mughara caves to be a sufficient refutation of Professor Picard's 'mistakes'; and so it proved to be.

Publication of Garrod's excavations on Mount Carmel had originally been planned for 1936, but the complexity of the work, particularly the problems that McCown and Keith faced with the human skeletons from Skhul, caused a delay. In the end it was decided to go ahead in 1937 with the first volume, the archaeology and palaeontology sections; the anthropology volume would follow in 1939. Garrod's report was the first part of the volume and included an account of the excavation of the caves, her determination of the unique Palaeolithic sequence that she had revealed in Palestine, and analysis of the tens of thousands of artefacts found, with pages of splendid illustrations. Part two of the first volume was Dorothea's analysis of and conclusions on the fossil fauna. McCown and Keith, to estimate the chronology of the remarkable human remains, had relied on Dorothea's 'definitive and brilliant studies', as they called her work, and it was truly pioneering.[16] Schematically, she described the changing fauna, climate and environment of the region over this vast period of time. In this she acknowledged the assistance of Dr Frederick Zeuner, a thirty-year-old refugee from Nazi Germany who was working as a research associate in palaeontology at the Museum.[17]*

Dorothea had identified, from the many thousands of bones, teeth, tusks, horns and antlers, fifty-two different species composed of an astonishing mixture of types. The collection included the first evidence of domestication (a subject that particularly interested her) in an animal from Palestine.† But that was simply the beginning of her work. During the Pleistocene, Palestine seemed to her to be 'a great highway of migration', through which North Africa received much of its Pleistocene mammal fauna, while successive waves of animals 'flowed in' from Asia.[18] Her purpose in this report was to endeavour to describe a detailed history of the fauna, which, she wrote, 'is inextricably woven with that of the

* The position was unpaid, as the Keeper of Geology, Dr Lang, told him: 'I am afraid that paid work in England for one of your attainments would be very difficult to obtain, even if you had the highest personal recommendations from anyone in England. The difficulty of getting even poorly paid posts for the most highly qualified man is very great indeed.' (Letter from Dr W. D. Lang to Dr F. Zeuner, 2 April 1934, DF 100/234/15.) Zeuner's talents, however, were recognized and he went on to specialize in past human environments, becoming Professor of Environmental Archaeology at London University.

† In her earlier Palestine reports, Dorothea had noted that it appeared very likely that the Natufians (the Mesolithic people who flourished about 12,000 years ago), whose culture Garrod had first discovered at Shukbah, practised a form of agriculture. 'It is now possible to announce,' she wrote in the *The Stone Age of Mount Carmel* [p. 154], 'the first evidence of the domestication of animals in the Natufian of Palestine.' The evidence was the skull of a dog. 'It is satisfactory to gain this further knowledge on the ways of the Natufians,' she continued, 'for it is well known that the early domestication of animals, which may take place during the nomad existence of a people, usually precedes the practice of agriculture.' She thought the dog may have been a watchdog or used for hunting. She found no evidence of the domestication of sheep or goats.

changing climatic and environmental conditions, a fact which makes it possible to picture in broad outline some of the varying aspects of the country during this time'.[19] This work was only possible because of the truly remarkable succession of deposits found in the neighbouring caves of Tabun and el-Wad, and Garrod's meticulous excavation and labelling of them, 'to which', Dorothea wrote, 'I cannot pay too high a tribute, and for which I cannot be too grateful'.

To begin to understand this mass of material, Dorothea devised two lines of attack. One was a special study of small mammals and the other was a study of the persistence and comparative frequency of the species. 'It is easy to understand,' she wrote, 'that small mammals may evolve rapidly, and that they may be so closely connected with conditions of soil and vegetation that they are usually more sensitive to climatic conditions and changes than many of the larger animals.' This was what Martin Hinton had discovered in his studies of small mammals and it demonstrated why the study of these animals was so important. She found primitive small species in the early levels, which appeared to be completely replaced by modern forms in the later ones. Larger species, such as the Persian fallow deer (*Dama mesopotamica*) and the red deer (*Cervus elaphus*), 'seem to survive almost throughout'. Among the larger species she found a greater quantity of the remains of the Persian fallow deer and of gazelle than those of any other species. Building on her preliminary work of 1932 on these two species, the one a woodland creature, the other a desert dweller, she ingeniously constructed a 'census' of these animals to attempt to establish the climate during the successive periods of Pleistocene time as represented at Tabun and el-Wad.[20] She tabulated the results into graphical form, showing layer by layer the climatic fluctuations that took place during the human occupation of the Palestine caves. Known as the Dama–Gazella graph, it is a masterpiece of clarity that is still used by prehistorians and archaeologists of the region.

Periodically, she discovered that there were suggestions of new faunal migrations from Asia. And then, in corresponding layers in the caves of both Tabun and el-Wad, she discovered something of 'primary importance'. It was nothing less than 'the dramatic, and apparently unmistakable, evidence of a great faunal break'.[21]* She believed that this was accom-

* The level indicating a drier climate was the Lower Levalloiso–Mousterian level of Tabun (Levels C and D) and that of the 'great faunal break' was the Upper Levalloiso–Mousterian

panied by greatly increased rainfall and perhaps a slight cooling of temperature. From this level onwards, Dorothea wrote, 'the fauna is of modern type.'[22] Species such as the hippopotamus and rhinoceros completely disappear. Her work on micromammals from the caves a few years later confirmed and strengthened the evidence of this great faunal change,[23] as, indeed, have modern studies.*

The book was received to wide acclaim, although some reviewers (perhaps understandably) appeared daunted by its technical detail. The prehistorian Abbé Breuil, who had taught Garrod in the 1920s, reviewed the work at length in *Nature*. He discussed Garrod's complex and fascinating archaeological findings, but when he came to Dorothea's report, he found that 'in spite of its great merit, it is too technical to be discussed at length. We must,' he wrote, 'confine ourselves to confirming what this distinguished palaeontologist says.'[24] The *American Journal of Physical Anthropology* called it 'an excellent publication in every respect . . . the work is of basic importance and will be indispensable to students of human prehistory'.[25]

Dr Clark reviewed the book enthusiastically for the Prehistoric Society, perhaps the least he could do in the circumstances; he, after all, had, published Picard's attack on Dorothea. He pays tribute to the presentation of the archaeological section and to Garrod's meticulous methods of working. Dorothea, he thought, had 'ably studied' the faunal remains. He found much of her writing 'primarily for the palaeontologist', but her more general remarks 'will be of the liveliest interest to prehistorians'. Of great interest to her must have been his unqualified support for her work on climate. 'By careful study of the fauna,' Clark wrote, with complete disregard for Picard's article in his own journal, 'Miss Bate has

---

level B of Tabun, and the corresponding level G of el-Wad. All the bones at the level of the faunal change were broken, some were crushed and she suggests that 'part of this deposit might have been accumulated in the form of carcasses violently washed in, or even . . . disturbed or redeposited.'

* In 'Dorothy Garrod's excavations in the Late Mousterian of Shukbah Cave in Palestine reconsidered', *PPS*, 70 (2004), 207–31, Jane Callander acknowledges how advanced was Dorothea's interpretation of the evidence: 'She also understood – and anticipates Tchernov's work to a surprising degree – that the microfauna were more "sensitive to climatic conditions and changes than are many of the larger animals".' (See Professor Eitan Tchernov, 'The Faunal Sequence of the Southwest Asian Middle Palaeolithic in Relation to Hominid Dispersal Events', in T. Akazawa, K. Aoki and O. Bar-Yosef (eds), *Neandertals and Modern Humans in Western Asia* (New York & London: Plenum Press, 1998), 77–90.)

demonstrated changes in the composition of the fauna, from which she is able to suggest important climatic fluctuations.' Furthermore he found 'particularly striking' her study of the deer and gazelle. 'No one reading this book,' he wrote, 'can fail to be impressed by the dispassionate and scientific atmosphere that pervades it, and, above all, by its essential modesty. The discoveries set out in its pages will everywhere be received with implicit confidence.' His only quibble was with the price: too much, he thought, at two guineas for a single volume, 'even if metaphorically speaking it be of pure gold'.[26]

The Stone Age of Mount Carmel led to Dorothy Garrod being awarded a Doctorate from Cambridge. Two years later, in 1939, she was appointed Disney Professor of Archaeology at Cambridge, the first woman to be appointed to a professorial chair by the university, and nine years before women were admitted as full members there. For Dorothea there was an enhanced reputation, but no academic recognition.

With the Mount Carmel work finally published, Dorothea could fully concentrate again on Bethlehem. Some of the vertebrate remains were exhibited in the Wellcome Museum, together with photographs of the dig. She and Elinor wrote up their preliminary finds on the bone beds and their report was published in *Nature* in September. For the first time, Dorothea revealed in print that the fauna was 'of earlier geological age than any known so far known from Palestine', and included a large cat (*Felis* size of *Panthera leo*), hippopotamus,* and an ox (*Bos*, which was subsequently re-identified as *Leptobos*, a type of extinct water buffalo[27]). She identified an antelope, a giraffoid, rhinoceros, *Elephas*, giant and tiny tortoises, and, of particular interest, *Hipparion* – the extinct early horse.[28] She suggests that the age of the fauna was not later than Early Pleistocene, her original theory, and that the species found indicated a warm climate 'with a more liberal supply of permanent water than is found in the country at the present day'. The position of the deposit and the fragility of the bones were, she believed, caused by upheavals of the earth, which

* A re-examination of the material in the 1950s by the palaeontologist Dirk Hooijer suggests that the fragment of molar that Dorothea thought was hippo in fact was a 'deceptive' *Archidiskodon* molar, an extinct species of elephant. He found no trace of hippopotamus in the collection. (D. A. Hooijer, 'An Early Pleistocene mammalian fauna from Bethlehem', *BBM(NH)G*, 3: 8 (1958), 265–92.)

subjected the remains to great pressure, and modern studies have confirmed this view.*

Gertrude Caton Thompson contributed to *Nature* a note on the flints, which did not take the question of their origin much further. 'The danger,' Caton Thompson wrote, 'of claiming humanly-fashioned tools from such an assemblage and so vast a selection is obvious. Nevertheless, we think that a certain number of pieces are difficult to explain in other terms.'[29] An anonymous reviewer in the same issue of *Nature* seized on this eagerly, at the same time warning that it would be 'premature to enlarge on this aspect before the specimens have received further expert consideration'.[30]

By the autumn of 1937, the security situation in Palestine had further deteriorated and the Wellcome Trustees began to discuss the consequences of allowing the excavations at Lachish and Bethlehem to continue. Their concern was not just of the possible danger to the archaeologists, but of the legal and financial burden that might fall on the Trust should anyone be shot or wounded. They concluded that they would not be liable and L. C. Bullock, writing on behalf of the Trustees, conveyed this to Starkey and to Sir Charles Marston. 'All these people,' Bullock wrote, 'are accepting employment with a full knowledge of the conditions existing in Palestine at the present time and so long as our officials take all reasonable precautions and are not negligent I am satisfied that no personal liability could arise.'[31]

Not three months later, James Starkey was shot dead on a lonely road near Hebron. The Trustees and Sir Charles Marston immediately reviewed their earlier decisions. Bullock wrote again to Marston. Choosing his words with the care that the emotional and legal situation demanded, he said that he felt there were serious concerns about possible claims for

* In 1988, Aharon Horowitz of Tel Aviv University described the Early Pleistocene geography of the country as being 'completely different' to the present. It was, he wrote, 'a rather vast flatland crossed by several river systems from the east to the Mediterranean'. The rivers were broad with several lakes and the climate subtropical. It was 'a rare combination of local tectonic activity and global climatic trends' that caused the great structural changes that raised the central hills and mountains of the country during the Pleistocene and raised the Bethlehem bone beds to their current elevated position. (A. Horowitz, 'The Quaternary Environments and Palaeography in Israel', in Y. Yom-Tov and E. Tchernov (eds), *The Zoogeography of Israel* (Dordrecht: Dr W. Junk Publishers, 1988), 53.) As with so much else, however, opinions differ, with some geologists placing the upheavals several million years earlier during the Pliocene, but even in a country as small as Israel, the upheavals could have happened in different areas at different times.

compensation, and although they had no legal liability to Mrs Starkey and her children, 'we may feel some moral obligation'. This, fairly swiftly, was translated into financial help. In the palpable shock of Starkey's death, no one seems quite to have known how to respond. The immediate concerns were how the expedition could continue 'in the absence'[32] of Starkey, and how to ensure the safety of Starkey's team. However, it was also thought, particularly by the British administration in Palestine, that it was most important for the morale of the whole district for excavations to continue.[33] It was, given the circumstances of the murder, an appalling dilemma. Starkey had been found, according to Sir Charles Marston's stepson, Holbrook Bonney, who was visiting Lachish, with 'two bullet holes in the chest and his head bashed in'.[34]

What had happened to him was pieced together by Olga Tufnell, Starkey's deputy at Lachish, and Charle Inge, an archaeology student at the British School of Archaeology in Jerusalem who was working with them. It was a report that must have been unbearably difficult to write. Starkey was being driven from Lachish to Jerusalem late on the afternoon of 10 January 1938. Near Hebron, on a bend so steep and narrow that escape was impossible, the road was blocked by a group of men armed with rifles. Starkey's driver, an Arab, explained to them who his English passenger was, that he had worked in the country for many years, and 'was a friend of the Arabs. They replied,' wrote Tufnell and Inge, 'that they did not believe him, that he was a Jew, and forced Starkey to get out of the car.' Contemporary newspaper reports suggested that Starkey's luxuriant beard had given him a Jewish appearance. The driver was ordered to proceed, but he had not driven twenty yards before he heard two shots and in his rear-view mirror saw Starkey fall to the ground. The driver, who was the source of this story, went straight to the police station. The locals were outraged at the murder and volunteered information to the police with the result that two members of the gang were captured and subsequently hanged.[35]

Hundreds of friends and well-wishers attended Starkey's funeral in Jerusalem. 'I was very touched,' Olga Tufnell wrote to Dorothea, 'to see Katherina [the housekeeper at Bethlehem] at the funeral among representatives of so many creeds and professions in Jerusalem. The list has just reached me of those who went to the Memorial Service [in London], I am glad you were able to be there – we all thought of you so much that day.'[36] Olga also told Dorothea that work at Lachish was to continue, with Gerald Lankester Harding, the Director of Antiquities for Transjordan, taking administrative charge: 'We have now heard from the Trustees that we are to carry on – as we had strongly urged; it would certainly have been Starks' wish and everyone in Jerusalem from the High Commissioner downwards is in favour of continuation.' Six policemen were to be based at Lachish, but Sir Charles Marston was concerned that the hanging of the two Arabs might lead to reprisals of such violence that the police guard would be of no use at all.[37] He accordingly sent a cable to Inge: 'If reaction from recent tragedy involves increased risks your lives must not be unduly endangered. Exercise your discretion. You have our authority to close down any moment.'[38]

With the future of the Bethlehem dig still unresolved, it was at this moment that the *Christian Science Monitor* chose to run an article about the Bethlehem excavations complete with official Wellcome photographs, on 19 January, just nine days after Starkey's murder. 'Research work in Palestine has been so bound up with archaeology,' the newspaper reported, 'that the discoveries in the bone-bearing beds at Bethlehem have been little noticed.' Extraordinarily, there is no reference to the murder of Starkey, but the report mentions both Elinor Gardner and 'Miss Dorothea M. Bate, who has been superintending the excavations on the palaeontological side'.[39] The article reads as if they were both interviewed, but there is no indication as to whether that was before or after Starkey's murder. Although Dorothea has kept the cutting in her files, she otherwise makes no reference to it. It is not impossible that, exasperated by the amount of newsprint devoted to Starkey and his work, Dorothea herself might for once have sought publicity; the newspaper was, after all, the journal of the Christian Science movement.

On 28 January, Dorothea wrote to Lankester Harding. 'I was just going to write to Mr Starkey when the tragic news of his death reached us – it was indeed a great shock, and I hate to think of what it must have been to all those who were working with him at Tel Duweir.' There were just six weeks before work was due to resume at Bethlehem, and a decision was urgently needed. 'I expect you have hardly had time to remember the Geological Section,' she wrote with some diffidence and a curious echo of the newspaper report,

> so I am writing to ask what you feel about the Bethlehem dig?
>
> No doubt we should be all right at Bethlehem, but I feel that all the pleasure in the work would be gone. The thought of the many journeyings that would be necessary from Tel Duweir to make arrangements for our housing . . . would make it difficult to preserve one's peace of mind.
>
> All this makes me wonder if it would be wise to postpone work at Bethlehem until next year when I hope the country will be in a happier state.[40]

Perhaps she was remembering the murder all those years ago of the abbot of Chrysostomo in Cyprus on the path 'where I pass nearly every day',[41] but her sentiments were disregarded by Lankester Harding. His concern was 'that we should endeavour to complete this season the work already begun breaking new ground'. To Dorothea's surprise – she scrawled in the margin of his letter, 'I have not heard of this' – he told her that Elinor Gardner would not be available for work in 1939, as she was committed to an excavation in Persia. 'If we do not work this year it would mean leaving it for two years which does not seem to me very wise. If what has already been started could be completed, we should at least be in a position to publish results so far achieved.' As far as the question of security in Bethlehem was concerned, they had been assured, he told Dorothea, 'that there is no danger whatever.'[42]

On 19 February she was told by Charles Inge that work was to continue and the Trustees would send Dorothea her usual cheque. The pit, he wrote, was in 'reasonable condition', Elinor Gardner would arrive in March and the housekeeper Katherina 'will be there before your arrival to get things ready'.[43] But it was not to be. The reports from Palestine of continuing violence, combined with the growing threat of war with Germany, made the Natural History Museum Trustees decide that her visit should be postponed. On 28 February 1938, Dorothea wrote to Elinor Gardner, who was excavating with Freya Stark and Gertrude Caton Thompson at Hadhraumat in Aden: 'I have sent you a cable to-day to this address in which I said "Not going Bethlehem. British Museum cancelled leave. Bate".' Outlining to Elinor the indecision of the past few weeks, she wrote that just about the time that the 'Wellcome people' had decided to go ahead with the work, 'the authorities here began asking questions about this matter and it was brought before our Trustees on Saturday (26th February). This morning the Director informed me that my leave of absence for going to Bethlehem had been cancelled.'[44] In spite of her earlier reservations about continuing work there, Dorothea was now prepared for it and it was with some distress that she wrote to Charles Inge: 'I only hope this may reach you before you have actually commenced the proposed preliminary work. I deeply regret that it is not possible for me to return to Bethlehem this year as Miss Gardner and I had hoped to add many treasures to our collection this season.'[45] Unlike so many of her letters to Starkey, this time she meant it.

Dorothea's distress was echoed by Charles Inge, who had just arrived in Bethlehem to prepare the accommodation and was about to begin the preliminary clearance of the pit. With considerable anger, he told the Wellcome Trustees that he had telegraphed the Museum asking for a reversal of their decision as he considered all fears for security in Bethlehem to be quite ungrounded. 'This has since been refused. We deeply regret,' he wrote, 'that Miss Bate will be unable to come out this year, but we hope that Miss Gardner will be able to find someone to collaborate with her in the removal of the bones for subsequent examination in London.'[46] Elinor, however, flatly refused to work with anyone other than Dorothea, although she did suggest that they might continue the work in the autumn. 'That,' wrote Inge to the Wellcome Trustees, 'will have to be decided in London, but we see no objections from this end. In any case we will leave the pit as it is now and will store the equipment in Bethlehem.'[47]

In spite of all their efforts, by August the volatile situation in Palestine and in Europe meant that Inge had to abandon the excavations 'for the present . . . It will be difficult,' he wrote to Dorothea, 'to say what will be possible for the future.' What he wanted to know immediately, however, was whether there was the possibility of publishing existing results. He also told her that the Palestine Museum was asking for a complete catalogue of the material so that a division could be made of the specimens and they wanted their share returned.[48] Dorothea did not reply until October and when she did the vagueness and uncertainty in her letter is quite unlike her usual direct approach. She could give him no idea of when a report might be prepared and also told him that, in order to undertake a report, she needed the total collection, and that made it impossible to return any of the specimens to Palestine before she had studied them.[49]

The unremitting pressure of these exhilarating, difficult and exhausting years was at last exacting a price, and it was her health. She was now nearly sixty. In addition to her major work on Mount Carmel, in the ten months between September 1937 and November 1938 she examined and reported on at least seventeen separate and substantial collections, including fauna from the famous dig at Megiddo, from Kent's Cavern in Devon (an ongoing excavation, which she had been dealing with for twenty years), and from two archaeological sites at Wadi Dhobai in Transjordan, excavated by John Waechter, a former pupil of Dorothy Garrod. She may have derived some satisfaction from Waechter's pointed remarks that 'The bone report supplied by Miss Bate is of great interest from a climatical point of view' when Leo Picard had supplied the geological information for this report.[50]

As if this were not enough, she had also been dealing with collections of faunal remains that excited the sensitivities of Alvan T. Marston, the discoverer of the Swanscombe skull. At the end of 1937, she had published a paper on an extinct species of deer, *Dama clactoniana* from London and Swanscombe. The antlers she examined had been discovered in 1934 in Swanscombe (before Marston found the human skull) by the young Louis Leakey, the great anthropologist who was to make astonishing hominid discoveries in East Africa. Marston took exception to some of Dorothea's conclusions and told her so in the spring of 1938.[51] Dorothea with great patience explained to him that the errors were in fact in his understanding of her report. Her charm and diplomacy were such that in the next few months he sent her three collections of antlers and other material for determination. The final collection she received in November 1938 was accompanied by a letter that she viewed with very mixed feelings. 'I am really hoping,' Marston enthused, 'that you will be able to find a progression through the various stages of the fauna and that is why you can count upon my sending every bone along that I get out.'[52] By 7 December she had given a preliminary examination to the collection and replied to Marston.

Two weeks later she collapsed. Wilfred Edwards, who had just succeeded Dr Lang as Keeper, at first thought she was suffering from no more than a cold that was also afflicting several other members of his department,[53] but Dorothea had been unwell since the autumn, hence the delay in her reply to Inge. Her illness was so severe she could not return to full-time work until the summer of 1939. Just how devastating it was is evident from photographs. In 1937, the Bethlehem pictures show a sturdy, apparently robust woman. A little more than a year later, in the photograph taken to mark the retirement of Dr Lang, she appears almost unrecognizably as a slim, slight figure, high-cheekboned and elegantly dressed, but whose expression reveals weariness and pain. Wilfred Edwards believed that she underwent two major operations – Christian Science discourages, but does not forbid, conventional medical treatment – but no records exist of these or the cause of her illness. Cancer is unlikely (although not impossible) given the high mortality rates for the disease in the 1930s. Dorothea herself does not reveal the problem.

Although she spent nearly every Christmas with her brother Thomas at Glanmonnow in Herefordshire, she chose not to stay with him to recuperate. Close though brother and sister were, her relations with her sister-in-law were strained. It was Ulrica, after all, who had deprived Dorothea of her inheritance. Although they were perfectly polite and charming to each other, David (Thomas and Ulrica's son) remembers vividly the undercurrent of tension between them. It was not until he brought his new wife to meet Dorothea that David learned for the first time something of what had transpired. To his amazement, Dorothea 'warned my wife of my mother's strong personality, though without mentioning how it had affected her personally. This was strong stuff for my aunt who was the soul of loyalty and had never before in my hearing breathed a word against my mother.'[54] Small wonder that Dorothea chose to be cared for by Leila at Waltons Park. In June, with gaining strength and with Leila for company, Dorothea went to Madeira, as a result of which, as Edwards wrote to Dr Lang at the beginning of July, she has 'much better, but has not yet returned to work'.[55]

Any possibility of producing a new report on Bethlehem was, for the foreseeable future, quite out of the question. Inge, however, was in ignorance of her ill health, and he wrote to Dorothea again in March 1939 to enquire after her report. 'I am afraid that I cannot give you any indication as to when the work on the fauna from Bethlehem will be completed,' she replied. 'The last few years have been such a wonderful harvest of palaeontological material from Palestine that its intensive study must occupy a considerable time.' Only then did she admit to her illness, adding, 'I shall not be returning to the Museum for some little time. This, as you will readily understand, has seriously delayed my research work.'[56] By August 1939, with war imminent, she was well enough to go occasionally to the Museum and was beginning to work through the faunal material from Tell ed Duweir. She was pursuing her interest in the domestication of animals and sent Inge a long series of questions concerning the skeleton of a dog found at the Tell, which she wanted to compare with that of the dog found at Mount Carmel. Her letter is much more like her old self, although there is an element of desperation in the list of questions, as if she feared her strength would desert her again before she had received the replies she needed.[57] Both Inge and Olga Tufnell were on holiday when her letter arrived. By the time they returned, the threat of war had become a reality.

CHAPTER 15

A Safe Place in the Country

From 1938 there had been just one priority for the Museum and that was to protect its collections in the event of war. Rooms that might provide gas- and splinter-proof shelters for staff were identified. The suggestion that a bombproof shelter might be built was rejected by the Treasury on the grounds that, if war came, all museums would be closed and their treasures removed.[1] For the Geology Department, Wilfred Edwards suggested that all type and figured specimens should be stored in the basement, although there was a still a shortage of cabinet and drawer space to put them in. The larger skeletons, which could not easily be removed, Edwards suggested protecting with sandbags. It would be, he warned the Trustees, the work of weeks, if not months.[2] In the summer of 1938 all personnel were sent on an anti-gas course, and training began in first aid and decontamination work.

That autumn, the Museum took possession of a bequest that was to become of enormous significance to Dorothea and which in part resolved the problem of where to send the collections. In the autumn of 1937, the zoologist *extraordinaire*, Walter, Lord Rothschild, had died and in his will left to the Natural History Museum his own exquisite Zoological Museum, which stood in the grounds of his home at Tring Park in Hertfordshire. For the casual visitor the Tring Museum was, and still is, a monument to taxidermy at its very best, with galleries of gleaming mahogany and glass cases filled with marvellous stuffed mammals and birds (more than four thousand of them mounted). Even today it is a place of wonder and delight, particularly for children. For zoologists and ornithologists there were thousands of mammal and bird skulls and skins to study, reptiles, amphibians and fish. The depression of the early 1930s had forced Rothschild to sell his unparalleled bird collection of around 300,000 specimens to the American Museum of Natural History in New York. At a total cost of $225,000, his sister Miriam wrote sadly, the birds were sold at 'a little less than a dollar a piece',[3] but there were still around fifty thousand birds' eggs, including the great auk and *Aepyornis*, the elephant bird. The entomology collection contained more than three million insects, including two million moths and butterflies and 100,000 fleas. In the grounds of Tring Park, Rothschild kept a small and idiosyncratic zoo, which included zebras, ponies, kangaroos, cassowaries, rheas, and the famous giant tortoises, on the backs of which Lord Rothschild would sit astride. It took a year for the negotiations to be completed, and an Act of Parliament had to be passed before the Tring Museum could be transferred to public ownership. It presented the perfect place to which at least some of the national collections could be evacuated.*

For the first half of 1939, Dorothea's poor health allowed her to attend the Museum only occasionally, although by July she seemed a little

* Walter Rothschild's nephew, Victor, who succeeded to the peerage on the death of his unmarried uncle, then offered in addition to present both Tring Mansion and Park to the Natural History Museum. This he did partly because he believed so great a space would provide opportunities for the study of zoology, but also because it could be used to store many more valuable specimens than the Museum. Martin Hinton, Keeper of Zoology, advocated acceptance of this magnificent gift, and involved the philanthropist Sir John Ellerman, who offered to pay for the running costs of house and grounds. In the end, agreement with Rothschild could not be reached and this great scheme lapsed. The house is now the home of the Arts Educational School. Ellerman himself was a shy and reclusive man, best known as a shipping magnate, although his business interests (and great wealth) spanned breweries and newspapers, including the *Daily Mirror*. He was a keen natural historian, whose great interest was small mammals, hence his friendship with Hinton, and he was engaged in writing the first volume of *The Families and Genera of Living Rodents* (published by BM (NH) Dept of Zoology (Mammals), 1940). He later corresponded with Dorothea on the subject, although he preferred to work with recent material, rather than extinct. His shyness was such that the *Daily Express* compared him to the reclusive millionaire Howard Hughes, while Dr William Stearn, the botanist and historian of the Natural History Museum, suggested that, like small, shy rodents, he kept out of sight as much as possible, and noted that few workers at the Museum 'connected this shy, tall, lean, middle-aged student of rodents with the fabulously wealthy director of the Ellerman Shipping Line'. (William T. Stearn, *The Natural History Museum at South Kensington: A History of the Museum, 1753–1980* (London: Natural History Museum, 1998), 189–90.)

stronger. That summer, visitors continued to come to the public galleries in their thousands, researchers could still be found examining the collections that had not been packed away and the work of the Museum, to some extent, continued. But, like a swan, serene on the surface, underneath there was frantic activity in every department. For the entire staff in the Geology Department, including Dorothea when she was well enough, their main occupation was listing and indexing type and figured specimens, packing these and the more fragile bones in special cardboard boxes, and creating vital lists of what collection was being sent where. As Dorothea told one anxious owner of bird remains, it was felt to be unwise to return his collection by sea. Instead, the case of specimens, 'duly labelled, has been sent to a safe place in the Country, together with some of our treasures'.[4] Photographs were taken of many of the more important specimens. Accession registers were also photographed and, even before the work was finished, over a mile of 35 mm film had been used and a million register entries photographed.[5] Edwards seized on this massive disruption to his department as an excuse to reorganize it, dispersing old collections, clearing overcrowded exhibition cases and planning their rearrangement.

To add to everyone's concerns, the IRA had started a bombing campaign on the mainland and 'unknown visitors' to the Museum had to have all their bags and packages searched.[6] On the outbreak of war at the beginning of September, the Natural History Museum, together with all museums and galleries, was closed to the public, although the staff continued their curatorial, and to some extent research, activities. They also started a satirical journal, *Tin Hat*, with such suggestions that beer would now be served to staff. It was a spirited and witty pastiche of a house magazine, with a range of humorous articles and jokes of this order: 'Ladies with laddered stockings will be requested to report to the Chief Fireman for fresh supplies of hose.'[7]

Against this background of disruption and uncertainty, Dorothea worked on those collections that were still available. One such was a collection of bones from north Syria, sent to her in 1938 by the archaeologist Max Mallowan, whose excavations at Ur in Iraq had resulted not only in splendid finds, but in his marriage to the crime writer Agatha Christie whom he met on her visit to the site. In January 1940, four months into the war, Mallowan wrote to Dorothea to thank her, some months late, for agreeing to examine his collection. He hoped to publish an account of his excavations by the end of February, he told her, 'in case I am called up for service in the East. At present,' he added, 'I am still cultivating archaeology and vegetables in the country.'[8] Dorothea had made a preliminary examination of the bones, but although Mallowan's labelling bore the suggested age of the material and was sufficient for her to follow the sequence, 'the dates convey very little to me,' she wrote to him, 'and I sometimes fancy that each archaeologist has own idea on the subject!'[9] It was a problem that she had constantly to confront. She sent him reports of Philip Guy's excavations at Megiddo and John Waechter's at Wadi Dhobai in Transjordan in the hope that he could correlate his proposed sequence of events with theirs.

It was Dorothea's meticulous examination of material and her pioneering interpretation of the climate and environment in which the fauna had once flourished that led, at last, to public recognition of her work. On 13 March 1940, at Burlington House in Piccadilly, the Geological Society of London awarded Dorothea the prestigious Wollaston Donation Fund for outstanding work. The award was presented to her in front of an invited audience by the President of the Society, Professor Henry Hurd Swinnerton. 'The Council of this Society,' he told her, 'desired to express its appreciation of the careful and detailed researches which you have been carrying on for so many years.' He outlined all her excavations in Pleistocene bone-bearing deposits and acknowledged that 'by careful and judicious study', she had 'extracted the maximum of information from the material so scrupulously collected by you'. He praised her 'admirable account' of the fossil fauna of the Wadi el-Mughara caves 'and of the climatic and environmental conditions they suggest'. He ended thus: 'Though work of this kind always brings its own reward, there is an added pleasure in knowing that your work has proved to be of value to others.'[10] The award was a cheque for the sum of £30. 4*s*. 7*d*. (about £770 in today's values).[11]* As Dorothea was still paid £250 a year by the Museum, as she had been for most of the previous decade, the award was rather more than one month's salary.† Five days later, as if no one had realized the

* The Wollaston Fund is still awarded today, but only to Fellows of the Society under the age of forty.

† In 1941 her salary was approximately £6,400 in today's terms, a substantial decrease from its equivalent value a decade earlier of approximately £8,600.

omission until that moment, she was nominated for a Fellowship of the Geological Society and was elected on 1 May.

There was a deceptive quiet in these early weeks of 1940. The expected bombing raids had not yet occurred and in February some of the galleries were reopened to the public. Instead of the usual huddle of children on Saturday mornings, however, those under twelve had now to be accompanied by an adult and everyone had to carry a gas mask. Although a total of thirty lorryloads of the most important and valuable specimens had been evacuated to Tring and elsewhere, much remained, packed into the basements or stacked away from windows. In June and July, Dorothea was well enough to be at work and was engaged in Edwards' reorganization of the geological collections.*

The future of the pit at Bethlehem had still not been satisfactorily resolved, to the concern of the Wellcome Trustees who were still paying out considerable sums in rent and compensation to the Hasbun brothers. Olga Tufnell feared that they would order the pit to be filled in and suggested as an alternative that they should find someone who could complete the work. The Department of Antiquities in Palestine, as Olga wrote to Elinor Gardner in January 1940, had mentioned Dr Moshe Stekelis, with whom Dorothea had collaborated on Jisr Banat Ya'qub. She added to Elinor that she had 'not heard anything of Miss Bate for some time and gather she is still not very fit after her illness'.[12] Elinor sent a copy of this letter to Dorothea, who reacted in a most healthy manner by firing off a letter to Olga Tufnell to ensure that any further specimens found would be sent to her at the Museum, under the terms of her original agreement with Starkey. Too practical to waste further time on anything other than the future of the pit, Dorothea sent to Olga the name of the best workman at Bethlehem for Dr Stekelis to contact. She also told her that she foresaw no difficulty now about being able to study the animal remains and prepare a report for publication, 'though not much can be done until the whole collection is available'.[13] To Dr Stekelis a few weeks later, she wrote wishing him 'much success' and briefed him on the state of the excavation. 'Most of the specimens,' she advised him, 'need a lot of attention before they can be extracted. If I can be of any

* Her task was weeding out specimens that were too fragmented to be determined or were of 'no value or interest', and in every case writing to the collector to ask whether the specimens should be returned or 'dealt with', a euphemism for destroyed.

help in telling you my method, please let me know.'[14] Her 'method' was the judicious use of celluloid and then plaster. Dr Stekelis, armed with quantities of the stuff, resumed the excavations in June 1940.

Over the course of the next four months, Stekelis excavated eighty-five specimens, including what he described as 'a wonderful elephant's lower jaw, with all the teeth in it' (which had needed 'kilograms' of celluloid solution to harden it), and also the flattened skull of a rhinoceros with all its teeth. These he packed into eleven boxes and dispatched them for store in the basement of the Palestine Archaeological Museum in Jerusalem. He sent to Dorothea a complete list of the bones he had found and added that he had refused a request by the Department of Antiquities to write a report for publication on his excavations. The reason he gave was 'my work was only a completion of yours and accordingly you were the real excavator', and he asked her to do it instead.[15] She found this incredible. 'I can't write an article on someone else's work!' she wrote vexedly to Elinor Gardner, and asked whether Elinor herself would write a brief report, but she was equally reluctant. Eventually Dorothea composed the briefest of notes for *Nature*, some four hundred words long, a quarter of which were acknowledgements to the sponsors of the expedition.[16] Her substantive report on the animal remains would have to await their arrival in London, and that was unlikely before war had ended.

In September 1940, the veneer of normality at the Museum was shattered. On successive nights it was struck by incendiary and high-explosive bombs. The Botany Department was badly damaged by fire caused by the incendiary bombs, while the Geology Department was struck by a high-explosive bomb that shattered nearly three hundred skylights and hundreds of panes of glass and internal partitions. Windows disappeared and subsequent rainstorms only added to the damage. Furthermore, the water used to extinguish fires in the Botany Department flooded the Geology Department, which took weeks to dry out. By October, twenty-eight bombs had landed on or near the Museum. Miraculously, injuries were only minor and the loss of specimens was surprisingly small; most were rescued from the wrecked rooms.

Dorothea missed it all. In late summer she was once more unwell and left London for Shepton Mallet, a pretty market town in Somerset. This was partly to recuperate but she also wanted to finish a report for publication: 'The fossil antelopes of Palestine in Natufian times'.[17] This was an analysis of material she had had insufficient time to study before the publication of *The Stone Age of Mount Carmel* and it included new antelope specimens from the Kebarah Cave, excavated by Francis Turville-Petre in 1931. Fundamental to the report was what the collection revealed about the changes in the Mesolithic climate and environment, and what might have caused the extinction of particular creatures. Drawing on a wide range of sources, including animals portrayed in rock art, she concluded that the extinctions and faunal change coincided with a much wetter phase in Palestine, which evidence from archaeological excavations showed had also affected Egypt, northwest and East Africa. Whether similar evidence existed in Iraq, Transjordan or Arabia was not known, and Waechter's work at the Wadi Dhobai in Transjordan, which might have provided some answers, had been interrupted by the war.

Needing some references checked for her report, she wrote to a colleague, Leslie Bairstow, a Cambridge academic who had joined the Geology Department in 1931. Her health was improving, she told him, and she had been out walking 'rain or shine every day for nearly 2 months and feel all the better for it'.[18] His wry and affectionate reply revealed conditions in the Museum that quite shocked her.

> I'm glad that you're flourishing in the fresh air, undaunted by the rain [he wrote]. I've been having some fresh air even here – on the roof putting temporary coverings over shattered skylights, sometimes during air raids and at least once in a thunderstorm. Have had several thorough soakings, so it isn't necessary to come to Somerset to get wet; not even to leave your own gallery, for after the Botany fire you have some very good showerbaths up there and I am mopping up at 3am during an air raid. I tried to look after your things and I don't think there was much damage.

Bairstow himself was less fortunate as 'about 200 of my books got wet'.[19]

With her report completed, and her health much improved, Dorothea returned to London, but not for long. She was now sixty-two, although as an unofficial scientific worker, the civil service rules of retirement at sixty did not apply to her. With the Geology Department so battered and the continuing threat of air raids, the Museum decided that, like the most valuable collections, Dorothea should be evacuated to the relative safety of Tring. Nothing could have suited her more, to exchange her urban environment and basement flat for the countryside of Hertfordshire and to continue her work in the congenial surroundings of Walter Rothschild's highly individual legacy. The drawback was her isolation from the department. She therefore made regular visits to the Museum on the Green Line bus (the cheapest form of transport) to consult the collections and to ensure she knew what was going on and was not forgotten.*

In April she returned, temporarily, to South Kensington. The severity of the blitz was such that it was now critical to move everything away from London that could possibly go. Tring could hold no more, so remote country houses were pressed into service to house specimens and books. It was a task of massive logistics. Collections from the Geology Department were to be sent to the fifteenth-century Tattershall Castle in Lincolnshire. The task of organizing the move fell to Dorothea and Arthur Hopwood, who had been appointed to the department on Charles Andrews' death. For days on end, in dim and dusty storerooms filled with great cabinets and hundreds of drawers, in the basement that Dr Lang had called 'that Augean mess',[20] Dorothea worked her way through the bird and mammal collections in her charge, noting every specimen to be moved on sheets of foolscap paper, with its name, number, whether it was a type or figured specimen, from which drawer or cabinet it had come, and its destination.[21] These lists were then typed and became more precious than gold; the loss of this information could cause unimaginable

* In February 1941 she had a 'hurried' conversation with Leslie Bairstow, in which she agreed to let him store some of his specimens in her room at the Museum, but she then became quite alarmed that it would lead to the removal of her cabinets from her room and possibly to the loss of the room altogether. It took all Bairstow's tact to soothe her. 'I hope you don't think in the hurry the other day I was trying to push you out of your gallery,' he wrote in a letter whose many emendations in draft are testimony to the care he took in phrasing it. 'What happened was that some weeks ago, without any suggestion on my part, the Keeper gave me instructions to stack some of my drawers there, and I thought it only courteous to discuss the matter with you before making any move. Naturally I shall be pleased if you can accommodate some drawers for me, as the back part of your gallery is one of the safest places in the department; but I don't want to displace your things or to obstruct or inconvenience you, and I propose simply to fill any space that you can spare and then to report to the Keeper for further instructions.' Bairstow had some two hundred drawers of specimens that were unsuitable for evacuation because they were either too heavy or too fragile to be moved, and many of these were valuable. Promising not to move any of her cabinets without 'further authorisation', Bairstow ended his letter chattily with news of the impending marriages of two of their colleagues. (Letter from Leslie Bairstow to Dorothea Bate, 24 February 1941, DF 100/80.)

chaos. Once listed, each specimen was carefully packed. By May, the collections were safely in Lincolnshire,[22] and Dorothea had returned to Tring. She at least could work in relative quiet, unlike her colleagues. In April a high-explosive bomb caused more damage to the Geology Department and blew out most of the windows that had just been replaced.

The war had all but put an end to the flow of fossil fauna coming into the Museum, but in July 1940, Dorothea had been sent a collection of fossil mammal bones of very great interest. They came from the Egyptian–Libyan desert, from Gebel Uweinat, famous for its palaeolithic rock drawings, and from the great sandstone plateau, Gilf Kebir, about seventy kilometres away. In one of those extraordinary moments of discovery, I found Dorothea's papers on Gilf Kebir as I was reading Michael Ondaatje's novel, *The English Patient*. Which was life and which art blurred together as I found brief notes from Dorothea referring to Uweinat, Gilf Kebir, and the Hungarian geographer Laszlo Almasy, after whom Ondaatje named his hero.

The fossil bones had been excavated by Oliver Myers, an archaeologist attached to an expedition to the region in 1938. The leader of the expedition was the explorer and scientist Ralph Bagnold, whose purpose was to investigate the physics of blown sand. Bagnold's extraordinary passion entailed waiting for sandstorms. To pass the time he explored Gilf Kebir. It was a plateau of truly vast proportions, roughly the size of Switzerland, with cliffs that rose up to 3,500 feet out of the desert. Bagnold and a companion drove along the foot of the Gilf and found 'a huge solitary sand drift that had spilled over from the plateau until its upper edge was level with the top'.[23] Not only was it firm enough to walk up, but on the fourth attempt they managed to drive their car onto the summit of the plateau itself. 'It seemed,' wrote Bagnold, 'we had strayed into a secret Stone Age World . . . A well-worn path led inland from the brink to successive factories of stone implements. Debris from the ancients' works lay strewn everywhere. It looked so fresh we half expected to come upon a group of uncouth artisans round the next corner. Yet we found no scrap of evidence that anyone had been here since prehistoric times.'[24]

Oliver Myers initially sent the faunal remains from Gebel Uweinat and Gilf Kebir to the palaeontologist and geologist John Wilfrid Jackson, who was based at the Manchester Museum.* Jackson and Dorothea had corresponded on palaeontological matters for over twenty years, exchanging information and seeking each other's advice. To Dorothea's considerable amusement, she received quite early in their acquaintance a letter from him in which he addressed her as 'Miss D. Bate MA', absent-mindedly confusing her initials with a degree.† Jackson, in this instance, confessed that the mammal remains from the Gilf Kebir had him stumped. He thought they were mostly gazelle, but preferred to send them to Dorothea for her opinion. 'Any remarks you care to make,' he added, 'would go under your name.'[25]

This was just the sort of material she had regretted not being available when she wrote the Natufian antelope report the previous year, and she wasted no time in examining the specimens. They were neither numerous nor well preserved, 'but', she wrote in her report for Myers, they were 'of particular interest since they are the first to be obtained in this area and they also help as an indicator of former climatic conditions, thus confirming the evidence previously only known from rock drawings'. As usual she had read widely before starting her report, referring to Almasy's discoveries of rock paintings at Uweinat and the importance of the animals represented in interpreting the remarkable changes in climate. The remains could not be definitely dated, but they were not later than 2500 BC. She identified at least nine animals, among them elephant, ox, antelope, gazelle, African jackal and African ostrich. None of them, apart from gazelles, are found in the region today. Where there is now burning, arid desert, just a few thousand years ago had been a green, warm and pleasant land with an abundant water supply. 'The fact that it would be impossible for *Bos* [ox] to live in the Gilf Kebir today,' Dorothea wrote, 'makes these animals of special interest.'[26] She sent the report to Jackson, who was delighted. 'It is very good of you to take all this trouble,' he

* Jackson had been taught his considerable skills by Professor Sir William Boyd Dawkins, whose renown extended from the earth sciences to archaeology and anthropology. As Dawkins had been born in 1837 and Jackson did not die until 1979, they together spanned an astonishing 142 years of geological and palaeontological expertise.

† In 2004, Dorothea's Museum keyring tag was found in a box in the palaeontology collections. It is small, made of brass and bears those unmistakable initials, 'D.M.A.B.' In all Dorothea's thousands of notes, letters and papers, I cannot think of one instance when 'M.A.' is omitted, whether in 'Dorothy M. A. Bate' to friends, 'Dorothea M. A. Bate' more formally, or even, in scribbled initials, always 'D.M.A.B.', never 'D.B.'

wrote. 'The report is splendid and should please Myers when I can reach him. I knew you were the right person to make anything of the fragments. I tried myself and got as far as Gazelle, Bos and jackal.'[27]

Jackson, based in Manchester, was seeing at first hand the misery of war. 'We had a bad blitz the other week,' he told Dorothea, '. . . and the loss of life is terrible . . . I am glad you are able to continue your work at Tring,' he added. 'It is so nice to keep on with the work one is fond of.'[28] It was even better than that, as Dorothea wrote to Dr Lang with whom she had remained in close touch after his retirement. 'I love being in the country more than ever, and it is also very satisfactory to be able to work out some collections which there never seemed time for in London.'[29] Some of these had awaited her attention for years. She wrote reports on the tiny remains of *Murinae* (rats and mice) from Palestine and Crete,[30] on shrews from the larger western Mediterranean islands,[31] and the Pleistocene mole she had discovered in Sardinia.[32] She examined in detail Dorothy Garrod's finds from the cave of Shukbah, which had been sent to her in 1928,[33] and a collection of small mammals from the 'Cedars of Lebanon', as she poetically wrote in her 'Collections in hand' notebook. These were not fossils, however, but recent remains mainly of voles and shrews, which arrived neatly packaged in pellets of the long-eared owl.[34] Altogether during these war years, she produced ten substantial reports for publication.

In the early years of the war, Dorothea was able to visit friends and family. In the summer she went to her brother Thomas in Herefordshire while Leila came over to Tring when she could obtain the petrol for the drive from Waltons. The RAF had commandeered her great house for the duration of the war and Leila was confined to 'one small room and a kitchenette'. The hall, where a hundred people had once danced, was now the officers' mess.[35] Dorothea spent weekends with the Langs at their home in Dorset, until the loss of all their domestic help to the war effort put an end to such visits. She went to Sussex to stay with the Smith Woodwards. Sir Arthur could no longer walk and had lost his sight, but with the help of his wife was completing the story of Piltdown Man. Dorothea found him frail, although his mind was as active as ever. Four months after her last visit, in September 1944, the day after he dictated the last words of his little book on Piltdown, *The Earliest Englishman*,[36] he died.

Much as Dorothea loved her life in the country, the death of her old friend emphasized a growing loneliness. The previous year she had lost another good friend, the distinguished geologist and palaeontologist Guy Pilgrim. They had corresponded regularly, particularly on the Bethlehem fauna. Since his retirement from the Geological Survey of India in 1930, he had worked on a temporary basis at the Natural History Museum, studying and writing on faunal deposits in India, Europe, and North America. For both of them the correspondence was as much an interchange of gossipy news as palaeontological information. He had sent Dorothea a paper of his on *Artiodactyla* (even-toed mammals such as antelopes) and her flattering response elicited this from him: 'You have the wonderful faculty,' Pilgrim wrote, 'of making the person with whom you are talking (or writing) feel that after all he is not such a poor worm as he had imagined, which is a blessed power in this censorious world.'[37] They exchanged ideas, facts, and references as well as friendship, and his death in 1943 left a great void. Even her visits to London became infrequent when the Green Line bus service was withdrawn. Contact with her colleagues at South Kensington diminished throughout the war as they were called up or, like Helen Muir-Wood, who went to work for the Admiralty in Bath,[38] were seconded to government departments.

Leslie Bairstow, who had managed to avoid military service for more than three years, had become her main source of news and gossip. When he was finally called up, his refusal to serve led to his immediate arrest, court martial and imprisonment for three months. On his release he was summoned before the conscientious objectors' tribunal and ordered to serve out the war either on the land or working in a hospital. He chose the latter, but Dorothea was deprived of one of her chief lines of communication with the department. In the summer of 1944, flying bombs caused horrendous destruction in London and considerable damage to the Museum. It was closed to the public and on her occasional visits there, Dorothea found a dark, depressing place, with windows boarded up and dusty silent galleries.* She sent Dr Errol White, the Deputy Keeper,

* At the western end of the Museum, however, the government had taken over three galleries and part of a corridor for the top-secret Special Operations Executive. Here the SOE's Camouflage Section had their Demonstration Room, where the most extraordinary and deadly devices were constructed for use by agents all over the world. The NHM's archivists, Susan Snell and Polly Tucker, have written a fascinating section on this work in their book *Life Through a Lens:*

a small parcel of specimens, not that she urgently needed his views on them, but 'because this makes an excuse for writing to you and I hope getting some news of you in return. I hope that all is well with you both and that you will think of coming down here when the days grow a bit longer and warmer.'[39] Although Tring gave her safety, she revealed sadly to Dr Lang that she went to occasional meetings, 'but otherwise do not meet many people'.[40]

She tried to keep in touch with friends abroad. In spite of his lapses as an archaeologist, she was fond of Joseph Baldacchino and had for years contributed to his department's section of the Malta Museum's annual report. The terrible bombing of Malta in 1943 so alarmed her that she wrote to the Malta Government Office in London to enquire after him, at the same time donating £1. 10*s.* (about £40 in today's values) to the Maltese Units Benevolent Fund. Baldacchino did not appear on any list of casualties, she was told, but more enquiries would be made. In January 1944, Baldacchino at last made contact with her. He and all her friends in Malta, he reported, were safe and in good health, although the Malta Museum had been partially demolished through enemy action and much of the zoology collections severely damaged or totally destroyed. The fossil fauna remains, however, collected from Ghar Dalam and other sites over so many years, were almost unharmed.

The one thing Dorothea could be sure of was that it would be many years before she was likely to receive further remains from Ghar Dalam. The cave had been requisitioned at the beginning of the war for refugees. 'Subsequently,' Baldacchino told her, 'it was taken over by the RAF for the storage of petrol.' While the excavated trenches were being filled in to level the floor, Baldacchino was dodging round the military, trying, 'as best I could', he told her, to protect sections of the deposit and the exposed examples of bone breccia with sandbags and sheets of corrugated iron. 'To reinstate the cavern,' he wrote mournfully, 'will undoubtedly take a long time and a lot of trouble.'[41] His reward, two years after the end of the war, was to be appointed Director of the Malta Museum.

In 1944 she began jotting down ideas for the Bethlehem report, reading, making copious notes and examining what specimens were available. She

---

Photographs from the Natural History Museum 1880 to 1950 (London: Natural History Museum, 2003).

sketched out paragraphs and a structure, and although the geological account of the site was to be written by Elinor Gardner, Dorothea was clearly fascinated by how the deposit came to be on a mountaintop and had her own ideas. The one she was most inclined towards was that the animal carcases or skeletons were washed into a narrow basin by periodic flooding on the surface, 'even though or perhaps because, this entails visualising a totally different topography to that of the present day'. She argued that it was well known that great changes took place on the earth's surface between Villafranchian (pre-Pleistocene) and Early Pleistocene times, 'and it seems not too much to suggest that the Bethlehem deposit was originally formed at no great height above the sea . . . and that subsequently this whole section of country was elevated to its present height'.[42] She amassed a great quantity of notes and paragraphs on the various fauna, particularly the elephant, but with the bulk of the collection still in Jerusalem, she could do no more.*

Dorothea also began work on another major report, this one on the fauna of the Mediterranean islands, much of which she herself had discovered. Again there are quantities of notes and almost countless lists of fossil Mediterranean fauna and comparative lists of Palestine mammals, and several times she has begun to construct a table of species. Occasionally she would draft a few paragraphs and sometimes even several pages on general themes; then she would stop as her narrative became a series of questions on what was not known rather than what was. At the heart of her problem was this question: 'What relationship is there between the Pleistocene faunas of the different Mediterranean islands?'[43] She had found different species of elephants and hippos on Cyprus and Crete, others were known from Malta and Sicily, while *Myotragus* was found only in the Balearics. Her studies were not helped, she wrote, by the methods of earlier collectors who were unaware of the need to note and record exactly the possible stratification of deposits, or to record the association of particular animals, while 'the collecting and preserving of

* In 1956, Dirk Hooijer examined the Bethlehem material at the NHM and confirmed her dating of the deposit to between Villafranchian and Early Pleistocene. (See D. A. Hooijer, 'An Early Pleistocene mammalian fauna from Bethlehem', *BBM(NH)G*, 3: 8 (1958), 265–92.) As for the flints, they were examined in 1961 by the anthropologist Professor John Desmond Clark who concluded finally that they were the work of nature, not man. (See Dr John Desmond Clark, 'Fractured chert specimens from the Lower Pleistocene Bethlehem beds, Israel', *BBM(NH)G*, 5: 4 (1961).)

the microfauna was almost entirely neglected.' In *The Stone Age of Mount Carmel*, she had noted that 'It must be remembered that almost everything remains to be discovered regarding the faunal sequence of the Pleistocene of the Mediterranean Islands.'[44] Since working with Dorothy Garrod, she had recognized that her own solitary, pioneering cave work in the Mediterranean islands would be considered now as test trenches, not full-scale excavations, remarkably successful though she had been.

What Dorothea had completed, and assumed had been published, was the report on Gilf Kebir and Uweinat that she had written in 1940 for Oliver Myers. It was not until after the war that she heard about it from John Wilfrid Jackson, who, to her surprise, told her that the war had held everything up. Myers, it appeared, had gone into some 'special government job' and had not published the report. Myers was, however, back at the Institute of Archaeology in London and was about to 'get going' on Dorothea's and other neglected reports.[45] It was only then that he discovered that Dorothea had not numbered the identifications, and he wrote not to Dorothea but to Jackson. When told of this by Jackson, she replied directly to Myers, her apparent charm hiding wrath:

> In a letter dated 21st March from Dr JW Jackson he tells me you are now assembling the material for the Report on your Gilf Kebir explorations.
>
> I also rather gather that you are not satisfied with my little Report on the animal remains! Will it not be much simpler if you communicate with me directly? I understand that you complain that I have not put numbers against the specimens referred to in my report – Well when I received the animal remains in 1940 I got practically no information regarding them and, with two exceptions, not a specimen was numbered.[46]

Nevertheless, she did suggest that Myers should come to see her at the Museum, as she told Jackson: 'It is always easier to talk things over and I could find out just what he requires.'[47]

Myers' reply to her could not have been more contrite. 'I wrote via Jackson as I thought it both more practical and polite as he had sent the stuff to you. He appears to have transformed my simple query about the numbers into a complaint. I am only too grateful,' he continued winningly, 'to all the kind people, yourself among them who are so good as to help us with all these technical reports at all.'[48]

Their meeting was a great success. Dorothea took Myers to lunch and they discussed the report, but when Myers looked at the collection it was apparent that Jackson had only sent part of it to Dorothea; the rest, they assumed, was still in Manchester. As Jackson by now had retired from the Manchester Museum, Myers left it to Dorothea to try to extract the rest of the Gilf Kebir collection from Jackson's successor and went off to excavate in the Sudan.

CHAPTER 16

Into Africa

The devastation of the war echoed through the first years of peace. The damage to the Natural History Museum was bad enough, but the shortage of builders and materials delayed repair work by months. The exhibition galleries, and then only a few, did not reopen until the end of 1946 and the evacuated collections had to remain stacked in boxes until space could be found for them. In a rare idle moment Dorothea sketched a list of things to be done 'Post War', most of which involved travel: to Basle and Bologna to visit their museums, and to Malta and Sicily to collect at known Pleistocene sites and perhaps to discover new ones. In the frantic months following the war, however, this could only be a 'list of desiderata'.[1]

With peace, a deluge of fossil fauna was sent to Dorothea for identification. It started reasonably enough: five boxes of mammal bones from Gibraltar, more owl pellets from Syria; John Wilfrid Jackson sent a box of bird bones from British archaeological sites, while small collections from enthusiastic amateurs also demanded her attention. In April and August 1946 she received several boxes of 'Mesolithic mammals etc' from Khartoum and 'still more March 1947',[2] sent by the archaeologist Anthony Arkell. To top them all, in June 1946 the eleven cases containing Stekelis's specimens from the Bethlehem elephant pit arrived. Chasing Oliver Myers' missing collection in Manchester and fulfilling anything on her 'list of desiderata' was simply impossible.

She was now sixty-eight and long past the official age of retirement, but as an unofficial worker, that was a decision, she was assured, that would be left to her. Furthermore, she did not have to return to London; as she told Dr Lang in December 1946, 'I feel very grateful to be still at Tring and hope very much to remain – when working at large specimens I go to London twice a week and hope to manage this way.' But in addition to all the work already piling up, she was faced with an area of research that for her was quite new. 'I am,' she continued in her letter to Lang, 'to be turned on to work at some of Dr Leakey's mammal collection.'[3] Louis Leakey's famed work was in East Africa, a region far outside her usual horizons. 'It is a big switch over,' she told Lang, 'and no light task to familiarise myself with the Recent and Pleistocene.'[4] That, for Dorothea, meant an exhaustive and exhausting study of all the available literature on the region and countless hours spent acquainting herself with the relevant collections in the Museum. Her nephew David, Thomas's son, visiting her on leave from his work as a barrister with the Colonial Legal Service in Nigeria, remembers the bookcases in her panelled office being filled with works on Africa – scientific, historical and travel.

The collection Leakey sent to her was a very large quantity of fossil pigs. The animals themselves were of considerable interest, but what made the work so significant was that their remains had been found in sites associated with early man. Pigs evolve relatively swiftly and Dorothea's study of them, Leakey hoped, would be critical in helping to date his spectacular discoveries.

For over twenty years, Louis Leakey had excavated in Kenya and Tanganyika (now Tanzania) in his search for the origins of man. Dorothea had known him for some time from his visits to the Museum and he was now the Curator of the Coryndon (later the National) Museum in Nairobi. Exceptionally, he addresses her in correspondence as 'Dorothy', while even some of her oldest friends still call her 'Miss Bate'. Leakey was a skilled publicist but really excelled himself when in 1946 he appeared on the new medium of television, holding up his latest great discovery. It was the jaw of a Miocene ape, later named *Proconsul nyanzae*, which he had discovered on Rusinga island in Lake Victoria.[5] By then, Leakey and his second wife Mary were already famous; their discoveries of Stone Age cultures, great extinct mammals, and apes millions of years old were of enormous interest to academics and newspapers alike.

The complexities of African prehistory were such that Leakey had become convinced of the need for prehistorians and palaeontologists of Africa to get together at the earliest moment to discuss 'mutual problems' on man's past on the continent. It was an enormous task but, as soon as the war ended, he convinced the Kenyan government of its importance and set about organizing the first Pan-African Congress on Prehistory. It was to be held in Nairobi in January 1947. Dorothea was invited by Leakey to attend and contribute, but by the autumn of 1946, only her colleague, Dr Kenneth Oakley, had been officially sanctioned. The problem for Dorothea lay not with the Museum, as she wrote to Dr Lang on New Year's Eve, just two weeks before the Congress was due to open, but with the Colonial Office, which 'doesn't see the importance of Natural Science and wouldn't put up the funds for fares'.[6] If the Colonial Office did not appreciate the importance of the Congress, others did. In the unlikely guise of a fairy godmother, the Kenyan government itself, encouraged by Leakey, issued her with a direct and 'special invitation'.[7] With scarcely time to pack, let alone buy clothes suitable for equatorial Africa, Dorothea literally took off for Kenya.

The Congress opened formally on Tuesday, 14 January. The dress code for women, noted Sonia Cole, a family friend of the Leakeys and biographer of Louis, was hat and gloves in spite of the heat. Outside the Town Hall in Nairobi where the main sessions of the Congress were to take place, flags of all twenty-six participating nations were flown, except those of Portugal and Egypt, which could not be found. When the Portuguese and Egyptian delegates discovered this, their protests, Sonia Cole recounts, led to all the flags being hauled down.[8]

Dorothea's main contribution to the formal part of Congress was to take place on the day after the congress opened. The audience reads like a mid-twentieth-century *Who's Who* of archaeology, palaeontology, anthropology, and anatomy. To find all those disciplines attending the same conference was extraordinary in itself, but it only emphasizes how great was the need for collaboration on the study of African prehistory. The delegates included Wilfred le Gros Clark, Professor of Anatomy at Oxford; the redoubtable archaeologist and prehistorian Abbé Breuil, who was elected president of the Congress; Camille Arambourg, the French palaeontologist whose later expedition to Abyssinia revealed spectacular hominid finds; the eighty-year-old doyen of palaeontology, Professor

Robert Broom; and Professor Raymond Dart, who discovered in South Africa the 'Taung baby', a tiny fossilized skull about two million years old. The assistant secretary to the Congress was the young John Desmond Clark, on the brink of a pioneering and celebrated career in archaeology and palaeoanthropology. He later became Professor of Anthropology at the University of California at Berkeley.

Dorothea's paper was titled 'The Pleistocene mammal faunas of Palestine and East Africa'. In this she drew brief comparisons and correlations between the fauna and concluded that Palestine was important to the knowledge of African faunas as it should 'supply a key to the dating and composition of the mammal migrations between Asia and Africa'.[9] The paper was well received, but Dorothea only learned that later. The arguments over who would fund her fare had so delayed her that she missed both the opening ceremony and reading her own paper. Kenneth Oakley read it in her place. She finally arrived on 17 January, to find herself surrounded by friends and acquaintances. There was of course Dr Oakley, who with others, including le Gros Clark, would not long afterwards reveal the Piltdown skull as a hoax. She had known le Gros Clark and the Abbé Breuil for years. Other friends included Professor Zeuner, who had assisted her Mount Carmel analysis; John Waechter, whose collections from Transjordan she had identified and Anthony Arkell, on whose large collections of faunal remains from a Mesolithic site at Khartoum in the Sudan she was currently working. Her reputation was high, Desmond Clark noted, 'due to her important work on the fauna from the Mount Carmel caves'.[10]

After an intense week of talk, study, debate and argument, the Leakeys had organized excursions for the delegates. Dorothea's handwriting in the little notebook she kept on this trip may be shaky at times, but she ignored her uncertain health and sixty-eight years; this was a totally new part of the world for her and she exploited it to the full. On the middle Sunday of the Congress, they all went off to Olorgesaile in the Kenya Rift Valley about thirty miles southwest of Nairobi.

In quavery pencil, in this case because of the movement of the car rather than any infirmity, Dorothea noted the creatures they passed: wildebeest, impala, a herd of giraffes, big bustard, white ant hills, zebra, roller, twenty-four Grants gazella, secretary bird and guinea fowl.[11] Olorgesaile was a place of spectacular scenery, a great extinct volcano with at its foot the site of a vast Pleistocene lake, now quite dry. In the early 1940s the Leakeys had found there an extraordinary concentration of Acheulian* flint implements of the Palaeolithic hunters whose home this had been around four hundred thousand years ago, as well as the fossil bones of the creatures they had hunted, giant pigs, baboons, hippos, and giraffes.† They decided to leave the flints in place where they found them and turn Olorgesaile into a great open-air museum. Huts were built so visitors could stay the night, shelters were erected over the excavations, and walkways suspended over the greatest concentration of hand-axes. It became a National Park and the delegates' visit was to mark its official opening. That took place at 1.30 p.m., under a burning sun. 'Speeches,' Kenneth Oakley noted privately, 'ought to have been banned!'[12] The proceedings, however, were considerably lightened when the Abbé Breuil, in the middle of his over-lengthy presidential address, found himself in the farcical situation of his trousers falling down. 'He was,' wrote Sonia Cole, 'wearing his safari outfit – khaki shirt, knickerbockers and outsize braces which broke under the post-prandial strain; the dignity of his address was completely spoilt by his efforts to hold up the knickerbockers, and by the attempts of his audience to control their laughter.'[13]

On the Saturday after the Congress ended, Dorothea and about fifty of the delegates set off on a 1,000-mile expedition to Tanganyika to visit three remarkable sites. They journeyed in a convoy of lorries and cars, with Dorothea determinedly noting the wildlife on the way in almost illegible writing as they jolted over potholes and up vertiginous mountain roads. Their first destination was 7,500 feet up at the Ngorongoro Crater. At twelve miles across and 2,000 feet deep, Dorothea could not have imagined anything like it. A shallow soda lake lay in the middle, pink with flamingos, while in the surrounding groves of yellow-barked acacias were herds of elephants, wildebeests, elands, and zebras. On the edge of the crater she observed two sunbirds, both metallic green, but one with a yellow upper breast, the other with a scarlet band round its lower throat. They were feeding in 'yellow pea shaped flowers', on a bush about ten

* Lower Palaeolithic culture, which preceded Mousterian, dated to about 1,500,000–150,000 years ago. Its name is derived from St Acheul in northern France where objects of this culture were first found.

† Louis Leakey thought then that the Olorgesaile site dated from about 125,000 years ago. It was not until the 1960s that its true age was established.

feet high from which she cut a specimen and pressed it into her notebook, where it still remains.[14] She also noted 'Rhino'. That evening a number of them invaded the expedition's food store, 'and ate all the pineapples', according to Mary Leakey, 'and most of the vegetables'.[15]

Dorothea and the more elderly scientists spent the night in log cabins; everyone else slept in tents. In the morning they began the descent from Ngorongoro to their next destination, the famous Olduvai Gorge, with its evidence of habitation by man more than a million and a half years ago. Here in the 1930s Leakey had begun to excavate the stone industry that he later called Olduwan and discovered a rich, varied and unique fossil fauna. It was a twenty-five-mile drive to Olduvai. They stopped first above the thirty-mile-long gorge to admire its massive 300-foot cliffs and geological sections. Down on the floor of the gorge in the sweltering heat, the Leakeys led the delegates round the excavation sites, showing them the beds of fossil fauna and stone implements, which they had left untouched as they had at Olorgesaile (the delegates were allowed to collect implements and bones), and explaining the geology of the gorge.

Early the next morning they left for the caves at Kisese. Dorothea, who had hoped to spend another night in a relatively comfortable log cabin, notes laconically and none too happily, 'arrived Kisese camp – tents'.[16] They had come to view the rock art of the caves and rock shelters, Palaeolithic paintings of ostriches and white rhinos, elands, giraffes, elephants and humans. One of the caves was a walk of two miles from the road, and then a climb of 200 feet up a steep cliff, all in the noonday heat. Leakey advised Dorothea and Professor Broom, who was twelve years older than her, not to attempt the climb, but they ignored him. Broom was, according to Leakey, an unforgettable sight in his customary garb of 'dark suit, wing collar and butterfly tie', while Dorothea, Leakey unsympathetically remarks, 'made very heavy weather of the ascent'.[17] She was not too exhausted, however, to note a bush buck and the tracks of antelopes in the undergrowth as they climbed towards the caves. That night as they sat round the camp fire, the inexhaustible Abbé Breuil gave a dissertation on the paintings, comparing them with those of Spain and of France, particularly the newly discovered cave paintings at Lascaux.[18]

On their return to Nairobi, Dorothea stayed first with the Leakeys, and then with Sonia Cole's mother-in-law, Lady Eleanor Cole, on whose land by Lake Elmentaita to the northeast of Nairobi was an Acheulian site with mammal remains. The lake at first light was glorious: 'Marvellous groups of flamingos rosy in early sunlight,' Dorothea noted as she sat in the cool of the verandah after breakfast, also 'other large groups of waders, smaller brown and white plovers . . . all a wonderful sight'.[19] After this brief rest, on 5 February, the Leakeys, Kenneth Oakley and John Waechter arrived in two cars, which also carried bedding and food, to take her off to Rusinga island in Lake Victoria where the *Proconsul* skull had been discovered.

It was a journey of two days to Lake Victoria, stopping at places of interest on the way including Songhor, a Miocene site from which Dorothea and Oakley collected rodent jaws, the tiny specimens difficult to spot in the glaring sun. A motor launch was waiting to take them across to Rusinga island on which they were to camp for five days. The boat would also take them round the island to save them walking great distances across it. At least, Dorothea must have thought, she was unlikely to be bothered by seasickness on a lake. They excavated at a number of sites, both Dorothea and Oakley collecting vertebrate and invertebrate fossils, and were taken to where the *Proconsul nyanzae* remains had been found. It was an amazing and exhausting experience. Dorothea's notes are a brief record of what they did. Kenneth Oakley records rather more, and it is apparent that the Leakeys' forceful personalities and energy caused discomfort to both. Mary Leakey 'scoffed' at one of Oakley's excavations on Rusinga, while (and this I found very hard to read) 'D. B', Oakley wrote, 'wept – overtired and feels L's are so exacting'.[20]

What happened next, Dorothea chose not to record anywhere. Kenneth Oakley and the Leakeys, however, did and at length. Late on the afternoon of 10 February, they were returning on the boat to their campsite when the helmsman sailed the boat 'gently but firmly', according to Mary Leakey, 'onto a submerged rock', where it stuck fast.[21] It began to bang against the rocks and Louis feared 'it would be smashed to smithereens. It was desperately important to get it off the reef as quickly as possible.'[22] The able-bodied in the party stripped off and went overboard to attempt to refloat the launch, but to no avail. Their predicament was spotted by local fisherman of the Luo tribe who were, according to the Leakeys' accounts, already completely naked. Louis, who evidently relished the scene, painted it vividly. 'Soon there were some thirty naked men and a few women milling round in the water, desperately trying to get the boat afloat. Meanwhile, poor Dorothea Bate, who was very Victorian in her outlook, kept her eyes tightly shut to avoid the sight of the naked men, while Mary periodically assured her that all would be well.'[23] What Mary and Louis did not know was that Dorothea was an appalling sailor and the violent movement of the boat as they tried to free it was probably much more the cause of her tightly closed eyes than were the supposedly naked Luo. Particularly as, according to Oakley, they were not. Far from their rescuers being naked fishermen, Oakley records that Leakey had gone ashore 'and collected football players!'[24] to try to free the boat. All the people in Oakley's photographs taken throughout the trip are visibly and modestly clothed.

By now it was dark, the launch could not be moved and sufficient men could not be enlisted to help until the morning. There was no alternative for the Leakeys but to take Dorothea and Kenneth Oakley back to the campsite, a nine-mile walk across the island in the dark. The launch was steadied; Oakley rolled up his trousers, clambered off the boat and waded ashore. That was not an option open to Dorothea. For Leakey, getting Dorothea off the boat was 'the most difficult manoeuvre' of all. As she could not wade ashore, she would have to be carried. 'The only men available to carry her were stark naked, and she simply *had* to open her eyes to climb onto their shoulders.'[25] Leakey is clearly hugely enjoying Dorothea's discomfort (whatever its cause), and he tells a good, if embellished story, but it was written, unlike Oakley's contemporaneous account, many years after the event.

Once ashore, the Leakeys' attention turned to Oakley who, Mary observed, 'had kept a reasonably stiff upper lip while being shipwrecked, [but] was very perturbed by the idea of walking across a bit of wild Africa in the dark . . . though we assured him that on the thickly populated Rusinga Island there were no particular perils'. Dorothea, 'freed from the embarrassment of the naked Luo', according to Mary, 'perked up and made light of the long walk'.[26] The one lantern they had was carried next to Dorothea, but somehow, Mary noted, 'Kenneth kept popping up between her and the light, where he felt a bit safer.'[27]* According to

* Mary's characterization of Oakley shocked his family as he had always regarded her as a friend. The weakness he apparently showed during that walk in the dark across Rusinga island is attributed by his son Giles to his lack of muscle strength and coordination, the legacy of a terrible illness he contracted as a young man. It was diagnosed then as polio, although in later life he developed multiple sclerosis.

Oakley, however, Dorothea was so exhausted that she nearly collapsed. Apart from blisters, neither Dorothea nor Oakley suffered any ill effects the next day. The launch had been refloated by a storm wave in the night and its passengers re-embarked with rather more dignity than they had left it.

Dorothea returned home with her little notebook full of requests from her conference colleagues. Many were for papers and casts for Leakey and the Coryndon Museum, but at least a dozen delegates asked her to find and send them various reports or enquired whether she could put them in touch with a particular scientist somewhere else in the world. Other delegates asked whether they might send collections to her for identification. In all there are nearly *forty* separate requests and against about three-quarters of them she has placed a tick. As a note to herself, she has jotted down '"Out of Africa" Karen Blixen', but with everything else demanding her attention, the book remained unticked.

Dr Lang was among the first to hear an account of her travels. 'The Pan African Congress was a grand success,' she wrote; '26 nationalities represented and it all went so smoothly and happily – I enjoyed it all more than I can say and received kindness and hospitality without measure. I visited all the intensely interesting sites and did a little collecting, camped on Rusinga Island, saw a lot of country, big game etc – all a marvellous experience.' There is, of course, no mention of naked tribesmen, footballers or even *mal de mer*. 'It was very difficult to tear myself away from Africa,' she told him, 'and I only reached this country again on 11 March.' The pressure of work was immediate. Not only had she brought back so many requests from the delegates, but 'mountains of correspondence'[28] were waiting for her at Tring. New collections arrived for her, including those from the Pan-African Congress delegates, from Syria, the Sudan and more from Jisr Banat Ya'qub. In May, John Desmond Clark, the assistant secretary of the Congress, brought in person part of his collection of fossil fauna from southern Somalia, scanty remains that he handed over with no identifications. With patience Dorothea explained the importance of labelling the bones with the site, locality, and level at which they were found. 'She was so interested,' wrote Desmond Clark; '. . . [she] was the kind of person who made you feel it was good to be with her.'[29] He sent the missing information on to her, with the promise of another three or four more bags of material, which he had still to label.

Most pressing of all were mammal bones from Lachish, which Olga Tufnell had asked her to determine. These arrived at the end of April 1947 in two tea chests and four assorted cases. For comparative purposes Dorothea asked Olga to send her from the Institute of Archaeology a collection from Tell Ajjul, a Bronze Age site in southern Palestine, excavated before the war but never previously examined. By the beginning of December she had completed reports on both collections, sent the Lachish report to Olga Tufnell and arranged to meet her at the beginning of January 1948. The illness of an aunt caused Olga to write to Dorothea to rearrange their meeting; she did not receive a reply for two months. John Desmond Clark also wrote to Dorothea in January as he had urgent need of her report on his collection for his thesis, which was due by April. He heard nothing. A winter cold had turned to flu and then pneumonia. This time Dorothea did not stay with Leila. Ill as she was, she managed to travel down to Westcliff-on-Sea in Essex to stay in the house of a Christian Science healer, Gertrude Gunson and her husband. Mrs Gunson had trained as a medical nurse before becoming a Christian Scientist, and both Gunsons are remembered today as warm and kindly people.[30] But Dorothea's recovery was slow and it was not until February that she was strong enough to write. Her notes on Desmond Clark's collection, which she had examined before Christmas, were sent to her, together with reference works. She wrote out her brief report in longhand, sending it back to the Museum to be typed.

She was still in Essex when she heard again from Oliver Myers. He wrote to her in great excitement from Khartoum. He had found three or four scraps of bone from his Gilf Kebir collection, which, he told her, had been tentatively identified 'as fish. Now if there were fish in the lakes of Gilf Kebir, it tells a very different story from what we supposed. I had thought purely temporary water as a result of perhaps two or three heavy successive rains, but fish, and apparently largish ones at that!' He could scarcely contain his enthusiasm. In the absence of any news about the missing bones in Manchester, he even suggested that as 'This seems such a vital clue . . . I am loath to abandon it even if it means going to Manchester myself to investigate. I could recognise the bones anyway by their colour and soil attachment if they can be found.'[31]

Dorothea returned to Tring before she replied to Myers, defusing his archaeological ardour with a cool lecture on the facts of the fish. 'It does not seem to be surprising,' she told him, 'to find fish remains at the Gilf or Uweinat.' She calmly quoted evidence which suggested the existence of a prehistoric lake with tropical vegetation; also dry lake beds had been discovered at the head of the wadis in the Gilf. 'I have mentioned these in my report (2nd ed) rewritten after I saw you in 1946, but I am not sure if you have had a copy of this.'[32] Myers' disappointment must have been acute, but Dorothea has merely noted that six weeks later she received a 'polite note' that he would come to the Museum one day. However, there is no record that he did.* As for the report that Dorothea had written for Myers on the fauna of Gilf Kebir and Gebel Uweinat, the manuscript remains in a file in the Natural History Museum, still unpublished.

* Nearly a year after this, Jackson wrote to tell her he had found some of the missing bones from the Gilf and proposed visiting her, bringing them with him. Dorothea was on leave and Jackson did not come. In 1970, Robert Savage, later Professor of Geology at Bristol University, came across a copy of Dorothea's manuscript, complete with bibliography and apparently ready for publication. He thought it still very useful and sent it to the distinguished archaeozoologist, Dr Juliet Clutton-Brock at the Natural History Museum, who was delighted the report had been found and thought it would be an excellent idea for it to be published. For whatever reason, by the 1980s still nothing had happened. Dr Clutton-Brock then showed it to a visitor to the Museum, Professor Joris Peters. He assumed Dorothea's report was never published because it was intended to go with Oliver Myers' archaeological findings, and he could find in the literature no evidence of a detailed report from Myers, only a preliminary one published in 1939. (*The Geographical Journal*, 93 (1939), 287–91.) He subsequently discovered that Myers had actually written three reports on the artefacts he discovered. 'These manuscripts,' Professor Peters wrote, 'are all undated and unpublished, and accompany the artefacts, housed in the Musée de l'Homme, Paris.' (Professor Joris Peters, personal communication.) Professor Peters managed to trace the information Dorothea lacked and restudied the faunal remains in the Natural History Museum. In 1987 he wrote a report on his findings. (Joris Peters, 'The faunal remains collected by the Bagnold-Mond Expedition in the Gilf Kebir and Gebel Uweinat in 1938', *Archéologie du Nil Moyen*, 2 (1987), 251–64.)

CHAPTER 17

Officer in Charge

Dorothea was now nearly seventy, and it would be reasonable to expect that her remaining years at Tring should have been those of quiet consolidation. It was not to be. Demand for her rare expertise was increasing from archaeologists and prehistorians internationally, while the Museum itself discovered it had need of more than her scientific skills. In mid 1948, the Officer in Charge at Tring was James Dandy, a botanist seconded to Tring from South Kensington. He had served four years there and was as anxious to return to South Kensington as his department was to have him back. He was later to become Keeper of the Botany Department. His predecessors since 1938, when Tring had been acquired by the Natural History Museum, were equally distinguished. The first Officer in Charge had been the zoologist Terence Morrison-Scott, later to become Director of the Natural History Museum. War service brought his tenure at Tring to an end in 1939 and he was replaced by John Norman, the Deputy Keeper of Zoology, whose ill health precluded military service. On his death at the age of forty-six in 1944, he was succeeded by Dandy. In 1948 – when the Museum was still struggling to return to normality after the war, with every member of staff overburdened with restoring collections and exhibition galleries and dealing with new acquisitions, there was no apparent successor to Dandy.

It may well have been the Keeper of Geology, Wilfred Edwards, who suggested to Norman Kinnear, the Director, that their next Officer in

Charge was already based at Tring, someone who knew the Museum intimately and who was a safe and enormously respected pair of hands. If there was a problem, it might be her age, which was a mere four months off her seventieth birthday, but, that aside, surely in Dorothea Bate there could be no better appointment. And so in July, to Dorothea's intense amusement, 'On the recommendation of the Director the Trustees decided that Miss DMA Bate (An Associate* of the Department of Geology) should take Mr Dandy's place as Officer in Charge of the Tring Museum.'[1] She was to become a Temporary Senior Scientific Officer and her salary was to be more than doubled to £525 per annum (about £12,300 in today's values), rising by £25 a year to £750 (Dandy had received over £1,000 p.a.). For the first time in her life she was to be given managerial responsibility.

The lengthy memo outlining these responsibilities was sent to her by the Museum Secretary, Thomas Wooddisse. She would, he informed her, be answerable to the Director of the Natural History Museum for all matters regarding the staff and exhibition galleries at Tring; she had to keep a record of visitor attendance to the galleries and of staff absences; was responsible for organizing maintenance of the buildings, and had to ensure that an accurate record of all financial matters was sent to the accountant in South Kensington on a monthly basis. She had the power to make minor changes to the running of the place and was told that any 'suggestions for the improvement of the administration of the Museum will be welcomed', but any major changes had to be referred either to the Director or to Wooddisse. She continued to be accountable to the Keeper of Geology for her own scientific work, but she was now also responsible for the curating and preservation of the zoology and entomology collections at Tring, and to the Keeper of Botany for any botanical work necessary.[2] She was expected to work forty-four hours a week and, in spite of her senior position, was permitted just forty-five minutes for lunch.

Dandy was to return to South Kensington on 15 November, although Dorothea was not informed of this until three days beforehand. In the three months since the announcement of her appointment, Dandy does

* Paid unofficial scientific workers like Dorothea had just been given the new title of Associates; those who were unpaid were Honorary Associates.

not appear to have shared any management information whatsoever with his successor, or even mentioned handing over his responsibilities to her. 'I am sure that Mr Dandy and I will be able to arrange matters harmoniously together,' Dorothea wrote to Wooddisse. However, 'I have been told nothing about the running of things! But I think all will go quite easily only please forgive any blunders I may start with! I shall of course keep the Director and yourself informed of all important happenings here.'[3]

Most of the 'happenings' appear to concern insufficient cabinets for the collections, and problems and delays over necessary renovations to the building, all of which had to be put out to competitive tendering, no matter how minor. She dealt with misdemeanours by the non-scientific staff: 'Mr Smit "on the mat" rather turbulent – has since attended to regulations.'[4] She put in hand a new telephone system, prepared for the regular visits by Norman Kinnear, the Director, to Tring, and in August 1949 allowed the press into the Tring Museum for a privileged view behind the scenes. The local paper, the *Buckinghamshire Herald*, described her as 'an elderly lady of brisk and business-like manner' who would not 'allow her "backroom boys" to be disturbed by people who have only their curiosity to satisfy. At the same time, Miss Bate is keen that the general public should have some idea of what the Museum's experts are doing. For that reason she gives a happy welcome to the Press – provided the reporter promises not to be "sensational" in what he writes. All scientific people,' he added, 'have a horror of cheapjack publicity.'[5]

Dorothea may have been seventy, but she was determined to demonstrate that neither her administrative responsibilities nor her science would suffer through her increased activities. She made the thirty-mile journey to London (once again on the bus) for her scientific work and meetings at least twice a week and often more. The Museum still could not afford typewriters for individual scientists, so she bought her own.

To the post-war palaeontologists and zoologists just starting their careers in the Museum, Dorothea was immensely kind and patient, sharing her encyclopedic knowledge of fossil mammals, and challenging them to use evidence and imagination. There was still so much that she would have liked to investigate, in Britain, the Mediterranean, the Near East and Africa, but now she simply did not have the time. What she could do, however, was encourage and support others, particularly the young scientists who joined the Museum after the war. 'She was,' the retired palaeontologist Dr Ellis Owen remembers, 'the most exciting person, always talking about something, making you think.'[6] She was seemingly full of energy, her slim, distinguished, white-haired figure forever 'rushing about' from one task to the next. Always neatly dressed in a well-tailored suit or a 'pretty blouse' and skirt, she charmed her colleagues and managed to get things done, or at least the most urgent. When colleagues disagreed and 'were hardly speaking, she would bring them together – she was a catalyst'.[7] Few knew she was a Christian Scientist; she would never talk about herself or her family, always turning the conversation away from anything personal, although she took great interest in the well-being of others. *In extremis*, if she had perhaps dropped something, she might be heard to utter, 'Oh damn it,' swiftly followed by, 'Oh, I shouldn't have said that, should I!' and a 'wonderful smile'.[8]

Collections continued to stream in, and she wrote short reports on material from Britain, Egypt, and Rhodesia, but she had so much work in hand that many boxes remained unopened. It was Leakey's collection from East Africa that presented her with the greatest problems. Fossil pigs had been discovered by a number of palaeontologists in South and East Africa. For years there had been confusion and argument over the nomenclature of some of the extinct species, not to mention the age of the sites they came from, and the arguments were very public, reverberating through the considerable literature and correspondence on fossil *Suidae*. Leakey himself did not help. Immensely charming, charismatic and utterly overwhelming, he wanted everything done at once and preferably his way.

The distinguished vertebrate palaeontologist, Professor Camille Arambourg, had published in 1947 a study of the *Suidae*[9] with which Leakey disagreed so fundamentally that, as he wrote to Dorothea in June the following year, he proposed to meet Arambourg to discuss the issue. He added, 'As you are now working on all this pig material, I would naturally like you to join us in this friendly discussion.'[10]* It would evidently be

* The heart of this disagreement was the description of a great quantity of fossil pigs, which Arambourg had discovered at Omo in Abyssinia and ascribed to the genus *Metridiochoerus*. Leakey was equally certain they should be ascribed to the species *Pronotochoerus jacksoni*. These were by no means the only species whose nomenclature was questioned and even a cursory study of more recent literature reveals the problems. (See for example, J. M. Harris and T. D. White, 'Evolution of the Plio-Pleistocene African *Suidae*', *Transactions of the American Philosophical Society*, 69: 2 (1979).)

anything but, and with some disbelief at this proposal, she scrawled on Leakey's letter, 'talk talk talk won't do it!' On 9 July she fired off to him a letter of her own so furious that, in the short term at least, it effectively silenced Leakey. When she needed advice from experts, she told him, she would ask for it; what she could not tolerate was anyone attempting to interfere with her work, particularly as, just three weeks previously, he had promised not to:

> As I told you, *I* do not wish to discuss the matter at this stage. Furthermore do you think that a man of Professor Arambourg's standing and experience will appreciate being told that he is all wrong and being asked to alter his considered opinions which have only just appeared in print? Should further material and study cause me to differ from some of Professor Arambourg's conclusions, I would of course let him know and give my reasons for so doing when publishing. You speak of clearing up the whole Pig question satisfactorily; nothing will do this but more complete material.

And then with a flash of her own considerable charm, she added, 'I am so sorry if I seem to be temporarily keeping you from playing with your favourite Pigs! But you pressed me to take on this work.'[11]

Just two weeks later she received a letter from Dorothy Garrod, who was digging at Angles-sur-l'Anglin near Vienne, asking her to examine another significant collection. A friend and colleague of Garrod's, Germaine Henri-Martin, was excavating a cave at Fontéchevade where she had found a human skull of considerable importance. The dating of it depended largely on the fauna that accompanied it, of which there was a very great deal. Germaine Henri-Martin, Garrod wrote persuasively, 'feels there is no one else who could really do the job'.[12] It would have been absorbing work with a major published report at the end of it, but it was a simply impossible task. 'I would gladly help her if it was possible,' Dorothea replied to Garrod,

> but it would be hardly fair to take on a job that I might have to postpone indefinitely.
>
> Since I was put on to work at some African mammal collections

> I seem to be busier than ever. My knowledge of African geography and zoology was weak and I had to start on these, now have at least three collections of fossils to deal with!
>
> The literature on French Cave faunas alone would be a big job – so I am afraid I must refuse – with regret.[13]

Apart from Leakey's favourite pigs, the two other African collections that so occupied Dorothea were Arkell's Khartoum finds and a collection from Cyrenaica in northeastern Libya, sent by the Cambridge archaeologist, Dr Charles McBurney.* During the war Anthony Arkell, the archaeological adviser to the Sudan government, had served with anti-aircraft forces and had spent much of his time sitting in a trench dug into a mound near the railway station in Khartoum. He had originally thought that the mound dated from the siege of Khartoum in 1885, but then realized that, in addition to recent material, it was largely formed of the debris from a Stone Age settlement: flint and bone tools and spearheads, fossil bones and pottery fragments.[14] He excavated the mound (which at the time of the early settlement had been a sandbank on the banks of the Blue Nile) in 1944 and 1945, and sent his entire collection of animal bones and fragments to Dorothea.†

Arkell published a preliminary report on his excavations in the journal *Antiquity* in December 1947. It is a straightforward, narrowly focused account of this Mesolithic site with little context. With Dorothea now involved with his substantive report, all that was to change. During the many hours that she spent discussing with Arkell the implications of his finds, she taught him to look not just at the individual artefacts and

* McBurney, a former pupil of Dorothy Garrod, was then a Fellow of King's College, Cambridge. He was appointed Professor of Quaternary Prehistory at Cambridge University in 1977 (a personal chair) and was to have the Geoarchaeology Laboratory there named after him.

† He initially sent a selection of the bones, animal and human, to be examined in Cairo by the man who had performed the autopsy on the mummified Tutankhamun in the early 1920s, Dr D. E. Derry. Although the tools seemed to indicate an early date for the site, Arkell thought it was not impossible that a people who could make and decorate pottery might have learned how to domesticate animals. At Mount Carmel, Dorothea and Dorothy Garrod had found evidence of domestication and the beginnings of agriculture but there was a total absence of pottery. Although Derry found no evidence of domestication in the animal bones (he identified crocodile, porcupine, hippopotamus, warthog, and buffalo), he was unable to determine all the specimens. There was only one person with the expertise to resolve the matter and Arkell accordingly sent the whole collection to Dorothea.

faunal remains, but at all relevant links and relationships across the region and beyond; northern Africa geographically is a bridge, linking Europe, western Asia, and southern Africa. She challenged Arkell in the same way that she challenged the young scientists in the Museum: to use both evidence *and* imagination.

Dorothea identified twenty-two mammals and five reptiles, including elephant, rhinoceros, river turtle, and the extinct reed-rat, *Thryonomys arkelli.** These twenty-seven species were not just an interesting list of long-dead fauna. She showed Arkell how they could be used as a key to understanding not just the habits and habitat of the early Khartoum settlers, but the peoples and climate of a great expanse of northern Africa. Sudan, the largest country in Africa, covers a million square miles, from Egypt in the north to Uganda in the south, with the great river Nile at its centre. Dorothea posed a series of questions that her examination of the fauna might answer, all of them fundamental to the discipline that is now called archaeozoology: what food did the settlers eat; was it procured close by or did they hunt far afield? What animals inhabited the region at the time; are they the same as today or similar to those in southern Sudan, or was this an altogether different and perhaps extinct fauna? What indications do they reveal of the environment and climate, of the age of the deposit or the interrelationships with other parts of Africa, and was there evidence of domestication? In the preface to *Early Khartoum*, Arkell acknowledges Dorothea's influence. 'It is thanks largely to her that it is now not the mere record of an excavation which had to be undertaken by the Sudan Government Antiquities Service, but a work that will have to be taken into consideration by prehistorians in north Africa and elsewhere, and one that I hope will encourage others . . . to elucidate still further the early connexions between the Nile Valley and lands far to the west of it.'[15] Dorothea's understanding and interpretation of the fauna

* She had identified this in her preliminary examination of the material in 1947 and published a report on it immediately. (Bate, 'An extinct reed-rat (*Thryonomys arkelli*) from the Sudan', *AMNH*, 11: 14 (1947), 65–71.) A strong swimmer and diver, the reed-rat needed a damp climate, reed-beds, swamps, and thick jungle. Its extinction, she noted, indicated great climate changes covering much of Saharan North Africa and spreading as far south as Khartoum. It also raised the possibility of a faunal change occurring since the habitation of the site. She discovered that it was related to a recent reed- or cane rat, which is now found across southern Africa from Angola to Mozambique, while other fossil species of *Thryonomys* had been found to the north in central and western Sahara.

caused Arkell, as he noted in *Early Khartoum*, to rewrite the summary and conclusions of his entire report.

Even before *Early Khartoum* was published, Arkell had returned to the field and was excavating two Pleistocene sites on the Blue Nile, Singa and Abu Hugar. All faunal remains from both of these went straight to Dorothea.*

In the midst of all this work, in the spring of 1949 she became ill once more, but this did not seem to be a recurrence of either malaria or pneumonia. She had time to organize in advance the smooth running of Tring and to leave detailed instructions for her staff before retreating to the ministrations of the Christian Science healer, Gertrude Gunson, in Westcliff-on-Sea. She was away for a fortnight but there is no record of what was wrong. On her return she resumed work with apparently unflagging enthusiasm, although now she used a walking stick and at times appeared to be in pain. She tried to pace herself, taking a few days' leave every month or so, but she was scarcely allowed to rest.

In May 1949, while she was still working her way through the Blue Nile material, Arkell sent her between two and three thousand specimens of fossil vertebrate remains from a site at Esh Shaheinab about forty miles from Khartoum. When Arkell came to write his archaeological report, Dorothea's influence is very clear. As he had learned from her with the Early Khartoum material, he used the material to look at the wider picture, at what the remains revealed of the relationship with other early settlement sites in North Africa and Egypt. For Dorothea the bones were of enormous interest. Esh Shaheinab was a Neolithic site; when she compared its fauna with the much older Mesolithic remains from Khartoum, she discovered 'a revolutionary change'. Of the twenty-four species that she identified from Esh Shaheinab, twelve had not occurred at Early

* He included with the material the 'beautiful', as Dorothea called it, 'and unusually well-preserved' skull of a buffalo. It had been recovered from a government office at Singa where for the past ten years it had been doing duty as a doorstop. This was in fact an extinct Pleistocene African buffalo, which in life would have had very long and widespread horns and whose relatives were portrayed in rock art in North Africa. Dorothea turned her study of the creature into a seminal work on the fossil buffalo of Africa and its relationship with Asiatic and recent African forms. She discovered that it was not of Asiatic origin as had previously been thought, but was related to the modern African buffalo. It was so distinct as to form a separate genus, which she called *Homoioceras singae* Bate. (Bate, 'The Pleistocene fauna of two Blue Nile sites: the mammals from Singa and Abu Hugar', *Fossil Mammals of Africa*, BMNH, 2 (1951).)

Khartoum,* and there was evidence of domestication of two species of goat, a small sheep and, perhaps making a fourth domesticated species, 'possibly' a dog. This, she remarked, 'was not a haphazard beginning of domestication of local animals, such as the Natufian dog of Palestine, but must have been a well-established custom'.[16] The two Sudanese sites may have been close geographically, but these faunal changes over time indicated great cultural and environmental differences. The hunter-fishermen of Early Khartoum had been superseded at Esh Shaheinab by a sophisticated people who had learned how to master both land and animals.

In her search through the literature, she came across a description of a dwarf goat that had been discovered at a site in Egypt, about thirty kilometres from Luxor. From drawings the creature appeared similar to one of the Esh Shaheinab specimens, but Dorothea could not be certain without close examination. The bones of the Egyptian dwarf goat were now in France, at the Musée des Sciences Naturelles in Lyons. Wilfred Edwards shared her interest and, with the blessing of the Trustees of the Museum, she was given leave to investigate the matter. The Keeper must also have thought that her health would greatly benefit from a break from that relentless workload.

On 24 August she caught the boat train to France and four days later was in the Musée des Sciences Naturelles in Lyons, examining the dwarf goat from Egypt. She found it to be quite different from published sources. The drawing she had seen had misleadingly foreshortened the length of the horn, which in reality she found to be long and slender. The horns from Esh Shaheinab were tiny, only six or seven centimetres long. She made notes and sketched the horns, and then escaped from the heat of Lyons to the village of La Clusaz, near Annecy, about three thousand feet up in the Aravis mountains. In 1949 it was a tiny place of pretty wooden chalets. For a few quiet days she stayed at a small, family-run hotel, spending her time walking through the cool forests and considering the implications of the Esh Shaheinab collection. Part of her report, she decided, would be a detailed comparison of the Shaheinab domestic fauna with that of sites of roughly similar age in the Nile valley, including

* They included lion, giraffe, buffalo, antelopes, 'and others which can be found in some part of the Sudan today'. There was no sign of swamp-loving animals and she surmised that the climate was evidently becoming drier.

Badari and the Fayum.[17] Furthermore, the great faunal record from all the different sites in Africa on which she was working would enable her to do further work on the degree of interchange that had taken place between the faunas of Africa and Asia. Her examination of Arkell's Sudan material was well advanced; now, refreshed by the mountain air, she planned also to begin work on the collections from Cyrenaica sent by Dr Charles McBurney.

CHAPTER 18

Swan Song

Charles McBurney, with the geologist Dr Richard Hey, had just spent two seasons excavating at three different sites of three different ages in Cyrenaica, a large promontory on the north coast of Libya. On her return from France, Dorothea arranged for these collections to be cleaned by the Museum's preparators. The established way was for the matrix to be chiselled away from the fossil with a variety of tools. However, a new method had just been developed, which dissolved the matrix by treating it with a solution of acetic acid, to reveal even the tiniest bones and teeth unharmed. Dorothea asked for the McBurney material to be so treated.*

While this was happening, and in between her work for Arkell and Leakey, somehow she found time to write articles for *Chamber's Encyclopaedia* on *Archaeopteryx*, the dodo and the moa. She also concerned herself with the plight of a friend whose archaeological journal was short of cash. To add to all this, in October 1949, eight cases of fossil vertebrate remains arrived from Ksâr 'Âkhil, the site in Syria from which Professor Alfred Day had sent collections to her for determination nearly thirty

* In this material she quickly identified as an 'outstanding' feature of the collection the remains of a Eurasiatic species of fossil vole, which she decribed as *Microtus cyrenae* Bate. (Bate, 'A fossil vole from Cyrenaica', *AMNH*, 12: 3 (1950), 981–5.) They had come from an Upper Palaeolithic cave deposit, the Hagfet ed Dabba or Cave of the Hyaena, and were new to science. Their presence through four successive layers of the cave would be important in helping Dorothea to determine the likely date of their immigration, the climatic conditions, and the probable age of the site.

years ago. New excavations since then by archaeologists from the Jesuit Fordham College in New York had revealed not only quantities of fossil fauna, but early human remains as well. Warning them that she had first to complete her African faunal reports, Dorothea agreed to take on the work. Early in 1950, Father J. Franklin Ewing sent her charts and photos of the site. 'It is indeed fortunate that you can do this study,' he wrote. 'I am jealous of the African faunas which will delay your attack on our material. I hope,' he added, 'that you are enjoying good health.'[1]

She was in fact far from well and it was with quite astonishing courage that she continued her work, determined not to reveal how ill she was. The clerical officer in the Geology Department of the Museum, Marjorie Firth, however, made a point of sending her daffodils ('they are a joy,' Dorothea told her) and butter, still rationed after the war.[2] In the middle of April 1950 she went to London to a conference organized by the Prehistoric Society at the Institute of Archaeology in Regent's Park. It was an opportunity to meet old friends. Perhaps she sensed it might be the last time she would see many of them. Professor Vaufrey presented a paper on Pleistocene elephants, Dorothy Garrod on the art of Lascaux, which she illustrated with a short film. Charles McBurney, Professor Zeuner and others took part in a discussion on older Palaeolithic cultures. It was a three-day conference, but Dorothea could only spare her time and energy for the first day.

At the end of April, she was told that the Trustees had purchased a small house just by the Tring Museum as a residence for the Officer in Charge at a reasonable rent. It needed work to be done on it, which, of course, would have to be put out to tender, but Dorothea had every hope of moving into it within a fairly short time. For so many years she had lived, as Wilfred Edwards described it, 'in the dreary discomfort of cold and inhospitable lodgings'.[3] Within a few months she would have a comfortable place to live just a few yards from the Tring Museum.

Dorothea had taken no leave since January and had been working under continuous pressure, both as administrator and scientist, and that was compounded by her frequent journeys to London. Rather than stop work for a meal, she often survived on endless cups of tea, biscuits, cake or tinned sardines, which she kept in a cupboard in her office. She would be seventy-two in November, and her exhaustion now was such that Wilfred Edwards insisted that she took a week's leave in early May, but it was not enough. A few weeks later she could no longer hide her ill health, and this time it took six weeks' rest by the sea with the Gunsons before she had the strength to return to work. She was back at Tring on 17 July, telling Wooddisse, the Museum Secretary, that she was 'nearly well again'.[4] Again she gives no indication of what was wrong, admitting only to 'rather a poor summer, sick leave etc'.[5] She talked to a local builder about her new house and threw herself back into work with an even greater intensity. In anticipation of a new home, she gave up her lodgings and moved to a local hotel.

For a visit to London in September 1950 she listed her tasks as follows: 'Cyrenaica Testudo remains to Dr Swinton . . . Register Cyrenaica voles. Finish sorting Cyrenaica coll[ection] . . . Sort out Shaheinab specimens to be identified and get numbers of bones of Hippo, Rhino, Giraffe, Antelopes Buff[alo].'[6] Even while she studied the Cyrenaica material, with a deadline of the end of the year for her report, during that autumn of 1950 she was also completing a substantial paper on fossil mammals from Arkell's two Blue Nile sites, and was drafting her full report for Arkell on Esh Shaheinab, which was to include comparisons with fauna from other Egyptian sites. In October she agreed to examine some gazelle remains from the Belt Cave in Iran, excavated by the American archaeologist Carleton S. Coon. 'You are the only person in the world who can identify these creatures,' he wrote to her. Knowing how hard it was for her to budget her health and time, he offered to bring them to her in February 1951, 'already cleaned and sorted'.[7] The strain she was under is reflected in her notes and official diary, her writing at times a loose scrawl as if the pen were too much for her to hold.

In October, her old friend, the former Keeper of Zoology Martin Hinton, wrote to her regarding a collection of rodents and small mammals from Essex that he had examined for her. In what seems now a curiously valedictory remark, he added this: 'I must take this opportunity of saying "Thank you" for all the interesting papers you have sent me in recent years. I have been very remiss in that matter but have always taken a very keen interest in your work.'[8]

That month Edwards again insisted that she took some leave, but even then she continued to work on her reports. On 13 November she completed the final typescript of her Cyrenaica report and sent it off to McBurney, and then drove herself even harder to complete the corrections of the proofs on the Blue Nile mammals. She planned to spend Christmas as usual with her brother Thomas in Herefordshire and wanted the proofs out of the way by then. Her new house was not yet ready, but she had no time to chivvy along the builders. The Tring Museum itself was undergoing major work, including rewiring and complete redecoration. As a result, Dorothea had to send brief progress reports to the Museum as well as field queries from the contractors and the Ministry of Works, which had to authorize all such projects.*

On 5 December, in spite of snow and bitter cold, she was in London, examining fauna from the Sudan. Her writing is a painful, disjointed scrawl, yet two days later she summoned the strength to host a dinner party.[9] On 20 December she noted in her official diary that an attendant at Tring was retiring. It is the last entry.[10] According to a letter that she wrote a few days later to Wooddisse, she 'got a sharp attack of influenza'. Her rooms were so uncomfortable that she did not dare go to bed, but managed instead to get herself to Westcliff-on-Sea, to her 'kind friends', the Gunsons.[11] Her brother Thomas, in Herefordshire, was possibly not aware of quite how ill she was. Leila, who throughout her life had leaned on Dorothea for love and support in times of ill health and trouble, was in Cyprus, recuperating from her own ailment. Nor were her friends aware of the gravity of her illness; at the Museum, the only one who might have realized was Helen Muir-Wood, but she was in the United States.

Dorothea asked for her notes for the Esh Shaheinab report to be sent to her and, when she was able, struggled to complete a cohesive sequence from the hundreds of notes and references on the fauna. The comparison of the domestic animals with fauna from sites in the Nile valley that she had so longed to do was now impossible. Even with her great determination and spirit, this time she no longer had the strength to fight. On 1 January 1951 she told Wooddisse, 'I think I may be here a bit.' She asked him to tell the Director that the building work 'was going on v. satisfactorily', and wished him a happy New Year. She could barely sign

* One such query concerned the enormous areas of plate glass in the public galleries, which (unsurprisingly) made the contractors anxious to know the terms of the Museum's insurance. Dorothea did not know and neither did the Ministry of Works. It was left to the Museum Secretary, Wooddisse, to inform the Ministry: 'The answer is of course that in accordance with general Government policy, nothing at all is insured at the Tring Museum.' (Wooddisse to H. F. King, Ministry of Works, 4 October 1950, DF 1004/719/6/271.)

the letter.[12] Ever practical, on 2 January she made a will, only to have to add a codicil to it three days later, having omitted, in her frailty, to bequeath her 'clothing, jewellery and personal effects' to Leila. Her Georgian silver tea service and £500 she left to Mrs Gunson.

Dorothea died on 13 January 1951. The doctor who certified her death knew nothing of her history and decided 'coronary thrombosis' was the cause. It was not that simple. According to her nephew, Sir David Bate, 'she, the most robust member of the family, died of cancer.'[13] The two major operations that Wilfred Edwards records her undergoing were unlikely to have been for that as they took place at least a decade earlier. It is more probable that when her cancer was diagnosed, it was either too far advanced or she could not bear the thought of further surgery and preferred to entrust her life to the healer, Mrs Gunson. Gertrude Caton Thompson certainly believed that, as she wrote much later, 'Because of her faith in Christian Science [she] refused the medical help which might have prolonged her life . . . [and] declared she would overcome the cancer if left alone in some undisclosed place. And there she died.'[14]

However kind and comforting Mrs Gunson was to Dorothea, she was a stranger. She did her best with the formalities of death, but the details recorded on Dorothea's death certificate are full of inaccuracies and omissions. Even her name is wrong. 'Minola', her second name, has become 'Mimosa'; her profession is given as archaeologist; her father's first name is 'unknown' and so is Henry's profession. Her address is given as 'The Royal Hotel, Tring Station'. Her family have no record of her funeral; neither has the Natural History Museum. She asked to be cremated, but where that took place and where her ashes were scattered, no one remembers either. It was not until the Spring of 2005 that I discovered that even these most final of arrangements had been made by the Gunsons. On 20 January 1951, Dorothea was cremated at Golders Green Crematorium in north London, her ashes scattered there on the quiet anonymity of the crocus lawn. And yet her death was a terrible blow to Thomas and to Leila, who wrote wretchedly to Dr Edwards: 'Thank you again for your sympathy. Dorothy's going is an irreparable loss to me – she was my "prop" in life and without her, many things are difficult.'[15]

In spite of Dorothea's poor health over so many years, few had appreciated how very ill she was. How could they, when she continued to produce an unceasing flow of reports, ideas and advice, and was so generous with her unique and phenomenal knowledge? In addition to the unpublished papers that she wrote on individual collections, of which there are well over a hundred, she published at least eighty-six scientific reports and reviews. Tributes to her were printed in the United States and France as well as in Britain where they appeared, among others, in *The Times*, *Nature*, and *Proceedings of the Geological Society of London*. Father Ewing in New York, who would now have to send his Ksâr 'Âkhil collection elsewhere, was 'much grieved';[16] the American ornithologist and museum curator Hildegard Howard, who knew her only from correspondence, 'felt the loss of a personal friend'.[17] Professor Vaufrey in Paris wrote the warmest of tributes to her work and to her constant vivacity, gaiety and charm, which she always showed in spite of her illness.[18] For Dorothy Garrod, Dorothea's death was 'a great loss to all of us, and to me personally means the disappearance of a friend of many years' standing'.[19] John Desmond Clark too told Dr Edwards, 'Her loss is a very great one and she was such a very likeable person.'[20] Just a few months before his own death in 2002, Professor Clark wrote that 'she was a most charming colleague whom I shall always remember with joy and respect.'[21]

Anthony Arkell, for whom she had laboured until a few days before her death to put into shape her Esh Shaheinab report, developed over three separate publications his own personal and professional tribute to her: 'Her widely lamented death,' he wrote in his book on Esh Shaheinab, has 'robbed archaeologists and prehistorians of a palaeontologist whose co-operation was invaluable and knowledge unique.'[22]

In the paper on the fauna of two Blue Nile sites, he went further.

> She had other collections in hand, and by her study of them, had she lived, more light would have been thrown on past climatic changes in North Africa, on the relations between the fauna of Africa and Asia, and on the origin of some domestic animals. Until someone arises to do Miss Bate the honour of following in her footsteps, the advance of prehistoric archaeology in North Africa and elsewhere must inevitably slow down.[23]

In the leading article in the *Archaeological News Letter* for May 1951, Arkell describes why her work, which he calls comparative zoology, was so vital for the prehistorian:

> It includes not only the identification of the various species represented but the comparative study of the various faunas which those species compose, and the deductions as to climate change, advances in domestication, etc to be drawn from them. It calls for a real interest in animals and their way of life, a wide knowledge of their teeth and bone structure, of the details in which species differ, and which thus render identification possible even from a comparatively small fragment, acquaintance with the literature, particularly with the reports on fauna from other excavations, and perhaps most important of all, acumen, persistence and patience.[24]

These were the qualities, he believed, that had characterized Dorothea all her life. 'Prehistorians in England, Europe and America,' he continued, revealing a profound and very moving grief, 'feel lost and helpless now that she has gone. There is so much waiting to be done in the field that she has made her own, and there is at present no one available to give this work the whole-time attention it needs.' What was required, he suggested, was for a fund to be endowed in her memory to 'enable promising students to carry on her important work'.[25] The Keeper of Geology, Wilfred Edwards, wrote to Arkell in total support of this article, which, he told Arkell, 'I ought to have written but which you wrote so much better than I should ever have done'.[26]

There was a brief flurry of interest in establishing such a fund. Professor Zeuner, Dorothy Garrod and Gertrude Caton Thompson promised letters of support. In July, it all looked very positive when Arkell heard from one of the Museum Trustees, Professor D. M. S. Watson, 'that the Trustees of the BM are taking steps to finance a student which is grand news'.[27] Edwards, from within the Museum, however, advised caution. The position, he told Arkell, was 'not yet absolutely clinched' and he urged Arkell to send the new Director of the Natural History Museum, Gavin Rylands de Beer, a copy of his article in the *Archaeological News Letter*.[28] De Beer, once he had seen the proposal, told Arkell that he thought the idea interesting but did not want any staff member of the Museum to write in support. Arkell then tried Dorothea's family, but Thomas's letter in reply, Arkell ruefully told Edwards, was 'not indicative of any active support from him!'[29] Arkell did not receive a reply from Leila, but during these weeks her daughter had been very ill in hospital.

Just how profound an effect the family's years of humiliating poverty had on Dorothea only now becomes clear. All her life she had spent as little as possible on herself and her comforts. Her aversion to owing anyone anything was evident all those years ago in Cyprus, when she tried to repay her father £10, and preferred to go hungry in Crete rather than accept supplies from the British archaeologists. That deeply moving letter from her father on his sixty-third birthday must have burned into her soul. According to Dr Edwards, she had latterly received a private income, although her family have no knowledge of it. The result of her economy, ironically, was to benefit Leila. In addition to the bequests stated in her will, Dorothea left to her sister, 'already', wrote Edwards, 'a wealthy woman', the very considerable sum of £15,000 (about £300,000 in today's values). Edwards' hope that some of it might find its way to science (as he wrote to Arkell, 'one never knows!'[30]) came to nothing. With momentum lost, in the climate of post-war England, the idea of a Dorothea Bate Studentship quietly disappeared. What Edwards did achieve was an additional staff post for a scientific officer to 'continue her work on Pleistocene mammals',[31] but that was not the same thing at all.

In the meantime, the papers she had been working on were published. The Esh Shaheinab report, which Dorothea had been writing at the time of her death, was published in March and the Singa mammals paper in May. Arkell was delighted with it, both for himself and Dorothea, as he told Dr Edwards. 'How pleased with it Dorothea would have been if she could have lived till today to see it. "Pleased" is hardly strong enough. She would have been thrilled.'[32]

The effect of her death was greatest on the Museum. The collections on which she had been working had to be taken on by her colleagues, all of whom were already overburdened themselves. In desperation, Wilfred Edwards even wrote to the shipping magnate and student of rodents, Sir John Ellerman, to ask him to take on some of Dorothea's work on small fossil mammals. 'She, as I know, in the past asked you whether you would consider extending your studies of the rodents to some of the fossil remains. I venture now to put this before you again with a much greater sense of urgency as Miss Bate has gone.' Ellerman still did not want to extend his work to fossils and could not help. As for Leakey's 'favourite pigs', it was hoped that Gavin Rylands de Beer might take them on, but it was too great a task to combine with directorship of the Museum.

In the end, using Dorothea's notes, Leakey wrote the paper himself.[33]

The Trustees of the Museum recorded their 'great regret' at her death and Professor D. M. S. Watson paid tribute to her qualities as a palaeontologist and her unique knowledge of Pleistocene mammals and birds.[34] It was Wilfred Edwards, however, who expressed the emotions of her colleagues. 'Dorothea's death is a very great loss to the scientific world in general and to the British Museum in particular. For myself,' he wrote to Leila, 'I have lost one of my best friends.'[35] It was his duty to write her obituary, and he carefully checked the facts with Thomas and Leila. From her diaries, in which she mentions her birthdays, Edwards had calculated that Dorothea had been born in 1878. Thomas, however, 'so far as he knew', believed that she had been born in 1879, which was the date Dorothea herself had given to the Museum accountant. Edwards, however, was right. Her birth certificate, incontrovertibly, records her birth as 8 November 1878. Neither Thomas nor Leila knew the extent of her achievements. 'I like so much what you have said about Dorothy,' Leila told Edwards. 'I think her family were really the last people to hear about all her work, and how much she had done. She was so very modest about it all, and I personally never heard about many of these things she did – which I regret so much now – and I wish I had been able to appreciate that part of her life.'[36]

Edwards had the task of sending Dorothea's belongings to her family and of asking for the return of Museum property. 'Dorothea had two pocket lenses and two small gauges for her use which were the property of the Trustees of the Museum,' he wrote to Leila. 'We have got all her other official instruments except these, and the lenses in particular would be quite likely to be in one of her personal handbags. If by any chance,' he continued with some embarrassment, 'you come across these we should be very glad if they could be returned.' The Museum, furthermore, was so short of equipment that Edwards even had to ask Leila this: 'I understand from Dr Muir-Wood that Dorothea had a typewriter of her own which presumably would have been at Tring. We are very short of typewriters in this department and they are not easy to obtain. If this instrument is still available, would it be possible for us to acquire it?'[37]

Unlike many women scientists, Dorothea's achievements were admired and respected by those she worked with in her lifetime, yet so little about her was remembered after her death. In the Natural History Museum today she is remembered as 'a great lady', although the extent of her knowledge and achievements has come as a surprise to many. According to the Curator of Quaternary Mammals, Andrew Currant, she was 'the spark that would ignite a project'.[38] He has lost count of the number of papers and books he has encountered in which the author acknowledges his debt to Dorothea, both in inspiring the work in the first place and then in guiding the author through it. 'We are especially indebted to Miss DMA Bate,' reads one such acknowledgement, 'who although herself fully engaged, has given unstintingly of her time and knowledge, and has directed and advised us on many problems.'[39] Arkell's great tribute to her was read, approved and then forgotten. Even if his appeal for funds for a studentship in 'comparative zoology' had succeeded, she in truth was the only one then with the ability to teach it.

The depth and breadth of her knowledge was exceptional, from the fossil fauna of British caves to that of East Africa, from the Tertiary period to recent. By compelling charm and innate ability she had attached herself to one of the world's greatest research museums, not in any overt attempt to storm a male bastion but simply to place herself in her chosen environment. Her exploration of Cyprus, Crete and the Balearics revealed for the first time the Pleistocene fauna of those islands as well as her own astonishing courage and spirit. As for her work in the Near East, perhaps the greatest tribute is that, some seventy years after her reports were published, they are still in use. Dorothea's lack of formal education enabled her to move effortlessly between academic disciplines, exploring and exploiting the fertile interface between the established earth and life sciences. She was in every respect a true pioneer, possessing in abundance that essential quality that characterizes our species, a ceaselessly enquiring mind.

As Edwards began the enormous task of sorting through Dorothea's papers – the collections on which she was working and those she had not yet begun – it became apparent that even if she had not been aware of imminent death, she had realized that her health would not permit her to continue her work for very much longer. On top of the neat pile of papers on which she was working, he found a scrappy sheet, torn from a flimsy notebook. It is headed 'Papers to Write'.[40] She has listed six papers that she was planning on specific fauna from individual Mediterranean islands, to be written when she had completed her African work. The last paper listed is 'On Pleistocene Mammals of Mediterranean Islands', the discoveries that had made her name as a young woman, travelling and excavating alone all those years ago. She had done so much work during the war on this report but had never had the time to complete it. Perhaps, as she wrote down the heading, she thought of Henry Woodward's marvellous conceit of her Mediterranean discoveries, when he had pictured 'the sensation you would make if you could walk down Piccadilly leading by a string your Pigmy Elephants, *Hippopotami*, *Myotragus*, Tortoises etc etc all in one long queue, the little elephant blowing his trumpet, and the Hippopotamus wagging its tail'.[41] It would have been, as Edwards notes, a report of the 'first importance'[42] and against it on this list she has written simply, 'Swan Song'.

What I found even more poignant is an entry in her 'collections in hand' notebook, which must have been made within a few days of her collapse. With a blue pencil, a firm double line has been drawn underneath the last collection listed and, in capitals, just as Dorothea had written at the end of her final work diary for the Balearics, there is just one word: 'FINIS.'

POSTSCRIPT

On a morning in late summer 2003, Dorothea Bate waited until the crowd had dispersed, then walked up the sweeping flight of steps to the entrance to the Natural History Museum. There she paused for photographs. Fifty-two years after her death, through the actress Jane Cartwright, the Museum had re-created her image as a Gallery Character. In the company of Charles Darwin, William Smith, Mary Anning and other select scientists, Dorothea now wanders the galleries of the Museum during school holidays, sharing stories of her life and her scientific achievements with visiting children – and adults – perhaps inspiring in at least some of them, as she did with young scientists so many years ago, that passion for science, discovery and exploration that formed her life.

Courtesy of the Natural History Museum.

NOTES

Abbreviations used in the Notes are as follows:

BM (NH)	British Museum (Natural History)
DMAB	Dorothea M. A. Bate
ESL	Earth Sciences Library, Natural History Museum
NHM	Natural History Museum
PEF	Palestine Exploration Fund Collections
PSMF Archives	Percy Sladen Memorial Fund Archives, deposited at the Linnean Society of London
TNA(PRO)	The National Archives (Public Record Office)
UPenn	Archives of University of Pennsylvania Museum of Archaeology and Anthropology
WA/HSW/AR/Lac	Reports and correspondence, Wellcome Archaeological Research Expedition to the Near East (WARENE), Wellcome Library of the History of Medicine

MSS HELD IN THE NATURAL HISTORY MUSEUM:

Accession no.	Uncatalogued MSS held in NHM Archives
DF	Departmental code for NHM Archives
P MSS BAT	Dorothea Bate MSS, Small Library, NHM Earth Sciences Library
Balearics III	Dorothea Bate, diary of third trip to Balearic islands, 1911, in P MSS BAT
Bethlehem 1935, 1936, Bethlehem 1937 I and II	Dorothea Bate's Bethlehem site notebooks in MSS Bethlehem excavations 1935–7, P MSS BAT
Crete	Dorothea Bate, diary of trip to Crete, 1904, in P MSS BAT
Cyprus I, II and III	Dorothea Bate, diary of trip to Cyprus, 1901–2, in P MSS BAT
Majorca II	Dorothea Bate, diary of second trip to Majorca, 1910, in P MSS BAT

P MSS LAN	Dr William Lang MSS, Small Library, NHM Earth Sciences Library
P MSS WOO	Sir Arthur Smith Woodward MSS, Small Library, NHM Earth Sciences Library

JOURNALS

AMNH	*Annals and Magazine of Natural History*
BAAS	*Report of the British Association for the Advancement of Science*
BASPR	*Bulletin of the American School of Prehistoric Research*
BBM(NH)G	*Bulletin of the British Museum (Natural History), Geology*
BJLS	*Biological Journal of the Linnean Society*
GM	*Geological Magazine*
ILN	*Illustrated London News*
JHE	*Journal of Human Evolution*
JRAI	*Journal of the Royal Anthropological Institute*
PEQ	*Palestine Exploration Quarterly*
PGA	*Proceedings of the Geologists' Association*
PPS	*Proceedings of the Prehistoric Society*
PRSE	*Proceedings of the Royal Society of Edinburgh*
PRSL	*Proceedings of the Royal Society of London*
PSAS	*Proceedings of the Society of Antiquaries of Scotland*
PBSS	*Proceedings of the University of Bristol Speleological Society*
PTRS	*Philosophical Transactions of the Royal Society of London*
PZSL	*Proceedings of the Zoological Society of London*
QJGS	*Quarterly Journal, Geological Society*

1: A Cathedral to Nature, pp. 5–13

1. Dorothea Bate's obituary by Wilfred Edwards, MSS for private circulation in the Department of Geology, 18 January 1951, Accession no. 2000/19.
2. Thomas Hearne, *An Epistolary Letter from T— H— to Sr H—s S—e who saved his life, and desired him to send over all the rarities he could find in his travels* (London, 1729).
3. Quoted in Sir William Henry Flower, *A General Guide to the British Museum (Natural History)* (London: BM (NH), 1896), 8.
4. Richard Bowdler Sharpe, Chapter 3: 'Birds', *History of the Collections* (London: BM (NH), 1906), 86.
5. Bowdler Sharpe, *History of the Collections*, 84.
6. Quoted in *The Times*, 18 April 1881.
7. Flower, *A General Guide to the British Museum (Natural History)*, 11.

8. 'The New Natural History Museum', *Nature*, 27 (1882), 55.
9. Flower, *A General Guide to the British Museum (Natural History)*, 20.
10. *Cassell's Saturday Journal*, 24 May 1893.
11. 'Ithuriel's Interviews', *London*, 30 August 1890.
12. *The Daily Telegraph*, 29 January 1884.

2: *This Noble Family, pp. 14–35*

1. *Wexford Independent*, 16 January 1839.
2. *Wexford Independent*, 2 February 1839.
3. TNA (PRO), WO 31/1102, 9 November 1855.
4. Henry Bate's personnel record, 24 December 1877, Dyfed–Powys Police Museum.
5. *Carmarthen Journal*, 4 January 1878.
6. Carmarthen quarter sessions Minute Books, 1866–91, Carmarthenshire Record Office.
7. Sir David Bate, personal communication.
8. *Cardigan and Tivy-side Advertiser*, 3 January 1890.
9. *Cardigan and Tivy-side Advertiser*, 5 September 1890.
10. *Cardigan and Tivy-side Advertiser*, 4 January 1889.
11. *Cardigan and Tivy-side Advertiser*, 4 January 1895.
12. *Cardigan and Tivy-side Advertiser*, 15 June 1888.
13. *Cardigan and Tivy-side Advertiser*, 11 November 1892.
14. *Cardigan and Tivy-side Advertiser*, 12 June 1891.
15. *Cardigan and Tivy-side Advertiser*, 12 August 1898.
16. *Cardigan and Tivy-side Advertiser*, 23 September 1898.
17. Sir David Bate, personal communication.
18. Letter from DMAB to Dr Richard Bowdler Sharpe, 15 October 1898, DF 230/3.
19. Letter from Dr Richard Bowdler Sharpe to the Director, 4 November 1902, DF 230/20/475.
20. DMAB's MSS notes etc on Syria and Palestine, P MSS BAT.
21. Notice of Henry Woodward's death, *GM*, 10: 58 (October 1921), 433.
22. Lady Woodward's biography of Sir Arthur Smith Woodward (draft notes), P MSS WOO, f. 449.
23. *The Church Family Newspaper*, 3 September 1897.
24. 'The retirement of Dr Henry Woodward', *GM*, 1: 9 (January 1902), 1.
25. Professor H. Alleyne Nicholson and R. Lydekker, *Manual of Palaeontology*, 3rd edn (Edinburgh and London: Wm Blackwood & Sons, 1889), i. 3.
26. Dr Henry Woodward, *Guide to the Exhibition Galleries of the Department of Geology and Palaeontology in the BM (NH)* (London: BM (NH), 1890), xii.
27. C. D. E. König, *Directions for Collecting Specimens of Geology and*

Mineralogy, for the British Museum (London: British Museum Department of Geology, 1837), 1.
28. König, *Directions for Collecting Specimens*, 2.
29. König, *Directions for Collecting Specimens*, 6.
30. König, *Directions for Collecting Specimens*, 5.
31. Dorothy [*sic*] M. A. Bate, 'A short account of a bone cave in the Carboniferous limestone of the Wye valley', *GM*, 4: 8 (1901), 102.
32. Ibid.
33. Andrew Currant, Curator of Quaternary Mammals, NHM, personal communication.
34. William T. Stearn, *The Natural History Museum at South Kensington* (London: NHM, 1998), 32.
35. Stearn, *The Natural History Museum at South Kensington*, 33.
36. DF 108/8, 27 September 1900.

3: *Leave to Collect, pp. 37–52*

1. Letter from Dr Charles Forsyth Major to Dr Henry Woodward, 2 February 1902, DF 100/160/9.
2. Cyprus I, 17 April 1901.
3. Cyprus I, 20 April 1901.
4. Cyprus I, 22 April 1901.
5. Jean Albert Gaudry, *Animaux fossiles et géologie de l'Attique: d'après les recherches faites en 1855–56, et en 1860 sous les auspices de l'Académie des sciences* (Paris, 1862–7), 14.
6. Arthur Smith Woodward, 'On the bone-beds of Pikermi, Attica and on similar deposits in northern Euboea', *GM,* 4: 8 (1901), 481–6.
7. Letter from Dorothea Bate to William Ogilvie-Grant, 13 May 1901, DF 230/22/139.
8. TNA (PRO), CO 67/128, Despatch 182, 7 October 1901.
9. Letter from Dorothea Bate to William Ogilvie-Grant, 13 May 1901, DF 230/22/139.
10. Cyprus I, 23 April 1901.
11. Cyprus I, 29 April 1901.
12. TNA (PRO), CO 67/131, Despatch 96, 15 May 1902.
13. Quoted in Jean Albert Gaudry, *Geology of the Island of Cyprus,* translated by F. Maurice (London: HMSO, 1878), 5.
14. Edward Vizetelly, *From Cyprus to Zanzibar by the Egyptian Delta: The Adventures of a Journalist* (London: C. Arthur Pearson, 1901), 26.
15. Vizetelly, *From Cyprus to Zanzibar,* 5–6.
16. Vizetelly, *From Cyprus to Zanzibar,* 10.
17. H. Rider Haggard, *A Winter Pilgrimage: Being an account of travels through*

Palestine, Italy, and the island of Cyprus, accomplished in the year 1900 (London: Longmans & Co., 1901), 64.

18. Cyprus I, 1 May 1901.
19. Cyprus I, 2 May 1901.
20. Bate, 'Field-notes on some of the birds of Cyprus', *Ibis* (October 1903), 571–81.
21. Mrs Lewis, *A Lady's Impression of Cyprus in 1893* (London: Remington & Co., 1894), 144.
22. TNA (PRO), CO 67/127, 23 April 1901.
23. Cyprus I, 6 May 1901.
24. George Chacalli (native of Cyprus), *Cyprus under British Rule* (printed at office of 'Phoni the Kyprou', Nicosia, 1902), 133.
25. Letter from DMAB to William Ogilvie-Grant, 13 May 1901, DF 230/22/139.
26. Cyprus I, 29 May 1901.
27. Cyprus I, 1 July 1901. In September 2003 I was shown this splendid specimen in the collections at the NHM's Zoological Museum at Tring. Until then, nothing was known about it to account for its condition apart from the fact that Dorothea had collected it in Cyprus and as Dr Jo Cooper, the palaeo-ornithologist, told me, it had puzzled them. It looks similar to Egyptian mummified bones, there was no preparation of it, soil is lodged in pockets of the skull and remains of ligaments can still be seen. Knowing that Dorothea had buried it solved the mystery.
28. Sir J. T. Hutchinson and Claude Delaval Cobham, *Handbook of Cyprus* (London: Waterlow & Sons Ltd, 1901), 60.
29. Cyprus I, 28 May 1901.
30. Cyprus I, 4 May 1901.
31. Anonymous letter, in Cyprus I.
32. Dr George Alexander Williamson, 'The Cyprus sphalangi and its connection with anthrax', *British Medical Journal* (1 September 1900).
33. Cyprus I, 11 June 1901.
34. Cyprus I, 6 May 1901.
35. Cyprus I, 20 June 1901.
36. Cyprus I, 10 May 1901.
37. Cyprus I, 6 July 1901.
38. Cyprus I, 18 July 1901.
39. Cyprus I, 20 July 1901.
40. Bate, 'Field-notes on some of the birds of Cyprus', *Ibis* (October 1903), 571–81.
41. Cyprus I, 9 September 1901.
42. Cyprus I, 22 September 1901.
43. *Handbook of Instructions for Collectors* (London: BM (NH), 1906), 22.
44. Cyprus I, 5 October 1901.

4: *In Search of Extinct Beasts, pp. 53–74*

1. Letter from Dr Charles Forsyth Major to Dr Henry Woodward, 2 February 1902, DF 100/160/9.
2. Cyprus II, 10 October 1901.
3. Cyprus II, 11 October 1901.
4. Cyprus II, 12 October 1901.
5. Bate, 'Further note on the remains of *Elephas cypriotes* from a cave-deposit in Cyprus', *PTRS*, series B, 197 (March 1905), 347–60.
6. Dr Henry Woodward (ed.), *Guide to the Exhibition Galleries of the Department of Geology and Palaeontology in the BM (NH)* (London: BM (NH), 1896).
7. Sir J. T. Hutchison and Claude Delaval Cobham, *Handbook of Cyprus* (London: Waterlow & Sons Ltd, 1901), 52.
8. Cyprus II, 13 October 1901.
9. Ibid.
10. Ibid.
11. *Handbook of Instructions for Collectors* (London: BM (NH), 1902), 129.
12. *Handbook of Instructions for Collectors* (London: BM (NH), 1921), 176.
13. W. H. Mallock, *In an Enchanted Island: or a Winter's Retreat in Cyprus* (London: Bentley & Son, 1889), 228.
14. Mrs Lewis, *A Lady's Impression of Cyprus in 1893* (London: Remington & Co., 1894), 187.
15. Cyprus II, 16 October 1901.
16. Kenneth J. Hsü, *The Mediterranean was a Desert: A Voyage of the* Glomar Challenger (Princeton, NJ: Princeton University Press, 1983), 178.
17. Cyprus II, 20 October 1901.
18. Mrs Lewis, *A Lady's Impression of Cyprus in 1893*, 245.
19. Cyprus II, 20 October 1901.
20. Ibid.
21. Cyprus II, 25 October 1901.
22. Cyprus II, 7 November 1901.
23. Letter from Dr Charles Forsyth Major to Dr Henry Woodward, 2 February 1902, DF 100/160/9.
24. Cyprus II, 16 November 1901.
25. Cyprus II, 17 November 1901.
26. Cyprus II, 18 November 1901.
27. Cyprus II, 19 November 1901.
28. Cyprus II, 21 November 1901.
29. TNA (PRO), CO 67/130, Confidential Despatch, 8 March 1902.
30. Cyprus II, 22 November 1901.
31. Cyprus II, 2 December 1901.

32. Cyprus II, 3 December 1901.
33. Cyprus II, 4 December 1901.
34. Cyprus II, 8 December 1901.
35. Letter from DMAB to William Ogilvie-Grant, 9 December 1901, DF 230/22/140.
36. Letter from Dr Charles Forsyth Major to Dr Henry Woodward, 2 February 1902, DF 100/160/9.
37. Ibid.
38. Cyprus II, 15 December 1901.
39. Cyprus II, 16 December 1901.
40. Cyprus II, 28 December 1901.
41. Letter from Dr Charles Forsyth Major to Dr Henry Woodward, 2 February 1902, DF 100/160/9.
42. George Chacalli, *Cyprus under British Rule* (Nicosia, 1902), 156.
43. Cyprus II, 6 January 1902.
44. Cyprus II, 7 January 1902.
45. Cyprus II, 10 January 1902.
46. Cyprus II, 14 January 1902.
47. Ibid.
48. Cyprus II, 28 January 1902.
49. Cyprus II, 30 January 1902.
50. Ibid.
51. Bate, 'The mammals of Cyprus', *PZSL*, 2 (1903), 342.
52. Letter from DMAB to Dr Henry Woodward, 4 February 1902, DF 100/34/77.
53. Cyprus II, 10 February 1902.
54. Cyprus II, 21 February 1902.
55. Letter from DMAB to Dr Arthur Smith Woodward, 22 February 1902, DF 100/34/79.
56. Cyprus II, 13 February 1902.
57. See Alan H. Simmons, *Faunal Extinction in an Island Society: Pygmy Hippopotamus Hunters of Cyprus* (New York and London: Kluwer Academic/Plenum Publishers, 1999); and David S. Reese, 'Tracking the extinct pygmy hippopotamus of Cyprus', *Field Museum of Natural History Bulletin*, 60: 2 (1989).
58. Letter from F. A. Jowle, Clerk to the Government Grant Committee, Royal Society, to Dr Henry Woodward, 18 February 1902, DF 100/34/78.
59. Cyprus II, 5 April 1902.
60. Unattributed newspaper cutting, in Gaze family papers.
61. Cyprus II, 8 April 1902.
62. Cyprus II, 7 April 1902.

5: The Most Unpromising Place, pp. 75–95

1. Cyprus III, 16 April 1902.
2. Cyprus III, 24 April 1902
3. Cyprus III, 25 April 1902.
4. Cyprus III, 19 April 1902.
5. Cyprus III, 30 April 1902.
6. Cyprus III, 1 May 1902.
7. Cyprus III, 3 May 1902.
8. Cyprus III, 4 May 1902.
9. Cyprus III, 5 May 1902.
10. Cyprus III, 6 May 1902.
11. Ibid.
12. C. I. Forsyth Major, 'On the pigmy hippopotamus from the Pleistocene of Cyprus', *PZSL*, 2 (1902), 107–12.
13. Ibid.
14. Letter from DMAB to Dr Arthur Smith Woodward, 29 May 1902, DF 100/34/80.
15. Ibid.
16. Cyprus III, 17 May 1902.
17. Cyprus III, 22 May 1902.
18. Cyprus III, 23 May 1902.
19. Cyprus III, 7 June 1902.
20. Cyprus III, 29 May 1902.
21. Cyprus III, 31 May 1902.
22. Cyprus III, 9 June 1902.
23. Letter from DMAB to Dr Arthur Smith Woodward, 7 February 1902, DF 100/34/76.
24. Cyprus III, 17 June 1902.
25. Bate, 'Preliminary note on the discovery of a pigmy elephant in the Pleistocene of Cyprus', *PRSL*, 71 (June 1903), 498.
26. Cyprus III, 18 June 1902.
27. Cyprus III, 19 June 1902.
28. Cyprus III, 23 June 1902.
29. Cyprus III, 28 June 1902.
30. Cyprus III, 4 July 1902.
31. Cyprus III, 10 July 1902.
32. Cyprus III, 15 July 1902.
33. Cyprus III, 23 July 1902.
34. Cyprus III, 15 July 1902.
35. In Cyprus III, letter from Francie Wodehouse to DMAB, 29 June 1902.
36. In Cyprus III, letter from Henry Bate to DMAB, 24 July 1902.

37. Cyprus III, 28 July 1902.
38. Cyprus III, 31 May 1902.
39. Bate, 'On the occurrence of *Acomys* in Cyprus', *AMNH*, 7: 11 (1903), 565.
40. Cyprus III, 16 September 1902.
41. Cyprus III, 9 August 1902.
42. Cyprus III, 25 August 1902.
43. Cyprus III, 26 August 1902.
44. Letter from DMAB to Dr Arthur Smith Woodward, 13 August 1902, DF 100/34/77.
45. Ibid.
46. Cyprus III, 7 October 1902.
47. Cyprus III, 9 October 1902.
48. Cyprus III, 10 October 1902.
49. Cyprus III, 11 October 1902.
50. Cyprus III, 12 October 1902.
51. Cyprus III, 13 October 1902.
52. Cyprus III, 14 October 1902.
53. Cyprus III, 17 October 1902.
54. Cyprus III, 23 October 1902.
55. Cyprus III, 30 October 1902.
56. Cyprus III, 2 November 1902.
57. Cyprus III, 8 November 1902.
58. Cyprus III, 16 November 1902.
59. Cyprus III, 17 November 1902.
60. Cyprus III, 18 November 1902.
61. Cyprus III, 19 November 1902.
62. *Annals of the Cyprus Natural History Society*, 3 (Nicosia: 1911), 5.
63. Cyprus III, 28 and 29 November 1902.
64. Cyprus III, 30 November 1902.
65. Cyprus III, 8 December 1902.
66. Cyprus III, 10 December 1902.

6: A bit of a shock to me!, pp. 96–117

1. Dr Ellis Owen, personal communication.
2. Sir Arthur Smith Woodward, 'Dr Charles William Andrews Obituary Notice', *PRSL*, series B, 100 (1926), i–iii.
3. Bate, 'On an extinct species of genet (*Genetta plesictoides* sp. n.) from the Pleistocene of Cyprus', *PZSL*, 2 (1903), 121–4.
4. N. D. Riley, *The Department of Entomology of the British Museum (Natural History) 1904–1964: A Brief Historical Sketch* (London: 12th International Congress of Entomology, 1964), 1.

5. Henry Woodward, 'Cave-hunting in Cyprus', *GM*, 4: 10 (June 1903), 241–6.
6. *History of the Collections contained in the Natural History Departments of the British Museum*, i (London: BM (NH), 1904).
7. DF 207/2, 4 August 1903.
8. Memo from G. F. Hampson to C. E. Fagan, 24 June 1903 (attached to letter from DMAB), DF 200/50.
9. Eileen Hawkins, Archivist of YWCA, personal communication.
10. Letter from DMAB to Michael Oldfield Thomas, 10 September 1903, DF 232/9/20.
11. Letter from DMAB to C. D. Sherborn, 31 August 1903, Handwriting Collection, NHM General Library.
12. Special report to Trustees, 1908, DF 104/52.
13. Report on progress, 1909–11, DF 104/52.
14. DF 102/9/14, December 1903.
15. Bate, 'On elephant remains from Crete, with description of *Elephas creticus* sp. n', *PZSL* (1907), 238–50.
16. Letter from DMAB to William Ogilvie-Grant, 29 December 1903, DF 230/22/142.
17. TNA (PRO), FO 421/207, Despatch 80.
18. Crete, 1 March 1904.
19. Crete, 28 February 1904.
20. E. S. Bosanquet, 'Bosanquet, Robert Carr', revised David Gill, *Oxford Dictionary of National Biography* (Oxford: OUP, 2004).
21. Crete, 3 March 1904.
22. Crete, 4 March 1904.
23. Mary Allsebrook, *Born to Rebel: The Life of Harriet Boyd Hawes* (Oxford: Oxbow Books, 2002), 85.
24. J. P. Droop, *Archaeological Excavation* (Cambridge: CUP, 1915), 64–5. It is known among archaeologists as 'Droop's 7th'.
25. Crete, 5 March 1904.
26. Charles Trick Currelly, *I Brought the Ages Home* (Toronto: Royal Ontario Museum, 1956), 64.
27. Vasso Fotou, *New Light on Gournia: Unknown Documents of the Excavations at Gournia and Other Sites on the Isthmus of Ierapetra by Harriet Ann Boyd*, Series Aegaeum 9 (Austin, Texas: University of Liège and University of Texas, 1993), 21–2.
28. Allsebrook, *Born to Rebel: The Life of Harriet Boyd Hawes*, 94.
29. *Illustrated London News*, 22 September 1906.
30. Crete, 6 March 1904.
31. Bate, 'The Caves of Crete', in Aubyn Trevor-Battye, *Camping in Crete* (London: Witherby & Co., 1913), 240.
32. Crete, 7 March 1904.
33. Crete, 14 March 1904.

34. Bate, 'Four and a half months in Crete in search of Pleistocene mammalian remains', *GM*, 5: 2 (1905), 193–202.
35. Crete, 8 March 1904.
36. Bate, 'The Caves of Crete', 240.
37. Crete, 8 March 1904.
38. Letter from DMAB to William Ogilvie-Grant, 3 April 1904, DF 230/22/141.
39. Crete, 9 March 1904.
40. Crete, 10 March 1904.
41. Crete, 11 March 1904.
42. Crete, 15 March 1904.
43. Crete, 16 March 1904.
44. T. A. B. Spratt, *Travels and Researches in Crete* (London, 1865), ii, 195.
45. Crete, 20 March 1904.
46. Crete, 21 March 1904.
47. Crete, 22 March 1904.
48. Bate, 'The Caves of Crete', 239.
49. Crete, 23 March 1904.
50. Ibid.
51. Crete, 24 March 1904.
52. Crete, 27 March 1904.
53. Ibid.
54. Crete, 29 March 1904.
55. Letter from DMAB to William Ogilvie-Grant, 3 April 1904, DF 230/22/141.
56. Crete, 1 April 1904.
57. Letter from DMAB to William Ogilvie-Grant, 3 April 1904, DF 230/22/141.
58. Crete, 3 April 1904.

7: Courageous as Ever, pp. 119–47

1. Crete, 7 April 1904.
2. Crete, 8 April 1904.
3. Crete, 9 April 1904.
4. Crete, 11 April 1904.
5. Crete, 14 April 1904.
6. Crete, 15 April 1904.
7. Elliot M. Lax, 'A Gazetteer of Cretan Palaeontological Localities', in David S. Reese (ed.), *Pleistocene and Holocene Fauna of Crete and its First Settlers* (Madison, Wisconsin: Prehistory Press, 1996), 4.
8. Crete, 23 April 1904.
9. Crete, 25 April 1904.
10. Crete, 27 April 1904.
11. Crete, 28 April 1904.

12. Crete, 29 April 1904.
13. Crete, 2 May 1904.
14. Letter from DMAB to Arthur Smith Woodward, 14 June 1904, DF 100/36/55.
15. T. A. B. Spratt, *Travels and Researches in Crete* (London, 1865), i, 137.
16. Crete, 16 May 1904.
17. Bate, 'Four and a half months in Crete in search of Pleistocene mammalian remains', *GM*, 5: 2 (1905), 201.
18. Crete, 16 May 1904.
19. Bate, 'Four and a half months in Crete', 200.
20. Bate, 'The Caves of Crete', in Aubyn Trevor-Battye, *Camping in Crete* (London, 1913), 252.
21. Crete, 23 May 1904.
22. Bate, 'Four and a half months in Crete', 201.
23. Crete, 27 May 1904.
24. *New York Times*, 25 September 1904.
25. Marshall J. Becker and Philip P. Betancourt, *Richard Berry Seager: Pioneer Archaeologist and Proper Gentleman* (Philadelphia: University of Pennsylvania Museum of Archaeology and Anthropology, 1997), 35.
26. Mary Allsebrook, *Born to Rebel: The Life of Harriet Boyd Hawes* (Oxford: Oxbow Books, 2002), 123.
27. Letter from Edith Hall to her parents, 18 January 1904, UPenn.
28. Allsebrook, *Born to Rebel*, 116.
29. Becker and Betancourt, *Richard Berry Seager*, 29–30.
30. Harriet Boyd Hawes, letter to Blanche Wheeler Williams, 10 July 1904, Box 4, Excavations – Crete, Gournia 1901–5, General Correspondence, Smith College Archives. Harriet calls the sarcophagi 'casellas' in both this correspondence and on site to Dorothea, who also used the word, but it was actually a term that Arthur Evans used for the underfloor storage cists that he discovered at Knossos. Harriet corrects this description in her later reports on her finds.
31. Crete, 28 May 1904.
32. Letter from J. L. Myres to Harriet Boyd, 27 August 1904, Harriet Boyd Hawes papers, Box 4, Folder 64, Smith College Archives.
33. Letter from Edith Hall to her parents, 29 May 1904, UPenn. These letters provide the only known surviving descriptions of Dorothea for her early life.
34. Spratt, *Travels and Researches in Crete*, i, 212.
35. Crete, 30 May 1904.
36. *Annual of the British School at Athens*, 10, Session 1903–4, 214.
37. Crete, 31 May 1904.
38. Letter from R. C. Bosanquet to Ellen Bosanquet, 23 April 1902, in Ellen S. Bosanquet (ed.), Robert Carr Bosanquet, *Letters and Light Verse* (Gloucester: John Bellows, 1938), 121.
39. Crete, 31 May 1904.

40. Crete, 1 June 1904.
41. Letter from Robert Carr Bosanquet to DMAB, 2 June 1904, attached to Crete diary.
42. Crete, 3 June 1904.
43. Crete, 4 June 1904.
44. Letter from Edith Hall to her parents, 5 June 1904, UPenn.
45. Letter from Edith Hall to her parents, 1 May 1904, UPenn.
46. Letter from Edith Hall to her sister, 13 April 1904, UPenn.
47. Letter from Edith Hall to her parents, 5 June 1904, UPenn.
48. Crete, 6 June 1904.
49. Crete, 8 June 1904.
50. Dr J. Arthur Evans, 'Excavations at Knossos, Crete 1904', *Report of the 74th Meeting of the British Association for the Advancement of Science* (Cambridge, 1904), 322–24.
51. Crete, 10 June 1904.
52. Crete, 15 June 1904.
53. Crete, 12 June 1904.
54. Charles Trick Currelly, *I Brought the Ages Home* (Toronto, 1956), 63.
55. Crete, 15 June 1904.
56. Crete, 16 June 1904.
57. Crete, 17 June 1904.
58. Letter from R. C. Bosanquet to Ellen Bosanquet, 19 June 1904, in Bosanquet (ed.), Robert Carr Bosanquet, *Letters and Light Verse,* 156.
59. Crete, 19 June 1904.
60. Crete, 20 June 1904.
61. Crete, 21 June 1904.
62. Crete, 24 June 1904.
63. Crete, 25 June 1904.
64. Crete, 26 June 1904.
65. Letter from Edith Hall to her sister, 8 July 1904, UPenn.
66. Crete, 27 June 1904.
67. Letter from Edith Hall to her father, 27 June 1904, UPenn.
68. Letter from Edith Hall to her mother and sister, 29 June 1904, UPenn.
69. Ibid.
70. Crete, 29 June 1904.
71. Letter from Edith Hall to her family, 7 July 1904, UPenn.
72. W. R. Halliday, 'Dawkins, Richard McGillivray', revised David Gill, *Oxford Dictionary of National Biography* (Oxford: OUP, 2004).
73. Letter from Edith Hall to her family, 7 July 1904, UPenn.
74. Crete, 1 July 1904.
75. Crete, 2 July 1904.
76. Letter from Edith Hall to her family, 7 July 1904, UPenn.
77. Crete, 2 July 1904.

78. Ibid.
79. Letter from Edith Hall to her family, 7 July 1904, UPenn.
80. Crete, 3 July 1904.
81. Letter from Edith Hall to her family, 7 July 1904, UPenn.
82. Crete, 3 July 1904.
83. Letter from Edith Hall to her family, 7 July 1904, UPenn.
84. Letter from Edith Hall to her sister, 8 July 1904, UPenn.
85. Crete, 15 July 1904.
86. Crete, 16 July 1904.
87. Crete, 20 July 1904.
88. Richard Pococke, 'Of the Island of Candia', in Pococke, *A Description of the East* (London, 1743), ii, Book IV.
89. Bate, 'Four and a half months in Crete', 193–202.
90. Crete, 21 July 1904.
91. Bate, 'Four and a half months in Crete', 193–202.
92. Letter from DMAB to Arthur Smith Woodward, 21 July 1904, DF 100/36/56.
93. Crete, 22 July 1904.
94. Crete, 23 July 1904.
95. Andries Spaan, '*Hippopotamus creutzburgi*: The Case of the Cretan Hippopotamus', in David S. Reese (ed.), *Pleistocene and Holocene Fauna of Crete and its First Settlers* (Madison, Wisconsin: Prehistory Press, 1996), 99.
96. Letter from DMAB to Edith Hall, 5 September 1904, UPenn.
97. Crete, 23 July 1904.
98. Crete, 25 July 1904.
99. Crete, 28 July 1904.
100. Crete, 30 July 1904.
101. Crete, 3 August 1904.
102. Crete, 5 August 1904.
103. Crete, 10 August 1904.
104. Becker and Betancourt, *Richard Berry Seager*, 50.
105. *New York Times*, 25 September 1904.
106. Letter from DMAB to Dr Arthur Smith Woodward, 21 July 1904, DF 100/36/56.
107. Marina Panagiotaki, 'Knossos objects: 1904 the first departure', in Gerald Cadogan, Eleni Hatzaki and Adonis Vasilakis (eds), *Knossos: Palace, City, State, Proceedings of the Conference in Herakleion organised by the British School at Athens and the 23rd Ephoreia of Prehistoric and Classical Antiquities at Herakleion in November 2000 for the Centenary of Sir Arthur Evans's Excavations at Knossos* (London: BSA Studies, 2004), xii, 565–79. For this information on Evans and his involvement in amending the archaeological laws, I am deeply grateful to Marina Panagiotaki for very generously sending me a proof copy of her article, and to Vasso Fotou for her unpublished material on Evans and the new law. See also Vasso Fotou,

'Harriet Boyd Hawes', in Getzel M. Cohen and Martha Sharp Joukowsky (eds), *Breaking Ground: Pioneering Women Archaeologists* (Ann Arbor: University of Michigan Press, 2004), 198–273.

108. Bate, 'Four and a half months in Crete', 193–202.

8: Exile in Ecclefechan, pp. 148–58

1. Letter from DMAB to Edith Hall, 5 September 1904, UPenn.
2. Letter from DMAB to Michael Oldfield Thomas, 14 November 1904, DF 232/10/27.
3. *The Pictorial Magazine*, 13 June 1903.
4. Letter from DMAB to Michael Oldfield Thomas, 14 November 1904, DF 232/10/27.
5. Tony Clubb, Verger of the Chapel Royal, Hampton Court Palace, personal communication.
6. *Dumfries and Galloway Standard*, 11 January 1905.
7. *The Westminster Gazette*, 3 May 1905.
8. *The Daily Graphic*, 5 May 1905.
9. 'Presentation of *Diplodocus carnegii*', transcript of speech, Newspaper Cuttings, Book II, NHM Archives.
10. Bate, 'On the mammals of Crete', *PZSL*, 2 (1905), 315–23.
11. N. D. Riley, *The Department of Entomology of the British Museum (Natural History) 1906–64: A Brief Historical Sketch* (London: 12th International Congress of Entomology, 1964).
12. Mary Allsebrook, *Born to Rebel: The Life of Harriet Boyd Hawes* (Oxford: Oxbow Books, 2002), 131.
13. Letter from Dr Arthur Evans to DMAB, 19 February 1906, in Cyprus [*sic*] III, 1902.
14. Bate, 'On a new genus of extinct muscardine rodent from the Balearic islands', *PZSL* (1918), 221.
15. John de Vos, *The Endemic Pleistocene Deer of Crete* (Amsterdam; Oxford: North-Holland Publishing, 1984).
16. Letter from DMAB to Dr Arthur Smith Woodward, 13 February 1906, DF 100/40/174.
17. Bate, 'The pygmy hippopotamus of Cyprus', *GM*, 5: 3 (1906), 241–55.
18. Letter from DMAB to Dr Arthur Smith Woodward, 15 August 1906, DF 100/40/175–6.
19. Letter from Dr Arthur Smith Woodward to DMAB, 15 August 1906, DF 100/40/175–6.
20. Letter from DMAB to Dr Arthur Smith Woodward, 20 August 1906, DF 100/40/177–8.
21. Standing Committee of Trustees, 27 October 1906, DF 900/5.

22. Letter from DMAB to Dr Arthur Smith Woodward, 20 August 1906, DF 100/40/177–8.
23. Bate, 'On elephant remains from Crete, with description of *Elephas creticus sp. n.*', *PZSL* (1907), 238–50.
24. Letter from DMAB to Michael Oldfield Thomas, 5 September 1906, DF 232/12.

9: A Needle in a Bundle of Hay, pp. 160–77

1. Bate, 'Preliminary note on a new artiodactyle from Majorca: *Myotragus balearicus*, gen. et sp. nov.', *GM*, 5: 6 (1909), 385–8.
2. Archduke Ludwig Salvator of Austria, *Die Balearen. Geschildert in Wort und Bild* (Leipzig; Würzburg, 1897).
3. Mary Stuart Boyd, 'Foreword and Forewarning' to *The Fortunate Isles: Life and Travel in Majorca, Minorca and Iviza* (London: Methuen & Co., 1911).
4. Letter from DMAB to Dr Arthur Smith Woodward, 12 April 1909, DF 100/47/53.
5. Bate, 'Preliminary note on a new artiodactyle from Majorca', 385–8.
6. J. A. Alcover, M. Llabrés and Ll. Moragues (eds), *Les Balears abans dels Humans*, (Palma de Mallorca: Societat d'Història Natural de les Balears, 2000), 45.
7. Bate, 'On the Pleistocene ossiferous deposits of the Balearic islands', *GM,* 6: 1 (1914), 337–45.
8. Bate, 'Preliminary note on a new artiodactyle from Majorca', 385–8.
9. Letter from DMAB to Dr Arthur Smith Woodward, 16 May 1909, DF 100/47/55.
10. Dr C. W. Andrews, 'A description of the skull and skeleton of a peculiarly modified rupicaprine antelope (*Myotragus balearicus* Bate)', *GM*, 6: 1 (1914), 378.
11. Bate, 'Preliminary note on a new artiodactyle from Majorca', 385–8.
12. Dr C. W. Andrews, 'A description of the skull and skeleton of a peculiarly modified rupicaprine antelope (*Myotragus balearicus* Bate) with a notice of a new variety, *M. balearicus* var. *major*', *PTRS*, series B, 206 (1915), 281–305.
13. Alcover, Llabrés and Moragues (eds), *Les Balears abans dels Humans*, 35.
14. Josep Antoni Alcover, Ramon Perez-Obiol, Errikarta-Imanol Yll and Pere Bover, 'The diet of *Myotragus balearicus* Bate 1909 *(Artiodactyla caprinae)*, an extinct bovid from the Balearic islands: evidence from coprolites', *BJLS*, 66: 1 (January 1999), 57–74.
15. Letter from DMAB to Dr Arthur Smith Woodward, 22 March 1910, DF 100/49/60.
16. Letter from DMAB to Dr Arthur Smith Woodward, 29 March 1910, DF 100/49/61.
17. Letter from DMAB to Dr Arthur Smith Woodward, 3 June 1910, DF 100/49/63.

18. *Annals of the Cyprus Natural History Society*, 1 (Nicosia, 1910).
19. Letter from DMAB to Dr Arthur Smith Woodward, 3 June 1910, DF 100/49/63.
20. Majorca II, 8 July 1910.
21. Majorca II, 9 July 1910.
22. Majorca II, 12 July 1910.
23. Majorca II, 14 July 1910.
24. Ibid.
25. Ibid.
26. Majorca II, 15 July 1910.
27. Majorca II, 16 July 1910.
28. C. W. Andrews, 'Note on a mounted skeleton of *Myotragus balearicus* Bate', *GM*, 6: 2 (1915), 337–9.
29. Pere Bover and Josep Antoni Alcover, 'The evolution and ontogeny of the dentition of *Myotragus balearicus* Bate 1909: evidence from new fossil data', *BJLS*, 68 (1999), 401–28.
30. Majorca II, 16 July 1910.
31. Majorca II, 19 July 1910.
32. Majorca II, 21 July 1910.
33. Majorca II, opp. p. 41.
34. Majorca II, 29 July 1910.
35. Majorca II, 31 July 1910.
36. Majorca II, 1 August 1910.
37. Bate, 'On the Pleistocene ossiferous deposits of the Balearic islands', 337–45.
38. Majorca II, 1 August 1910.
39. Majorca II, 2 August 1910.
40. Majorca II, 4 August 1910.
41. Bate, 'On the Pleistocene ossiferous deposits of the Balearic islands', 337–45.
42. Majorca II, 7 August 1910.
43. Majorca II, 8 August 1910.
44. Majorca II, 11 August 1910.
45. Frederick Chamberlin, *The Balearics and their Peoples* (London: John Lane the Bodley Head, 1927), 44.
46. Majorca II, 12 August 1910.
47. Majorca II, 13 August 1910.
48. Majorca II, 14 August 1910.
49. Majorca II, 16 August 1910.

10: The Fates Seem against Me, pp. 179–95

1. Letter from DMAB to Dr Arthur Smith Woodward, 10 December 1910, DF 100/49/65.
2. Letter from DMAB to Michael Oldfield Thomas, 18 March 1911, DF 232/17/18.

3. Bate, 'On the Pleistocene ossiferous deposits of the Balearic islands', *GM*, 6: 1 (1914), 337–45.
4. Application form for Percy Sladen Memorial Fund, 11 April 1911, PSMF Archives, DMAB file.
5. Letter from Dr Henry Woodward to Mr Bailey Saunders, Chairman of Trustees of Percy Sladen Memorial Fund, 17 April 1911, PSMF Archives.
6. Balearics III, 27 September 1911.
7. Gaston Vuillier, *The Forgotten Isles: Impressions of Travel in the Balearic Isles, Corsica and Sardinia*, translated by Frederic Breton (London: Hutchinson, 1897), 110.
8. Balearics III, 6 October 1911.
9. Ibid.
10. Balearics III, 5 October 1911.
11. Balearics III, 2 October 1911.
12. Balearics III, 3 October 1911.
13. Bate, 'On the Pleistocene ossiferous deposits of the Balearic islands', 337–45.
14. Josep Antoni Alcover, 'Mammals of the Pityusic Islands', in H. Kuhbier, J. A. Alcover and C. Guerau d'Arellano Tur (eds), *Biogeography and Ecology of the Pityusic Islands* (The Hague: Dr W. Junk, 1984), 456.
15. Paul Y. Sondaar, Miguel McMinn, Bartomeu Seguí and Josep Antoni Alcover, 'Palaeontological interest of Karstic deposits from the Gymnesic and Pityusic islands', in A. and J. Ginés (eds), *Karst and Caves in Mallorca* (Palma de Mallorca: Federació Balear d'Espeleologia, 1995), 155.
16. J. A. Alcover, M. McMinn and C. R. Altaba, 'Eivissa: a Pleistocene oceanic-like island in the Mediterranean', *National Geographic Research and Exploration*, 10: 2 (1994), 236–8.
17. Balearics III, 7 October 1911.
18. Balearics III, 11 October 1911.
19. Balearics III, 12 October 1911.
20. Ibid.
21. Balearics III, 13 October 1911.
22. Balearics III, 14 October 1911.
23. Arthur Keith, *Ancient Types of Man* (London: Harper & Brothers, 1911), 118.
24. Keith, *Ancient Types of Man*, xii.
25. Frederick Chamberlin, *The Balearics and their Peoples* (London, 1927), 233.
26. Balearics III, 14 October 1911.
27. Majorca II, 7 August 1910.
28. Damià Ramis and Josep Antoni Alcover, 'Revisiting the earliest human presence in Mallorca', *PPS*, 67 (2001), 261–9.
29. Balearics III, 14 October 1911.
30. Balearics III, 15 October 1911.
31. Balearics III, 17 October 1911.
32. Balearics III, 20 October 1911.

33. Bate, 'On the Pleistocene ossiferous deposits of the Balearic islands', 337–45.
34. Balearics III, 21 October 1911.
35. Ibid.
36. Balearics III, 26 October 1911.
37. Balearics III, 28 October 1911.
38. Balearics III, 1 November 1911.
39. Balearics III, 3 November 1911.
40. Balearics III, 9 November 1911.
41. Balearics III, 13 November 1911.
42. Ibid.
43. Balearics III, 14 November 1911.
44. Balearics III, 16 November 1911.
45. Balearics III, 18 November 1911.
46. Balearics III, 20 November 1911.
47. Balearics III, 25 November 1911.
48. Balearics III, 28 November 1911.
49. Balearics III, 30 November 1911.
50. C. W. Andrews, 'Note on a mounted skeleton of *Myotragus balearicus* Bate', *GM*, 6: 2 (1915), 337–9.
51. Juan Flaquer y Fabregues, 'Dorothea M. A. Bate', *Revista de Menorca*, Año XXIV (Quinta Época), tomo XV, Mahon (1920).
52. J. A. Alcover, S. Moyà-Solà, J. Pons-Moyà, *Les Quimeres del Passat* (Mallorca: Ciutat de Mallorca, 1981).
53. *Naturalesa i Societat*, Circular de la Societat d'Història Natural de les Balears, 25 (October 2001).

11: *War and Other Reasons, pp. 196–212*

1. Letter from DMAB to Dr Arthur Smith Woodward 20 December 1911, DF 100/51/45.
2. Standing Committee of Trustees, 24 February 1912, DF 900/7.
3. Standing Committee of Trustees, 22 February 1913, DF 900/8.
4. *The Times,* 31 January 1913.
5. Juan Flaquer y Fabregues, 'Dorothea M. A. Bate', *Revista de Menorca*, Año XXIV (Quinta Época), Tomo XV, Mahon (1920).
6. Sir David Bate, personal communication.
7. Letter from DMAB to William Ogilvie-Grant, 23 July 1912, DF 230/22/143.
8. Sir David Bate, personal communication.
9. Letter from DMAB to Henry Woodward, 15 October 1912, DMAB file, PSMF Archives.
10. Bate, 'The Caves of Crete', in Aubyn Trevor-Battye, *Camping in Crete* (London, 1913), 239.

11. Dr Kenneth P. Oakley lecture notes, DF 116/17/83–95.
12. *The Times,* 4 December 1953.
13. Professor Chris Stringer and Andrew Currant, Pfizer Lecture, NHM, 25 November 2003.
14. Record of Dr Kenneth P. Oakley's interview with Mr Albert Victor Eade, 16 August 1969, DF 116/17/164–6.
15. Special report to Trustees, 1908, DF 104/52.
16. Standing Committee of Trustees, 22 February 1913, DF 900/8.
17. Dr C. W. Andrews, 'A description of the skull and skeleton of a peculiarly modified rupicaprine antelope (*Myotragus balearicus* Bate) with a notice of a new variety, *M. balearicus* var. *major*', *PTRS*, series B, 206 (October 1915), 281–305.
18. Sir David Bate, personal communication.
19. Geology Department account book, 1902–37, DF 102/9.
20. Handwritten note in 'Malta envelope', Malta file, P MSS BAT.
21. Bate, 'On a small collection of vertebrate remains from the Har [*sic*] Dalam cavern, Malta; with note on a new species of the genus *Cygnus*', *PZSL*, 2 (1916), 421–30.
22. Letter from Guiseppe Despott to DMAB, 3 August 1916, DF 100/109/14.
23. G. Despott, 'Archaeological investigations in Malta II: Tal Herba and Burmeghez fissures', *BAAS* (1915), 213–14.
24. G. Despott, 'Excavations conducted at Ghar Dalam (Malta) in the summer of 1917', *JRAI*, NS, 48 (1918), 214–221.
25. Sir Arthur Keith, 'Discovery of Neanderthal Man in Malta', *Nature*, 101 (1918), 404.
26. *The Star,* 24 January 1916.
27. *Evening News,* 29 January 1916.
28. Standing Committee of Trustees, 24 October 1914, DF 900/8.
29. Letter from Henry Woodward to Dr Sherborn, 3 October 1917, Handwriting Collection, NHM General Library.
30. B. N. Peach and J. Horne, 'On a bone cave in the Cambrian limestone in Assynt, Sutherlandshire', *BAAS* (1892), 720–1. The bones were identified by the palaeontologist and Fellow of the Royal Society, Edwin Tulley Newton. Peach and Horne published a further report on the cave in 1917.
31. Letter from Henry Woodward to DMAB, 26 July 1915, Accession no. 2000/19.
32. Note to the Percy Sladen Memorial Fund, 21 February 1934, DMAB file, PSMF Archives.
33. *Christian Science Monitor,* Saturday, 2 June 1915.
34. Sir David Bate, personal communication.
35. W. N. P. Barbellion [Bruce Frederick Cummings], *The Journal of a Disappointed Man* (London, 1923; reprinted Gloucester, 1984), 199.
36. Barbellion, *The Journal of a Disappointed Man*, 74.
37. Standing Committee of Trustees, 23 October 1915, DF 900/9.

38. Standing Committee of Trustees, 24 February 1917, DF 900/9.
39. *The Daily Telegraph*, 28 May 1915.
40. Bruce Frederick Cummings, *The Louse and its Relation to Disease* (London: BM (NH) Economic Series 2, 1915).

12: *Man and Mammals, pp. 213–37*

1. Sir Alan Muir-Wood, personal communication.
2. *The Times*, 1 December 1919.
3. *The Museums Journal*, May 1928.
4. Letter from Leila Luddington to Dr W. N. Edwards, dated 'Tuesday', probably in June 1951, Accession no. 2000/19.
5. Letter from Professor H. Obermaier to DMAB, 16 January 1922, File 6, MSS on Malta, Gibraltar, Balearics, etc., P MSS BAT.
6. Gertrude Caton Thompson, *Mixed Memoirs* (Gateshead: Paradigm Press, 1983).
7. M. A. Murray, *Excavations in Malta*, with a chapter by Gertrude Caton Thompson (London: Bernard Quaritch, 1923), part I. 6–7.
8. M. A. Murray, *Excavations in Malta* (London: Bernard Quaritch, 1925), part II. 10.
9. Letter from DMAB to Professor Raymond Vaufrey, 11 January 1929, DF 100/218/13.
10. Chris Stringer and Clive Gamble, *In Search of the Neanderthals* (London: Thames & Hudson, 1993), 78.
11. Kenneth P. Oakley, 'Ghar Dalam', in Kenneth P. Oakley, Bernard Grant Campbell and Theya Ivitsky Molleson (eds), *Catalogue of Fossil Hominids, Part II: Europe,* (London: BM (NH), 1971), 264.
12. These include C. Savona-Ventura, 'Palaeolithic Man in Malta', *The Sunday Times*, 20 July 1997 and 3 August 1997.
13. Letter from Charles Andrews to Dr Arthur Smith Woodward, 20 May 1924, MS ADD 242/3, University College London, Library Services.
14. Letter from F. A. Bather to Guy Pilgrim, 7 August 1925, DF 100/178/1.
15. Keeper's report to the Sub-Committee on Geology etc, 19 January 1925, DF 104/55.
16. Ibid.
17. Letter from DMAB to Dr William Lang, 4 January 1929, DF 100/80/42.
18. Keeper's report to the Sub-Committee on Geology etc, 1924, DF 104/55.
19. Letter from Professor Alfred Ely Day to F. A. Bather, 24 August 1924, DF 100/108/10.
20. Letter from Professor Alfred Ely Day to DMAB, 25 July 1925, DF 100/108.
21. F. Turville-Petre, *Researches in Prehistoric Galilee 1925–1926* (London: British School of Archaeology in Jerusalem, 1927), 18.
22. P. L. O. Guy, 'Prehistoric Man in Palestine', *The Times*, 24 August 1925.

23. *The Times*, 25 August 1925.
24. Letter from DMAB to F. A. Bather, 15 September 1925, 'Notes on the Palestine caves', P MSS BAT.
25. Letter from F. A. Bather to DMAB, 8 October 1925, 'Notes on the Palestine caves', P MSS BAT.
26. Bate, MSS list of birds headed: 'Lake Huleh', 21 September 1925, in Wonnacott Collection, Birds, Box 4, P MSS BAT.
27. Letter from F. A. Bather to Sir Arthur Keith, 30 October 1925, 'Notes on the Palestine caves', P MSS BAT.
28. Letter from Sir Arthur Keith to F. A. Bather, 6 November 1925, 'Notes on the Palestine caves', P MSS BAT.
29. Letter from Francis Turville-Petre to DMAB, 28 August 1926, DF 100/216/14.
30. Letter from DMAB to Professor John Garstang, undated, DF 100/120/18.
31. Bate, 'On the Animal Remains Obtained from the Mugharet-el-Zuttiyeh in 1926', in Turville-Petre, *Researches in Prehistoric Galilee 1925–1926*, 48.
32. Bate, 'On the Animal Remains Obtained from the Mugharet-el-Zuttiyeh in 1926', in Turville-Petre, *Researches in Prehistoric Galilee 1925–1926*, 47.
33. MSS notes for monthly report, July 1928, reports file 1934, in Reports 1922–50, P MSS BAT.
34. Martin A. C. Hinton, 'Preliminary note upon the mammalian remains from Merlin's Cave', *PBSS*, 2: 2 (1926), 156.
35. Meare Lake Village Reports 1924–36, in Reports 1922–50, P MSS BAT.
36. A. P. Currant, 'The late glacial mammal fauna of Gough's Cave, Cheddar, Somerset', *PBSS*, 17: 3 (1986), 286–304.
37. Letter from DMAB to R. F. Parry, 5 April 1928, Gough's Cave correspondence, Accession no. 2004/87.
38. Letter from DMAB to R. F. Parry, 8 August 1928, Gough's Cave correspondence, Accession no. 2004/87.
39. Currant, 'The late glacial mammal fauna of Gough's Cave, Cheddar, Somerset', 286–304.
40. Sylvia Benton, 'Excavation of the Sculptor's Cave, Covesea, Morayshire', *PSAS*, 65 (12 January 1931), 177–216.
41. Letter from DMAB to Sylvia Benton, 15 July 1930, DF 100/82.
42. Letter from Sylvia Benton to DMAB, 16 November 1930, DF 100/82.
43. Letter from Sylvia Benton to DMAB, 15 January 1931, DF 100/82.
44. Letter from Sylvia Benton to DMAB, 21 January 1931, DF 100/82.
45. Letter from DMAB to Sylvia Benton, 31 January 1931, DF 100/82.
46. Letter from DMAB to Sylvia Benton, 15 January 1932, DF 100/82.
47. Letter from Sylvia Benton to DMAB, 28 January 1932, DF 100/82.
48. Keeper's Report to the Sub-Committee on Geology etc., 1928, DF 104/58.
49. Sir Mortimer Wheeler, *Still Digging* (London: Pan Books, 1958), 64.
50. Letter from Dorothy Garrod to Dr Arthur Smith Woodward, 10 November 1923, DF 100/120/17.

51. Ofer Bar-Yosef and Jane Callander, 'A forgotten archaeologist: the life of Francis Turville-Petre', *PEQ*, 129 (1997), 2–18.
52. M. A. C. Hinton, *Monograph of the voles and lemmings (Microtinae) living and extinct* (London: BM (NH), 1926).
53. Bate, 'The animal remains', 92, in Garrod *et al.*, 'Excavation of a Mousterian rock shelter at Devil's Tower, Gibraltar', *JRAI*, 58 (1928), 33–113.
54. Joanne Henrietta Cooper, 'Late Pleistocene Avifaunas of Gibraltar and their Palaeoenvironmental Significance', 1999, unpublished Ph.D. thesis, University of London.
55. PEF/DA/BSAJ (UK), AGM, 18 November 1927.
56. D. A. E. Garrod and D. M. A. Bate, 'Excavations at the Cave of Shukbah, Palestine, 1928', *PPS*, 8 (1942), 1–20. The spelling of 'Shukbah' varies: today it is more usually found as 'Shukba' or 'Shuqba'.
57. D. A. E. Garrod, 'Excavations of a Palaeolithic cave in western Judaea', *BAAS* (1928).
58. D. A. E. Garrod, 'A new Mesolithic industry: the Natufian of Palestine', *JRAI*, 62 (1932), 257–69.
59. Letter from DMAB to John Crowfoot, undated, 'Notes on the Palestine caves', P MSS BAT.
60. Letter from DMAB to John Crowfoot, 3 September 1928, 'Notes on the Palestine caves', P MSS BAT.
61. Letter from John Crowfoot to DMAB, 4 October 1928. 'Notes on the Palestine caves', P MSS BAT.
62. Garrod, 'Excavations at the Cave of Shukbah, Palestine 1928', 1–20.
63. Brian Boyd and Zoë Crossland, 'New fieldwork at Shuqba cave and in Wadi en-Natuf, Western Judea', *Antiquity*, 74: 286 (December 2000), 755–6.
64. Letter from Dorothy Garrod to DMAB, 16 June 1929, DF 100/120.
65. Bate, untitled MSS report on the Mugharet el-Wad, 1 September 1930, 'Notes on the Palestine caves', P MSS BAT.
66. Garrod, 'A new Mesolithic industry: the Natufian of Palestine', 257–69.
67. Letter from Dorothy Garrod to DMAB, 16 June 1929, DF 100/120.
68. Garrod, 'A new Mesolithic industry: the Natufian of Palestine', 257–69.
69. Bate, 'Note on the fauna of the Athlit caves', *JRAI*, 62 (1932), 277–9.
70. Ibid.
71. T. D. McCown, 'Appendix VI: Excavation of the Human Burials of the Mugharet es-Skhul', in D. A. E. Garrod and D. M. A. Bate, *The Stone Age of Mount Carmel* (Oxford: Clarendon Press, 1937), i. 132–3.
72. T. D. McCown and Sir Arthur Keith, *The Stone Age of Mount Carmel: The Fossil Human Remains from the Levalloiso-Mousterian* (Oxford: Clarendon Press, 1939), ii, Preface.
73. T. D. McCown, 'Fossil men of the Mugharet es-Sukhul, 1932', *BASPR*, 9 (May 1933), 15.

74. Sir Arthur Keith, note in T. D. McCown, 'The oldest complete skeletons of man', *BASPR*, 10 (May 1934), 19.
75. D. A. E. Garrod, 'Excavations at the Wady el-Mughara, Palestine (1932–33)', *BASPR*, 10 (May 1934), 8.
76. Ibid.

13: *The Bone Beasts of Bethlehem, pp. 238–59*

1. Letter from DMAB to Sir Themistocle Zammit, 12 February 1934, Malta file, folder II, P MSS BAT.
2. 'Report on the Museums of Malta, Cyprus and Gibraltar, by Alderman Chas. Squire and D. W. Herdman, to the Carnegie Corporation of New York, 1932', quoted in *Report on the Working of the Museum Department* (Malta, 1934–5).
3. Dr J. G. Baldacchino, 'Report on the Natural History Section', in *Report on the Working of the Museum Department* (Malta, 1934–5).
4. Application for grant, Percy Sladen Memorial Fund, 21 February 1934, DMAB file, PSMF Archives.
5. Letter from Dorothy Garrod to the Percy Sladen Memorial Fund, 20 February 1934, DMAB file, PSMF Archives.
6. Letter from DMAB to Spencer Savage, 28 February 1934, DMAB file, PSMF Archives.
7. Bate, 'Two new mammals from the Pleistocene of Malta', *PZSL*, 2 (1935), 247–64.
8. Letter from DMAB to Elinor Gardner, 17 August 1934, DF 100/224/7.
9. Note in Palaeontological notes and references folder, 'Bethlehem Excavations 1935–7', P MSS BAT.
10. Undated aide-mémoire, DF 100/224.
11. Letter from Dorothy Garrod to DMAB, 26 November 1932, DF 100/120.
12. Pamela Jane Smith, 'Dorothy Garrod, first woman Professor at Cambridge', *Antiquity*, 74: 283 (March 2000), 131–6.
13. D. A. E. Garrod, 'Part I: Description and Archaeology', in Garrod and D. M. A. Bate, *The Stone Age of Mount Carmel* (Oxford: Clarendon Press 1937), i. 57–8.
14. Tabun notes in Mammal notes, Equus folder, in Mammal Notes Box, P MSS BAT.
15. Jane Callander and Ofer Bar-Yosef, 'Saving Mount Carmel Caves', *PEQ*, 132 (2000).
16. Garrod, 'Part I: Description and Archaeology', in Garrod and Bate, *The Stone Age of Mount Carmel*, i.
17. Garrod, 'Part I: Description and Archaeology', in Garrod and Bate, *The Stone Age of Mount Carmel*, i. 61 fn.

18. MSS notes on Palestine birds 1934, Palestine box file, P MSS BAT.
19. Jane Callander, personal communication.
20. Letter from Dorothy Garrod to DMAB, 22 June 1934, DF 100/120.
21. Gertrude Caton Thompson, *Mixed Memoirs* (Gateshead: Paradigm Press, 1983), 158.
22. Caton Thompson, *Mixed Memoirs*, 235.
23. Standing Committee of Trustees, 23 June 1934, DF 900/13.
24. Letter from Dorothy Garrod to DMAB, 22 June 1934, DF 100/120.
25. Bate, 'Two additions to the Pleistocene cave-fauna of Palestine (*Trionyx* and *Crocodilus*)', *AMNH*, 10: 14 (1934), 474–8.
26. Bate, 'A fossil wart-hog from Palestine', *AMNH*, 10: 13 (1934), 120–9.
27. Letter from James Starkey to DMAB, 3 July 1934, DF 100/224.
28. Letter from DMAB to Olga Tufnell, 23 January 1940, DF 100/224.
29. Letter from James Starkey to DMAB, 3 July 1934, DF 100/224.
30. Letter from Sir Henry Wellcome to DMAB, 16 October 1934, DF 100/224.
31. Letter from DMAB to the editor of *Nature*, 3 August 1934, Reports 1934, in Reports 1922–50, P MSS BAT.
32. Bate, 'Discovery of a fossil elephant in Palestine', *Nature*, 134: 1 (1934), 219.
33. Letter from James Starkey to DMAB, 2 October 1934, DF 100/224.
34. Letter from James Starkey to DMAB, 4 October 1934, DF 100/224.
35. Letter from DMAB to James Starkey, 5 October 1934, DF 100/224.
36. Letter from Sir Henry Wellcome to DMAB, 16 October 1934, DF 100/224.
37. WA/HSW/AR/Lac/A.7, Geological Section, first report, 28 October–30 November 1934.
38. Ibid.
39. WA/HSW/AR/Lac/A.7, Geological Section, first report, 28 October–30 November 1934, f. 4.
40. Letter from DMAB to James Starkey, 26 December 1934, DF 100/224.
41. WA/HSW/AR/Lac/A.7, Geological Section, second report, 1 December 1934–31 March 1935.
42. WA/HSW/AR/Lac/A.7, Geological Section, second report, 1 December 1934–31 March 1935, f. 1.
43. Register of specimens, MSS Bethlehem excavations, P MSS BAT.
44. Bethlehem maps and plans, 1935–7, P MSS BAT.
45. WA/HSW/AR/Lac/A.7, Geological Section, third report, 1–30 April 1935, f. 10.
46. WA/HSW/AR/Lac/A.7, Geological Section, first report, 28 October–30 November 1934, f. 1.
47. WA/HSW/AR/Lac/A.7, Geological Section, first field report, 9 March–1 April 1935, f. 2.
48. Bethlehem 1935.
49. Although this is from an undated and unaddressed partial draft of a letter, in 1940 she was to offer this information to the archaeologist Dr Moshe

Stekelis. Malta Bos Equus folder, in Malta, Gibraltar, Corsica, Baleares, Sardinia Box, P MSS BAT.

50. WA/HSW/AR/Lac/A.7, Geological Section, Bethlehem, third report, 1–30 April 1935, f. 11.
51. WA/HSW/AR/Lac/A.7, Geological Section, Bethlehem, third report, 1–30 April 1935, f. 4.
52. Letter from DMAB to James Starkey, 18 July 1935, DF 100/224.
53. Letter from DMAB to Dr Kalman Lambrecht, 9 June 1931, DF 100/151/22.
54. Henry Fairfield Osborn, *Proboscidea: A Monograph of the Discovery, Evolution, Migration and Extinction of the Mastodonts and Elephants of the World*, 2 vols (New York, 1936 and 1942), ii. 1257.
55. Letter from Sir Arthur Smith Woodward to DMAB, 29 August 1935, DF 116/36/3–4.
56. Gertrude Caton Thompson, *Mixed Memoirs* (Gateshead: Paradigm Press, 1983), 160.
57. P. R. Lowe, 'Struthious remains from northern China and Mongolia: with descriptions of *Struthio wimani*, *Struthio andersoni* and *Struthio mongolicus*, with a note by D. M. A. Bate on remains of carinate birds', *Palaeontologia Sinica*, series C, 6, fasc. 4 (1931), 1–47.
58. Letter from Professor Carl Wiman to DMAB, 26 November 1935, DF 100/228.
59. Letter from Professor Carl Wiman to DMAB, 30 November 1935, DF 100/228.
60. WA/HSW/AR/Lac/A.4, Geological Section, first report of the expedition [Gardner], 13–28 March 1936, f. 1.
61. Sir Harry Luke and Edward Keith-Roach (eds), *The Handbook of Palestine and Trans-Jordan*, advertising section (London: Macmillan & Co., 1934), 11.
62. Luke and Keith-Roach (eds), *The Handbook of Palestine and Trans-Jordan*, 161.
63. Luke and Keith-Roach (eds), *The Handbook of Palestine and Trans-Jordan*, 162.
64. WA/HSW/AR/Lac/A.4, Geological Section, first report of the expedition [Gardner], 13–28 March 1936, f. 2.
65. WA/HSW/AR/Lac/A.4, Geological Section, first report of the expedition [Starkey], 13–28 March 1936, f. 1.
66. WA/HSW/AR/Lac/A.4, Geological Section, first report of the expedition [Gardner], 13–28 March 1936, f. 1.
67. Bethlehem 1936.
68. Luke and Keith-Roach (eds), *The Handbook of Palestine and Trans-Jordan*, 156.
69. Bethlehem 1936.
70. Ibid.
71. Ibid.
72. 'Zerka Main', 30 April 1936, MSS Bethlehem Excavations 1935–7, P MSS BAT.
73. Ibid.

74. Bethlehem 1936.
75. Ibid.
76. Letter from C. T. Trechmann to DMAB, 17 October 1936, Misc. letters file, Misc. Palestine Box, P MSS BAT.
77. Letter from DMAB to C. T. Trechmann, 22 October 1936, Misc. letters file, Misc. Palestine Box, P MSS BAT.
78. Letter from DMAB to J. Baldacchino, 10 July 1936, Malta file, folder II, P MSS BAT.
79. Letter from G. Hudson Lyall to Sir Charles Marston, 11 August 1936, WA/HSW/AR/Lac/B11 [Letters file].
80. Letter from G. Hudson Lyall to G. E. Pearson, 12 August 1936, WA/HSW/AR/Lac/B11 [Letters file].
81. Letter from J. L. Starkey to G. Hudson Lyall, 13 October 1936, WA/HSW/AR/Lac/B11 [Letters file].
82. Letter from J. L. Starkey to G. Hudson Lyall, 12 December 1936, WA/HSW/AR/Lac/B11 [Letters file].
83. Letter from J. L. Starkey to G. Hudson Lyall, 11 February 1937, WA/HSW/AR/Lac/B11 [Letters file].
84. Letter from G. Hudson Lyall to Martin Price, 17 February 1937, WA/HSW/AR/Lac/B11 [Letters file].

14: The Stone Age of Mount Carmel, pp. 260–78

1. Bethlehem 1937 I.
2. Ibid.
3. Bethlehem 1937 II.
4. WA/HSW/AR/Lac/A.5, Geological Section, first report of the expedition [Gardner], 16 March–15 April 1937, f. 2.
5. Bethlehem 1937 II.
6. Gertrude Caton Thompson, *Mixed Memoirs* (Gateshead: Paradigm Press, 1983), 176.
7. Bethlehem 1937 II.
8. Ibid.
9. WA/HSW/AR/Lac/A.5, Geological Section, second report of the expedition [Gardner], 16–28 April 1937, f. 7.
10. WA/HSW/AR/Lac/A.5, Geological Section, second report of the expedition [Gardner], 16–28 April 1937, f. 9.
11. Letter from Dorothy Garrod to DMAB, 21 August 1937, DF 100/120.
12. Letter from DMAB to J. G. D. Clark, 29 June 1937, Misc. Palestine mammals envelope, Misc. Palestine file, P MSS BAT.
13. Bate, 'Note on the Pleistocene mammalian fauna of Palestine', 27 June 1937, Misc. Palestine mammals envelope, Misc. Palestine file, P MSS BAT.

14. Letter from DMAB to Dr J. G. D. Clark, 30 June 1937, Misc. Palestine mammals envelope, Misc. Palestine file, P MSS BAT.
15. Bate, 'Palaeontology: The Fossil Fauna of the Wady el-Mughara Caves', in Dorothy A. E. Garrod and Dorothea M. A. Bate, *The Stone Age of Mount Carmel* (Oxford: Clarendon Press, 1937), i. Part II. 139.
16. Theodore McCown and Sir Arthur Keith, *The Stone Age of Mount Carmel: The Fossil Human Remains from the Levalloiso-Mousterian* (Oxford: Clarendon Press, 1939), ii. viii.
17. Obituary of Professor Frederick Zeuner, *PGA*, 75: 1 (1964), 117–20.
18. Bate, 'Palaeontology: The Fossil Fauna of the Wady el-Mughara Caves', in Garrod and Bate, *The Stone Age of Mount Carmel*, i. Part II. 226.
19. Bate, 'Palaeontology: The Fossil Fauna of the Wady el-Mughara Caves', in Garrod and Bate, *The Stone Age of Mount Carmel*, i. Part II. 139.
20. Bate, 'Palaeontology: The Fossil Fauna of the Wady el-Mughara Caves', in Garrod and Bate, *The Stone Age of Mount Carmel*, i. Part II. 141, fig. 1.
21. Bate, 'Palaeontology: The Fossil Fauna of the Wady el-Mughara Caves', in Garrod and Bate, *The Stone Age of Mount Carmel*, i. Part II. 150.
22. Bate, 'Palaeontology: The Fossil Fauna of the Wady el-Mughara Caves', in Garrod and Bate, *The Stone Age of Mount Carmel*, i. Part II. 149.
23. Bate, 'Pleistocene *Murinae* from Palestine', *AMNH,* 11th series, 9: 55 (July 1942), 465–86.
24. Abbé Henri Breuil, review of Garrod and Bate, *The Stone Age of Mount Carmel*, i, *Nature*, 141: 1 (1938), 304–6.
25. Aleš Hrdlička, review of Garrod and Bate, *The Stone Age of Mount Carmel*, i, *American Journal of Physical Anthropology*, 23: 4 (1938), 468.
26. Dr J. G. D. Clark, review of Garrod and Bate, *The Stone Age of Mount Carmel*, i, *PPS*, NS, 3 (1937), 486–8.
27. D. A. Hooijer, 'An Early Pleistocene mammalian fauna from Bethlehem', *BBM(NH)G*, 3: 8 (1958), 265–92.
28. Elinor W. Gardner and Dorothea M. A. Bate, 'The Bone-bearing Beds of Bethlehem: their fauna and industry', *Nature*, 140 (4 September 1937), 431–3.
29. Ibid.
30. Anon., review, 'Ancient fauna and early man at Bethlehem', of Gardner and Bate, 'The Bone-bearing Beds of Bethlehem', *Nature*, 140 (4 September 1937), 381.
31. Letter from L. C. Bullock to Sir Charles Marston, 28 October 1937, WA/HSW/AR/Lac/B11 [Letters file].
32. Letter from L. C. Bullock to Sir Charles Marston, 11 January 1938, WA/HSW/AR/Lac/B11 [Letters file].
33. Note on telephone call from Mr Harding, Director of Antiquities, Transjordan, 11 January 1938, WA/HSW/AR/Lac/B11 [Letters file].
34. Letter from Holbrook Bonney to Sir Charles Marston, 24 January 1938, WA/HSW/AR/Lac/B13.

35. Charles Inge and Olga Tufnell, 'Extraordinary report II, 31 January 1938, re murder of J. L. Starkey on 10 January 1938', WA/HSW/AR/Lac/B13.
36. Letter from Olga Tufnell to DMAB, 28 January 1938, DF 100/224.
37. Letter from L. C. Bullock to Sir Henry Hallett Dale, 24 January 1938, WA/HSW/AR/Lac/B11 [Letters file].
38. Cable from Sir Charles Marston to Charles Inge, 24 January 1938, WA/HSW/AR/Lac/B13.
39. *Christian Science Monitor,* 19 January 1938.
40. Letter from DMAB to Gerald Lankester Harding, 28 January 1938, DF 100/224.
41. Cyprus II, 21 November 1902.
42. Letter from Gerald Lankester Harding to DMAB, 9 February 1938, DF 100/224.
43. Letter from Charles Inge to DMAB, 19 February 1938, DF 100/224.
44. Letter from DMAB to Elinor Gardner, 28 February 1938, DF 100/224.
45. Letter from DMAB to Charles Inge, 1 March 1938, DF 100/224.
46. WA/HSW/AR/Lac/A.6, Geological Section, 7th field report, 1–15 March 1938.
47. WA/HSW/AR/Lac/A.6, Geological Section, 8th field report, 1–31 March 1938, f. 56.
48. Letter from Charles Inge to DMAB, 22 August 1938, DF 100/224.
49. Letter from DMAB to Charles Inge, 6 October 1938, DF 100/224.
50. J. d'A. Waechter and V. M. Seton-Williams, 'The excavations at the Wadi Dhobai 1937–38 and the Dhobaian industry' [vertebrate report by D. M. A. Bate and geological report by L. Picard], *Journal of the Palestine Oriental Society,* 18 (1938), Conclusion, 22.
51. Letter from A. T. Marston to DMAB, 28 March 1938, DF 100/162/8.
52. Letter from A. T. Marston to DMAB, 17 November 1938, DF 100/162.
53. Letter from W. N. Edwards to W. D. Lang, 3 January 1939, DF 100/152/8.
54. Sir David Lindsay Bate, personal communication.
55. Letter from W. N. Edwards to W. D. Lang, 7 July 1939, DF 100/152/8.
56. Letter from DMAB to Charles Inge, 14 March 1939, DF 100/224.
57. Letter from DMAB to Charles Inge, 3 August 1939, DF 100/224.

15: A Safe Place in the Country, pp. 279–94

1. Standing Committee of Trustees, 26 February 1938, DF 900/14.
2. Standing Committee of Trustees, 26 March 1938, DF 900/14.
3. Miriam Rothschild, *Dear Lord Rothschild* (London: Hutchinson, 1983), 302.
4. Letter from DMAB to Dr W. R. B. Oliver, 12 October 1939, DF 100/172/13.
5. Keeper's report to the Sub-Committee on Geology etc, 12 April 1940, DF 104/69.

6. Standing Committee of Trustees, 25 February 1939, DF 900/14.
7. *Tin Hat,* BM (NH) (1939–42), 1 (30 September 1939).
8. Letter from M. E. L. Mallowan to DMAB, 2 January 1940, Mallowan file in MSS notes, Syria and Palestine file, P MSS BAT.
9. Letter from DMAB to M. E. L. Mallowan, 10 January 1940, Mallowan file in MSS notes, Syria and Palestine file, P MSS BAT.
10. Professor H. H. Swinnerton's 'Addresses to medallists etc', Proceedings of the Geological Society of London, session 1939–40, *QJGS*, 96 (1940), lvi–lvii.
11. *QJGS*, 97 (1941), xviii.
12. Letter from Olga Tufnell to Elinor Gardner, 1 January 1940, DF 100/224.
13. Letter from DMAB to Olga Tufnell, 23 January 1940, DF 100/224.
14. Letter from DMAB to Dr Moshe Stekelis, 17 April 1940, DF 100/224.
15. Letter from Dr Moshe Stekelis to DMAB, 25 October 1940, MSS Bethlehem Excavations 1935–7, P MSS BAT.
16. Bate, 'The bone-bearing beds of Bethlehem', *Nature*, 147: 2 (June 1941), 783.
17. Bate, 'The fossil antelopes of Palestine in Natufian times', *GM*, 77: 6 (1940), 418–43.
18. Letter from DMAB to Leslie Bairstow, 24 November 1940, DF 100/80.
19. Letter from Leslie Bairstow to DMAB, 30 November 1940, DF 100/80.
20. Letter from W. D. Lang to W. N. Edwards, 10 December 1939, DF 100/152/8.
21. Second World War evacuation list of fossil mammal specimens sent to Tattershall Castle, Lincolnshire, April/May 1941, P MSS BAT.
22. Standing Committee of Trustees, 23 May 1941, DF 900/14.
23. Ralph A. Bagnold, *Sand, Wind and War: Memoirs of a Desert Explorer* (Tucson, Arizona: University of Arizona Press, 1990), 115.
24. Bagnold, *Sand, Wind and War,* 117.
25. Letter from John Wilfrid Jackson to DMAB, 13 June 1940, NHM, Zoology Department, Mammal Group.
26. Bate, MS report, 'Mammalian remains from the Gilf Kebir and Uwainat', NHM, Zoology Department, Mammal Group.
27. Letter from John Wilfrid Jackson to DMAB, 12 January 1941, NHM, Zoology Department, Mammal Group.
28. Letter from John Wilfrid Jackson to DMAB, 4 January 1941, NHM, Zoology Department, Mammal Group.
29. Letter from DMAB to Dr W. D. Lang, 11 August 1943, P MSS LAN.
30. Bate, 'New Pleistocene *Murinae* from Crete', *AMNH*, 11: 9 (1942), 41–9; and Bate, 'Pleistocene *Murinae* from Palestine', *AMNH*, 11: 9 (1942), 465–86.
31. Bate, 'Pleistocene shrews from the larger western Mediterranean islands', *AMNH*, 11: 11 (1944), 738–69.
32. Bate, 'The Pleistocene mole of Sardinia', *AMNH*, 11: 12 (1945), 448–61.

33. D. A. E. Garrod and D. M. A. Bate, 'Excavations at the cave of Shukbah, Palestine, 1928', with Bate, 'Appendix on the fossil mammals of Shukbah', *PPS*, 8 (1942), 15–20.
34. Bate, 'Note on small mammals from the Lebanon mountains, Syria', *AMNH*, 11: 12 (1945), 141–58.
35. Robert Gibson, *The Annals of Ashdon: No Ordinary Village* (Essex Record Office, 1988), 310.
36. Sir Arthur Smith Woodward, *The Earliest Englishman* (London: Watts & Co., 1948).
37. Letter from Guy Pilgrim to DMAB, 6 August 1941, MSS Bethlehem Excavations 1935–7, P MSS BAT.
38. Sir Alan Muir-Wood, personal communication.
39. Letter from DMAB to Dr Errol White, 12 December 1944, DF 100/80/42.
40. Letter from DMAB to Dr W. D. Lang, 5 November 1944, P MSS LAN.
41. Letter from J. G. Baldacchino to DMAB, 26 July 1945, DF 100/79/35.
42. Notes for 'Bethlehem report', 11 March 1944, MSS Bethlehem Excavations 1935–7, P MSS BAT.
43. Bate, undated notes, from MSS notes on Mediterranean island faunas, P MSS BAT.
44. Bate, 'Palaeontology: The Fossil Fauna of the Wady el-Mughara Caves', in D. A. E. Garrod and D. M. A. Bate, *The Stone Age of Mount Carmel* (Oxford: Clarendon Press, 1937), i. Part II. 155.
45. Letter from John Wilfrid Jackson to DMAB, 28 February 1946, NHM Zoology Department, Mammal Group.
46. Letter from DMAB to Oliver Myers, 27 March 1946, NHM Zoology Department, Mammal Group.
47. Letter from DMAB to J. Wilfred Jackson, 27 March 1946, NHM Zoology Department, Mammal Group.
48. Letter from Oliver Myers to DMAB, 1 April 1946, NHM Zoology Department, Mammal Group.

16: Into Africa, pp. 295–305

1. List headed 'Post War', in 'Notebook of collections in hand 1928–50', P MSS BAT.
2. 'Notebook of collections in hand 1928–50', P MSS BAT.
3. Letter from DMAB to Dr W. D. Lang, 31 December 1946, P MSS LAN.
4. Letter from DMAB to Dr W. D. Lang, 31 May 1948, P MSS LAN.
5. Sonia Cole, *Leakey's Luck* (London: Collins, 1975), 165.
6. Letter from DMAB to Dr W. D. Lang, 31 December 1946, P MSS LAN.
7. Standing Committee of Trustees, 17 May 1947, DF 900/15.
8. Cole, *Leakey's Luck*, 153.

9. Bate, 'The Pleistocene mammal faunas of Palestine and East Africa', in L. S. B. Leakey and Sonia Cole (eds), *Proceedings of the Pan-African Congress on Prehistory, Nairobi, 1947* (Oxford: Basil Blackwell, 1952), 38.
10. Professor J. Desmond Clark, personal communication.
11. Africa 1947 notebook, in Misc. Palestine Box, P MSS BAT.
12. 19 January 1947, Oakley papers.
13. Cole, *Leakey's Luck*, 155.
14. Africa 1947 notebook, in Misc. Palestine Box, P MSS BAT.
15. Mary Leakey, *Disclosing the Past* (London: Weidenfeld & Nicolson, 1984), 93.
16. Africa 1947 notebook, in Misc. Palestine Box, P MSS BAT.
17. L. S. B. Leakey, *By the Evidence: Memoirs 1932–1951* (London and New York: Harcourt Brace Jovanovich, 1974), 207.
18. Africa 1947 notebook, in Misc. Palestine Box, P MSS BAT.
19. Ibid.
20. 9 February 1947, Oakley papers.
21. Leakey, *Disclosing the Past*, 94.
22. Leakey, *By the Evidence*, 209.
23. Leakey, *By the Evidence*, 210.
24. 10 February 1947, Oakley papers.
25. Leakey, *By the Evidence*, 210.
26. Leakey, *Disclosing the Past*, 94.
27. Leakey, *Disclosing the Past*, 94–5.
28. Letter from DMAB to Dr W. D. Lang, 28 March 1947, P MSS LAN.
29. Professor J. Desmond Clark, personal communication.
30. Mrs Muriel Scutt, and Mrs Jasmine Kyne, Clerk of Southend Christian Science Society, personal communication.
31. Letter from Oliver Myers to DMAB, 7 May 1948, NHM Zoology Department, Mammal Group.
32. Letter from DMAB to Oliver Myers, 4 June 1948, NHM Zoology Department, Mammal Group.

17: Officer in Charge, pp. 306–15

1. Standing Committee of Trustees, 24 July 1948, DF 900/15.
2. Memo to DMAB from Thomas Wooddisse, 25 August 1948, DF 1004/719/6.
3. Letter from DMAB to Thomas Wooddisse, 13 November 1948, DF 1004/719/6.
4. Tring Museum Administration diary, 2 February 1949, NHM, Tring Museum Archive.
5. *Buckinghamshire Herald*, 19 August 1949.
6. Dr Ellis Owen, personal communication.

7. Dr Ellis Owen and Dr Jean Ingles, personal communication.
8. Dr Ellis Owen, personal communication.
9. Camille Arambourg, *Contribution à l'étude géologique et paléontologique du bassin de lac Rodolphe et de la basse vallée de l'Omo, II: Paléontologie*, Mission Scientifique de l'Omo, 1932–3 (Paris, 1947), 232–562.
10. Letter from Dr Louis Leakey to DMAB, 21 June 1948, in 'Miss Bate, Pig Notes: E A Pigs', Accession no. 2002/23.
11. Letter from DMAB to Dr Louis Leakey, 9 July 1948, in 'Miss Bate, Pig Notes: E A Pigs', Accession no. 2002/23.
12. Letter from Dorothy Garrod to DMAB, 18 July 1948, DF 100/120.
13. Letter from DMAB to Dorothy Garrod, 2 August 1948, DF 100/120.
14. A. J. Arkell, *Early Khartoum* (Oxford: OUP, 1949), 1.
15. Arkell, *Early Khartoum*, vii.
16. Bate, 'The fauna of Esh Shaheinab', *The Archaeological News Letter*, 2: 8 (January 1950), 128–9.
17. A. J. Arkell, note on the fauna, in Arkell, *Shaheinab* (London: OUP, 1953), Chap. 3.

18: Swan Song, pp. 316–26

1. Letter from Father J. Franklin Ewing to DMAB, 18 February 1950, Ksâr 'Âkhil file, photographs box, P MSS BAT.
2. Letter from DMAB to Mrs Marjorie Firth, 2 March 1950, DF 5012/42.
3. W. N. Edwards, Dorothea Bate obituary, for private circulation in the Department of Geology, 18 January 1951, Accession no. 2000/19.
4. Letter from DMAB to Thomas Wooddisse, 17 July 1950, DF 1004/719/6.
5. Letter from DMAB to S. Hazzeldine-Warren, 5 October 1950, Nazeing file, Mammal Box, P MSS BAT.
6. MS note headed 'London', in Sudan–Egypt file, P MSS BAT.
7. Letter from Carleton S. Coon to DMAB, 19 October 1950, 'Coon letters to Bate' envelope, P MSS BAT.
8. Letter from Martin Hinton to DMAB, 29 October 1950, Nazeing file, Mammal notes Box, P MSS BAT.
9. Letter from DMAB to Mrs Marjorie Firth, 7 December 1950, DF 5012/42.
10. Tring Museum Administration diary, 20 December 1950, NHM, Tring Museum Archive.
11. Letter from DMAB to Thomas Wooddisse, 1 January 1951, DF 1004/719/5/257.
12. Ibid.
13. Sir David Bate, personal communication.
14. Gertrude Caton Thompson, *Mixed Memoirs* (Gateshead: Paradigm Press, 1983), 235.

15. Letter from Leila Luddington to W. N. Edwards, undated, Accession no. 2000/19.
16. Letter from Dorothy Garrod to Kenneth Oakley, 6 March 1951, Garrod Box 1951, Cambridge University Museum of Archaeology and Anthropology Archives.
17. Hildegard Howard, 'Tribute to Dorothea Bate', *The Auk*, 69 (August 1952), 491.
18. Professor Raymond Vaufrey, 'Tribute to Dorothea Bate', *L'Anthropologie*, 55 (1951), 555.
19. Letter from Dorothy Garrod to Father J. Franklin Ewing, 12 March 1951, Garrod Box 1951, Cambridge University Museum of Archaeology and Anthropology Archives.
20. Letter from J. Desmond Clark to W. N. Edwards, 12 November 1951, Cyrenaica file, P MSS BAT.
21. Professor J. Desmond Clark, personal communication.
22. A. J. Arkell, 'The Fauna', in *Shaheinab* (Oxford: OUP, 1953), 11.
23. A. J. Arkell, 'Foreword' to D. M. A. Bate, 'The Pleistocene fauna of two Blue Nile sites: the mammals from Singa and Abu Hugar', *Fossil Mammals of Africa*, BM (NH) 2, (1951), vi.
24. A. J. Arkell, 'Zoology and prehistoric archaeology: an opportunity', *The Archaeological News Letter*, 3: 11 (May 1951).
25. Ibid.
26. Letter from W. N. Edwards to A. J. Arkell, 26 July 1951, DF 100/78/1.
27. Letter from A. J. Arkell to W. N. Edwards, 24 July 1951, DF 100/78.
28. Letter from W. N. Edwards to A. J. Arkell, 26 July 1951, DF 100/78.
29. Letter from A. J. Arkell to W. N. Edwards, 13 August 1951, DF 100/78.
30. Letter from W. N. Edwards to A. J. Arkell, 26 July 1951, DF 100/78.
31. Keeper's report to the Subcommittee on Geology, etc., 10 May 1951, DF 103/124.
32. Letter from A. J. Arkell to W. N. Edwards, 1 May 1951, DF 100/78.
33. L. S. B. Leakey, 'Some East African Pleistocene *Suidae*', *Fossil Mammals of Africa*, 14 BM (NH), Department of Geology [Mammals], 1958.
34. Standing Committee of Trustees, 27 January 1951, DF 900/15.
35. Letter from W. N. Edwards to Leila Luddington, 22 May 1951, Accession no. 2000/19.
36. Letter from Leila Luddington to W. N. Edwards, 14 November 1951, Accession no. 2000/19.
37. Letter from W. N. Edwards to Leila Luddington, 22 May 1951, Accession no. 2000/19.
38. Andrew Currant, Curator of Quaternary Mammals, NHM, personal communication.
39. E. E. Allen and J. G. Rutter, *A Survey of the Gower Caves* (Swansea: Vaughan Thomas, 1948), 5.

40. MS note, in 'collections in hand notebook', P MSS BAT.
41. Letter from Henry Woodward to DMAB, 26 July 1915, Accession no. 2000/19.
42. W. N. Edwards, 'Obituary of Dorothea Bate', *Proceedings of the Geological Society of London* (1951), lvi–lviii.

SELECT BIBLIOGRAPHY

A. Primary Sources

I. MANUSCRIPT SOURCES

Cambridge University Museum of Archaeology and Anthropology Archives
Dorothy Garrod letters: Box 1951

Carmarthenshire Record Office
Carmarthen Quarter Sessions Minute Books, 1866–1891

Dyfed-Powys Police Museum
Superintendent Henry Reginald Bate, personnel record, 24 December 1877

Gaze family papers
Papers and photographs relating to Wodehouse family (private collection)

Linnean Society of London
Dorothea Bate file: Percy Sladen Memorial Fund Archives, deposited at the Linnean Society of London

The National Archives (Public Record Office)
War Office: Office of the Commander-in-chief: memoranda and papers, Viscount Hardinge, WO 31/1102
Commissary General of Musters Office and successors: General Muster Books and Pay Lists, 77th Foot 1st Battalion, WO 12
Colonial Office: Cyprus, Original Correspondence, Despatches 1901–2, CO 67
Foreign Office: Affairs of Crete. Further Correspondence 1904, FO 421, FO 195

Kenneth P. Oakley papers
MS notes on visit to Egypt and East Africa, 1947: in possession of Giles Oakley (private collection)

Palestine Exploration Fund Collections
Archive of the British School of Archaeology in Jerusalem: PEF/DA/BSAJ(UK)

Smith College Archives
Harriet Boyd Hawes papers: Box 4

University of London
Sir Arthur Smith Woodward papers: MS ADD 242, Library Services, University College London
Dr Joanne Henrietta Cooper, 'Late Pleistocene Avifaunas of Gibraltar and their Palaeoenvironmental Significance', unpublished PhD thesis 1999

Archives of University of Pennsylvania Museum of Archaeology and Anthropology
Edith Hall Dohan papers – Crete
Richard Berry Seager papers – Crete

Wellcome Library of the History of Medicine
Wellcome Archaeological Research Expedition to the Near East (WARENE), reports and correspondence: WA/HSW/AR/LAC/A series; and WA/HSW/AR/LAC/B11 and /B13

The Natural History Museum

All material from the Natural History Museum Library and Archives collections is used here by permission of the Trustees of the Natural History Museum.

Natural History Museum Archives:
Central Administration:
DF1004 Director's case and policy files
Board of Trustees:
DF900 Minutes of the Standing Committee
DF901 Minutes of General and Sub-committee meetings
Department of Palaeontology (formerly Geology):
DF100 Departmental correspondence
DF102 Departmental finance and accounts
DF103 Reports to Trustees
DF104 Reports of progress
DF108 Departmental visitors' books
DF116 Correspondence and papers on Piltdown Man
Department of Zoology:
DF200 Keeper's correspondence

DF207Departmental finance and accounts
DF230 Bird Section correspondence
DF232 Mammal Section correspondence
Accession files
No. 2000/19: Dorothea Bate obituaries and letters relating to her death
No. 2002/23: 'Miss Bate's Pig Notes: E.A. Pigs'
No. 2004/87: Gough's Cave correspondence

Natural History Museum, Earth Sciences Library:
Dorothea M. A. Bate MSS: Small Library P MSS BAT
Dr William Dixon Lang MSS: Small Library P MSS LAN
Sir Arthur Smith Woodward MSS: Small Library P MSS WOO

Natural History Museum, General Library:
Handwriting Collection
Charles Davies Sherborn, *Index Animalium* MS notes: MSS SHE A

Natural History Museum, Zoology Department, Mammal Group
Dorothea Bate, general letters file

Natural History Museum, Tring Museum Archive
Administration Diaries 1949 and 1950

II. PRINTED SOURCES

Books

Dates are of editions consulted, not necessarily of first publication

Alcover, Josep Antoni, 'Mammals of the Pityusic Islands', in H. Kuhbier, J. A. Alcover and C. Guerau d'Arellano Tur (eds), *Biogeography and Ecology of the Pityusic Islands*, The Hague: Dr W. Junk, 1984
—, S. Moyà-Solà and J. Pons-Moyà, *Les Quimeres del Passat*, Mallorca: Ciutat de Mallorca, 1981
—, M. Llabrés and Ll. Moragues (eds), *Les Balears abans dels Humans*, Palma de Mallorca: Societat d'Història Natural de les Balears, 2000
Allen, E. E. and J. G. Rutter, *A Survey of the Gower Caves*, Swansea: Vaughan Thomas, 1948
Allsebrook, Mary, *Born to Rebel: The Life of Harriet Boyd Hawes*, Oxford: Oxbow Books, 2002
Annual of the British School at Athens, 10, Session 1903–4, London, 1904
Annual Report on the working of the Museum Department 1933–4, Malta, 1934
Arambourg, Camille, *Contribution à l'étude géologique et paléontologique du*

bassin de lac Rodolphe et de la basse vallée de l'Omo. Part II: *Paléontologie*, Mission Scientifique de l'Omo, 1932–3, Paris (1947), 232–562
Arkell, A. J., *Early Khartoum*, Oxford: OUP, 1949
—, *Shaheinab*, Oxford: OUP, 1953
Bagnold, Ralph A., *Sand, Wind and War: Memoirs of a Desert Explorer*, Tucson, Arizona: University of Arizona Press, 1990
Barbellion, W. N. P. (Bruce Frederick Cummings), *The Journal of a Disappointed Man,* London: Chatto & Windus, 1923; reprinted Gloucester: Hogarth Press, 1984
Becker, Marshall J. and Philip P. Betancourt, *Richard Berry Seager: Pioneer Archaeologist and Proper Gentleman*, Philadelphia: University of Pennsylvania, Museum of Archaeology and Anthropology, 1997
Boekschoten, G. J. and P. Y. Sondaar, *The Pleistocene of the Katharo Basin (Crete) and its Hippopotamus,* Amsterdam: University of Amsterdam, 1966
Bosanquet, Ellen S. (ed.), Robert Carr Bosanquet, *Letters and Light Verse,* Gloucester: John Bellows, 1938
Boyd, Mary Stuart, *The Fortunate Isles: Life and Travel in Majorca, Minorca and Iviza,* London: Methuen & Co., 1911
Caton Thompson, Gertrude, *Mixed Memoirs,* Gateshead: Paradigm Press, 1983
Chacalli, George (native of Cyprus), *Cyprus under British rule,* printed at office of 'Phoni the Kyprou', Nicosia, 1902
Chamberlin, Frederick, *The Balearics and their Peoples,* London: John Lane the Bodley Head, 1927
Cole, Sonia, *Leakey's Luck*, London: Collins, 1975
Cohen, Getzel M. and Martha Sharp Joukowsky (eds), *Breaking Ground: Pioneering Women Archaeologists,* Ann Arbor: University of Michigan Press, 2004
Cummings, Bruce Frederick, *The Louse and its Relation to Disease: Its Life-history and Habits, and How to Deal with it,* London: BM (NH), Economic Series 2, 1915
Currelly, Charles Trick, *I Brought the Ages Home*, Toronto: Royal Ontario Museum, 1956
Cuvier, Baron Georges, *Recherches sur les Ossemens Fossiles de Quadrupèdes*, Paris, 1812
Davies, William and Ruth Charles, (eds), *Dorothy Garrod and the Progress of the Palaeolithic: Studies in the Prehistoric Archaeology of the Near East and Europe*, Oxford: Oxbow Books, 1999
Droop, J. P., *Archaeological Excavation*, Cambridge: CUP, 1915
Falconer, H., *Palaeontological Memoirs and Notes of H. Falconer, with a Biographical Sketch of the Author,* C. Murchison, *et al.* (eds), London: Robert Harwicke, 1868, 2 vols
Flitch, J. E. Crawford, *Mediterranean Moods: Footnotes of travel in the islands of Mallorca, Menorca, Ibiza and Sardinia*, London: Grant Richards, 1911

Flower, Sir William Henry, *A General Guide to the British Museum (Natural History)*, London: BM (NH), 1896

Fotou, Vasso, *New Light on Gournia. Unknown Documents of the Excavations at Gournia and other sites on the Isthmus of Ierapetra by Harriet Ann Boyd*, Series Aegaeum 9, Austin, Texas: University of Liège and University of Texas, 1993

Garrod, D. A. E. and D. M. A. Bate, *The Stone Age of Mount Carmel*, i, Oxford: Clarendon Press, 1937

Gaudry, Jean Albert, *Animaux fossiles et géologie de l'Attique: d'après les recherches faites en 1855–56, et en 1860 sous les auspices de l'Académie des sciences*, Paris: F. Savy, 1862–7

—, *Geology of the Island of Cyprus*, translated by F. Maurice, London: HMSO, 1878

Gibson, Robert, *The Annals of Ashdon: No Ordinary Village*, Essex Record Office, 1988

Girouard, Mark, *Alfred Waterhouse and The Natural History Museum*, London: Natural History Museum, 1999

Green, Col. A. O., *Cyprus, a short account of its history and present state*, Kilmacolm, Scotland: M. G. Collart, 1914

Gurney, Gerald, *Table Tennis: The Early Years*, London: International Table Tennis Federation, 1989

Haggard, H. Rider, *A Winter Pilgrimage. Being an account of travels through Palestine, Italy, and the island of Cyprus, accomplished in the year 1900*, London: Longmans & Co., 1901

Handbook of Instructions for Collectors, London: BM (NH), 1902, 1906 and 1921 edns

Hearne, Thomas, *An Epistolary Letter from T— H— to Sr H—s S—e who saved his life, and desired him to send over all the rarities he could find in his travels*, London, 1729

Hinton, Martin A. C., *Monograph of the voles and lemmings (Microtinae) living and extinct*, London: BM (NH), 1926

History of the Collections contained in the Natural History Departments of the British Museum, London: BM (NH), 1904 and 1906 edns

Horowitz, A., 'The Quaternary Environments and Palaeogeography in Israel', in Y. Yom-Tov and E. Tchernov (eds), *Zoogeography of Israel*, Dordrecht: Dr W. Junk, 1988

Howard, Esmé William, Baron Howard of Penrith, *Theatre of Life*, London: Hodder & Stoughton, 1935

Hsü, Kenneth J., *The Mediterranean was a Desert: A Voyage of the* Glomar Challenger, Princeton, NJ: Princeton University Press, 1983

Hutchinson, Sir J. T. and Claude Delaval Cobham, *Handbook of Cyprus*, London: Waterlow & Sons Ltd., 1901

Keith, Sir Arthur, *Ancient Types of Man*, London: Harper & Brothers, 1911

König, C. D. E., *Directions for Collecting Specimens of Geology and Mineralogy, for The British Museum*, London: British Museum, Department of Geology, 1837

Lawson, T. J., (ed.), *The Quaternary of Assynt and Coigach: Field Guide*, Cambridge: Quaternary Research Association, 1995

Lax, Elliot M., 'A Gazeteer of Cretan Palaeontological Localities', in David S. Reese (ed.), *Pleistocene and Holocene Fauna of Crete and its First Settlers*, Madison, Wisconsin: Prehistory Press, 1996

Leakey, Louis S. B., 'Some East African Pleistocene *Suidae*', *Fossil Mammals of Africa* 14, London: BM (NH), 1958

—, *By the Evidence: Memoirs 1932–1951*, London & New York: Harcourt Brace Jovanovich, 1974

Leakey, Mary, *Disclosing the Past*, London: Weidenfeld & Nicolson, 1984

Lewis, Mrs, *A Lady's Impression of Cyprus in 1893*, London: Remington & Co., 1894

Luke, Sir Harry and Edward Keith-Roach (eds), *The Handbook of Palestine and Trans-Jordan*, London: Macmillan & Co., 1934

McBurney, Charles B. M., *The Stone Age of Northern Africa*, Harmondsworth, Middlesex: Penguin Books, 1960

— and Richard W. Hey, *Prehistory and Pleistocene Geology in Cyrenaican Libya: A Record of Two Seasons' Geological and Archaeological Fieldwork in the Gebel Akhdar Hills*, Cambridge: CUP, 1955

McCown, Theodore D. and Sir Arthur Keith, *The Stone Age of Mount Carmel: The Fossil Human Remains from the Levalloiso-Mousterian*, ii, Oxford: Clarendon Press, 1939

Mallock, W. H., *In an Enchanted Island: or a Winter's Retreat in Cyprus*, London: Bentley & Son, 1889

Mercer, Roger, with contributions by Richard Tipping *et al.*, *Kirkpatrick Fleming, Dumfriesshire: An Anatomy of a Parish in Southwest Scotland*, Dumfriesshire and Galloway Natural History and Antiquarian Society, 1997

Minnit, Stephen and John Coles, *The Lake Villages of Somerset*, Glastonbury Antiquarian Society, 1996

Murray, Margaret Alice, *Excavations in Malta*, with a chapter by Gertrude Caton Thompson, London: Bernard Quaritch, 1923–5

Nicholson, Professor H. Alleyne and R. Lydekker, *Manual of Palaeontology*, i, 3rd edn, Edinburgh and London: William Blackwood, 1889

Oakley, Kenneth P., 'Ghar Dalam', in Kenneth P. Oakley, Bernard Grant Campbell and Theya Ivitsky Molleson (eds), *Catalogue of Fossil Hominids, Part II: Europe*, London: BM (NH), 1971

Osborn, Henry Fairfield, *Proboscidea : A Monograph of the Discovery, Evolution, Migration and Extinction of the Mastodonts and Elephants of the World*, 2 vols, New York: American Museum of Natural History, 1936 & 1942

Panagiotaki, Marina, 'Knossos objects: 1904 the first departure', in Gerald Cadogan, Eleni Hatzaki and Adonis Vasilakis (eds), *Knossos: Palace, City, State, Proceedings of the Conference in Herakleion organised by the British School at Athens and the 23rd Ephoreia of Prehistoric and Classical Antiquities at Herakleion in November 2000 for the Centenary of Sir Arthur Evans's Excavations at Knossos*, London: BSA Studies, 2004, xii, 565–579

Peach, B. N. and J. Horne, *Guide to the Geological Model of the Assynt Mountains*, London: HMSO, 1914

Pococke, Richard, *A description of the East, and some other Countries*, London, 1743–45, 2 vols

Report on the Working of the Museum Department, Malta: Malta Museum, 1934–5

Riley, Norman Denbigh, *The Department of Entomology of the British Museum (Natural History) 1904–1964: A Brief Historical Sketch*, London: 12th International Congress of Entomology, 1964

Rixon, A.E., *Fossil Animal Remains: Their Preparation and Conservation*, London: Athlone Press, 1976

Rothschild, Miriam, *Dear Lord Rothschild*, London: Hutchinson, 1983

Salvator, Archduke Ludwig of Austria, *Die Balearen. Geschildert in Wort und Bild*, Würzburg; Leipzig, 1897

Simmons, Alan H., *Faunal Extinction in an Island Society: Pygmy Hippopotamus Hunters of Cyprus*, New York and London: Kluwer Academic/ Plenum Publishers, 1999

Simonelli, Vittorio, *Candia*, Parma, 1897

Snell, Susan and Polly Tucker, *Life Through a Lens: Photographs from the Natural History Museum 1880 to 1950*, London: Natural History Museum, 2003

Sondaar, Paul Y., Miguel McMinn, Bartomeu Seguí and Josep Antoni Alcover, 'Palaeontological Interest of Karstic Deposits from the Gymnesic and Pityusic Islands', in A. and J. Ginés (eds), *Karst and Caves in Mallorca*, Palma de Mallorca: Federació Balear d'Espeleologia, 1995

Spaan, Andries, '*Hippopotamus creutzburgi*: The Case of the Cretan Hippopotamus', in David S. Reese (ed.), *Pleistocene and Holocene Fauna of Crete and its First Settlers*, Madison, Wisconsin: Prehistory Press, 1996

Spratt, Thomas Abel Brimage, *Travels and researches in Crete*, London: John van Voorst, 1865, 2 vols

Stearn, William T., *The Natural History Museum at South Kensington: a history of the Museum, 1753–1980*, London: Natural History Museum, 1998

Stringer, Chris and Clive Gamble, *In Search of the Neanderthals*, London: Thames & Hudson, 1993

Tchernov, Professor Eitan, 'The Faunal Sequence of the Southwest Asian

Middle Palaeolithic in Relation to Hominid Dispersal Events', in T. Akazawa, K. Aoki and O. Bar-Yosef (eds), *Neandertals and Modern Humans in Western Asia,* New York and London: Plenum Press, 1998

Trevor-Battye, Aubyn, *Camping in Crete. With notes upon the animal and plant life of the island . . . Including a description of certain caves and their ancient deposits by Dorothea M. A. Bate,* London: Witherby & Co., 1913

Tufnell, Olga, *et al.*, *Lachish (Tell ed-Duweir),* iv, The Wellcome-Marston Archaeological Research Expedition to the Near East, London, 1958

Turville-Petre, Francis, *Researches in Prehistoric Galilee 1925–1926, with sections by D. M. A. Bate and C. Baynes, and 'A Report on the Galilee Skull,' by Sir A. Keith, et al.*, London: British School of Archaeology in Jerusalem, 1927

Vizetelly, Edward, *From Cyprus to Zanzibar by the Egyptian Delta: The Adventures of a Journalist,* London: C. Arthur Pearson, 1901

Vos, John de, *The Endemic Pleistocene Deer of Crete,* Amsterdam; Oxford: North-Holland Publishing, 1984

Vuillier, Gaston, *The Forgotten Isles: Impressions of Travel in the Balearic Isles, Corsica and Sardinia,* translated by Frederic Breton, London: Hutchinson, 1897

Wheeler, Sir Mortimer, *Still Digging,* London: Pan Books, 1958

Woodward, Sir Arthur Smith (ed.), *Guide to the Fossil Mammals and Birds in the Department of Geology and Palaeontology,* London: BM (NH), 1904

—, *The Earliest Englishman,* London: Watts & Co., 1948

Woodward, Dr Henry (ed.), *Guide to the Exhibition Galleries of the Department of Geology and Palaeontology in the British Museum (Natural History),* London: BM(NH), 1896

Articles

Alcover, Josep Antoni, Miquel McMinn and Cristian Ruiz Altaba, 'Eivissa: A Pleistocene oceanic-like island in the Mediterranean', *National Geographic Research & Exploration,* 10: 2 (1994), 236–8

—, Ramon Perez-Obiol, Errikarta-Imanol Yll and Pere Bover, 'The diet of *Myotragus balearicus* Bate 1909 (*Artiodactyla caprinae*), an extinct bovid from the Balearic islands: evidence from coprolites', *Biological Journal of the Linnean Society,* 66: 1 (1999), 57–74

Anon., review: 'Ancient fauna and early man at Bethlehem', of E. W. Gardner and D. M. A. Bate, 'The bone-bearing beds of Bethlehem: their fauna and industry', *Nature,* 140, (September 1937), 381

Andrews, Dr C. W., 'Some suggestions on extinction', *Geological Magazine,* 4: 10 (1903), 1–2

—, 'A description of the skull and skeleton of a peculiarly modified rupicaprine antelope (*Myotragus balearicus* Bate)', *Geological Magazine,* 6: 1 (1914)

—, 'A description of the skull and skeleton of a peculiarly modified rupicaprine antelope (*Myotragus balearicus* Bate), with a notice of a new variety, *M. balearicus* var. *major*', *Philosophical Transactions of the Royal Society of London*, series B, 206 (October 1915), 281–305

—, 'Note on a mounted skeleton of *Myotragus balearicus* Bate' *Geological Magazine*, 6: 2 (1915), 337–9

—, 'Obituary notice of Dr Charles Immanuel Forsyth Major', *Proceedings of the Royal Society of London*, series B, 95 (1924), liv-lv

Annals of the Cyprus Natural History Society, 1 (Nicosia: 1910) and 3 (Nicosia: 1911)

Arkell, Anthony J., 'Zoology and prehistoric archaeology: an opportunity', *The Archaeological News Letter*, 3: 11 (1951)

Bar-Yosef, Ofer and Jane Callander, 'A forgotten archaeologist: the life of Francis Turville-Petre', *Palestine Exploration Quarterly*, 129 (1997), 2–18

—, 'The woman from Tabun: Garrod's doubts in historical perspective', *Journal of Human Evolution*, 37 (1999), 879–85

Benton, Sylvia, 'Excavation of the Sculptor's Cave, Covesea, Morayshire', *Proceedings of the Society of Antiquaries of Scotland*, 65, (12 January 1931), 177–216

Bosanquet, E. S., 'Bosanquet, Robert Carr (1871–1935)', revised David Gill, *Oxford Dictionary of National Biography*, Oxford: OUP, 2004

Bover, Pere and Josep Antoni Alcover, 'The evolution and ontogeny of the dentition of *Myotragus balearicus* Bate, 1909: evidence from new fossil data', *Biological Journal of the Linnean Society*, 68 (1999), 401–28

Boyd, Brian and Zoë Crossland, 'New fieldwork at Shuqba cave and in Wadi en-Natuf, Western Judea', *Antiquity*, 74: 286 (December 2000), 755–6

Breuil, Abbé Henri, review of *The Stone Age of Mount Carmel*, i, *Nature*, 141: 1 (1938), 304–6

Callander, Jane, 'Dorothy Garrod's Excavations in the Late Mousterian of Shukbah Cave in Palestine reconsidered', *Proceedings of the Prehistoric Society*, 70 (2004) 207–31

— and Ofer Bar-Yosef, 'Saving Mount Carmel Caves', *Palestine Exploration Quarterly*, 132 (2000), 94–112

Clark, Dr John Desmond, 'Fractured chert specimens from the Lower Pleistocene Bethlehem beds, Israel', *Bulletin of the British Museum Natural History (Geology)*, 5: 4 (1961)

Clark, Dr. John Grahame Douglas, review of Garrod and Bate, *The Stone Age of Mount Carmel*, i, *Proceedings of the Prehistoric Society*, NS, 3 (1937), 486–8

Currant, Andrew P., 'The late glacial mammal fauna of Gough's Cave, Cheddar, Somerset', *Proceedings of the University of Bristol Speleological Society*, 17: 3 (1986), 286–304

Despott, Guiseppe, 'Archaeological investigations in Malta. II: Tal Herba and Burmeghez fissures', *Report of the British Association for the Advancement of Science* (1915), 213–14

—, 'Excavations conducted at Ghar Dalam (Malta) in the summer of 1917', *Journal of the Royal Anthropological Institute*, NS, 48 (1918), 214–21

Edwards, W. N., 'Obituary notice of Dorothea Bate', *Proceedings of the Geological Society of London* (1951), lvi–lviii

Evans, Sir J. Arthur, 'Excavations at Knossos, Crete 1904,' *Report of the 74th Meeting of the British Association for the Advancement of Science* (1904), 322–4

Flaquer y Fabregues, Juan, 'Dorothea MA Bate', *Revista de Menorca*, Año XXIV, (Quinta Época) Tomo XV, Mahon (1920)

Garrod, D. A. E., 'Excavation of a Mousterian rock-shelter at Devil's Tower, Gibraltar', with sections by L. H. D. Buxton, G. Elliott Smith and D. M. A. Bate, *Journal of the Royal Anthropological Institute*, 58 (1928), 33–113

—, 'Excavations of a Palaeolithic cave in Western Judaea', *Report of the British Association for the Advancement of Science*, 1928

—, 'A new Mesolithic industry: the Natufian of Palestine', *Journal of the Royal Anthropological Institute*, 62 (1932), 257–69

—, 'Excavations at the Wady el-Mughara, Palestine (1932–33)', *Bulletin of the American School of Prehistoric Research*, 10 (May 1934), 7–11

— and D. M. A. Bate, 'Excavations at the Cave of Shukbah, Palestine, 1928', *Proceedings of the Prehistoric Society*, 8 (1942), 1–20

Guy, P. L. O., 'Prehistoric Man in Palestine', *The Times*, 24 August 1925

Halliday, W. R., 'Dawkins, Richard MacGillivray (1871–1955)', revised David Gill, *Oxford Dictionary of National Biography*, Oxford: OUP, 2004

Harris, J. M. and T. D. White, 'Evolution of the Plio-Pleistocene African *Suidae*', *Transactions of the American Philosophical Society*, 69: 2 (1979)

Hill, J. E., 'A memoir and bibliography of Michael Rogers Oldfield Thomas, FRS', *Bulletin of the British Museum Natural History (Historical Series)*, 18: 1 (May 1990), 25–113

Hinton, Martin A. C., 'Preliminary Note upon the Mammalian Remains from Merlin's Cave', *Proceedings of the University of Bristol Speleological Society*, 2: 2 (1926), 156–8

Hooijer, D. A. 'An early Pleistocene mammalian fauna from Bethlehem', *Bulletin of the British Museum (Natural History) Geology*, 3: 8 (1958), 265–92

Howard, Hildegard, 'Tribute to Dorothea Bate', *The Auk*, 69 (August 1952), 491

Hrdlička, Aleš, review of *The Stone Age of Mount Carmel*, i, *American Journal of Physical Anthropology*, 23: 4 (1938), 467–8

Irwin, D. J., 'The Exploration of Gough's cave and its development as a show cave', *Proceedings of the University of Bristol Speleological Society*, 17: 2 (1985)

Keith, Sir Arthur, 'Discovery of Neanderthal Man in Malta', *Nature*, 101 (1918), 404

Lowe, P. R. and D. M. A. Bate, 'Struthious remains from northern China and Mongolia; with descriptions of *Struthio wimani*, *Struthio andersoni* and

Struthio mongolicus. With a note by D. M. A. Bate on remains of carinate birds', *Palaeontologia Sinica*, series C, 6, fasc. 4 (1931), 1–47
McCown, Theodore T., 'Fossil men of the Mugharet es-Sukhul', *Bulletin of the American School of Prehistoric Research*, 9 (May 1933), 9–15
—, 'The oldest complete skeletons of man', with note by Sir Arthur Keith, *Bulletin of the American School of Prehistoric Research*, 10 (May 1934), 13–19
Major, Charles Immanuel Forsyth, 'On the pigmy hippopotamus from the Pleistocene of Cyprus', *Proceedings of the Zoological Society of London*, 2 (1902), 107–12
Martel, E. A., 'Les cavernes de Majorque', *Spelunca, Bulletin & Mémoires de la Société de Spéléologie*, 32 (February 1903)
Naturalesa i Societat, Circular de la Societat d'Història Natural de les Balears, 25 (October 2001)
'The New Natural History Museum', *Nature*, 27 (1882), 55
Obituary of Guy Ellcock Pilgrim, *Obituary Notices of Fellows of the Royal Society*, 4: 13 (1944), 577–60
Obituary of Professor Frederick Zeuner, *Proceedings of the Geologists' Association*, 75: 1 (1964), 117–20
Peach, B. N. and J. Horne with E. T. Newton, 'On a bone cave in the Cambrian limestone in Assynt, Sutherlandshire', *Report of the British Association* (1892), 720–1
— and J. Horne, 'Bone cave in the valley of Allt Nan Uamh (Burn of the caves), near Inchnadamff, Assynt, Sutherlandshire, with notes on the bones found in the cave by E. T. Newton FRS', *Proceedings of the Royal Society of Edinburgh*, 37 (1917), 327–48
Peters, Joris, 'The faunal remains collected by the Bagnold–Mond expedition in the Gilf Kebir and Gebel Uweinat in 1938', *Archéologie du Nil Moyen*, 2 (1987), 251–64
Picard, Professor Leo, 'Inferences on the problem of the Pleistocene climate of Palestine and Syria drawn from flora, fauna and stratigraphy', *Proceedings of the Prehistoric Society*, 5 (1937), 58–70
Ramis, Damià and Pere Bover, 'A review of the evidence for domestication of *Myotragus balearicus* Bate 1909 (artiodactyla, caprinae) in the Balearic islands, *Journal of Archaeological Science*, 28 (2001), 265–82
— and Josep Antoni Alcover, 'Revisiting the earliest human presence in Mallorca', *Proceedings of the Prehistoric Society*, 67 (2001), 261–9
Reese, David S., 'Tracking the extinct pygmy hippopotamus of Cyprus', *Field Museum of Natural History Bulletin*, 60: 2 (1989)
—, 'Cypriot hippo hunters no myth', *Journal of Mediterranean Archaeology*, 9: 1 (1996), 107–12
—, 'The extinct pygmy hippos of Cyprus', *Sunjet* [Cyprus Airways in-flight magazine], 9: 2 (1996)
'The retirement of Dr Henry Woodward', *Geological Magazine*, 1: 9 (1902), 1

Savona-Ventura, C., 'Palaeolithic Man in Malta', *The Sunday Times*, 20 July 1997 and 3 August 1997
Sherborn, Dr Charles Davies, Obituary, *Nature*, 150 (1942), 146–7
Shindler, Karolyn, 'Bate, Dorothea Minola Alice (1878–1951)', *Oxford Dictionary of National Biography*, Oxford: OUP, 2004
Smith, Pamela Jane, 'Dorothy Garrod, first woman Professor at Cambridge', *Antiquity*, 74: 283 (March 2000), 131–6
—, Jane Callander, Paul G. Bahn and Geneviève Pinçon, 'Dorothy Garrod in words and pictures', *Antiquity*, 71: 272 (1997), 265–70
Swinnerton, Professor H. H., 'Addresses to medallists etc.', Proceedings of the Geological Society of London, session 1939–40, *Quarterly Journal of the Geological Society*, 96 (1940), lvi–lvii
Turville-Petre, Francis, 'Excavations in the Mugharet el-Kebarah, *Journal of the Royal Anthropological Institute*, 62 (1932)
Waechter, J. d'A., V. M. Seton-Williams, D. M. A. Bate and L. Picard, 'The excavations at the Wadi Dhobai 1937–38 and the Dhobaian industry', [vertebrate report by DMAB and geological report by L.Picard], *Journal of the Palestine Oriental Society*, 18 (1938)
Williamson, Dr George Alexander, 'The Cyprus sphalangi and its connection with anthrax', *British Medical Journal* (1 September 1900)
Woodward, Sir Arthur Smith, 'On the bone-beds of Pikermi, Attica and on similar deposits in Northern Euboea', *Geological Magazine*, 4: 8 (1901), 481–6
—, 'Dr Charles William Andrews Obituary Notice', *Proceedings of the Royal Society of London*, series B, 100 (1926), i–iii
Woodward, Dr Henry, 'Cave-Hunting in Cyprus,' *Geological Magazine*, 4: 10 (June 1903), 241–6
Vaufrey, Professor Raymond, 'Tribute to Dorothea Bate', *L'Anthropologie*, 55 (1951), 555

Dorothea Bate's principal published reports (in chronological order)

The full bibliography is held in the Earth Sciences Library, Natural History Museum.

Bate, D. M. A., 'A short account of a bone cave in the Carboniferous limestone of the Wye valley', *Geological Magazine*, 4: 8 (1901), 101–6
—, 'Field-notes on some of the birds of Cyprus', *Ibis* (October 1903), 571–81
—, 'On a new species of wren from Cyprus', *Bulletin of British Ornithologists' Club*, 8 (1903), 51–2
—, 'A preliminary note on the discovery of a pigmy elephant in the Pleistocene

of Cyprus', *Proceedings of the Royal Society of London*, 71 (June 1903), 498–500
—, 'On an extinct species of genet (*Genetta plesictoides* sp. n.) from the Pleistocene of Cyprus', *Proceedings of the Zoological Society of London*, 2 (1903), 121–4
—, 'The mammals of Cyprus', *Proceedings of the Zoological Society of London*, 2 (1903), 341–8
—, 'On the occurrence of *Acomys* in Cyprus', *Annals and Magazine of Natural History*, 7: 11 (1903), 565–7
—, 'On the ossiferous cave deposits of Cyprus', *Geological Magazine*, 5: 1 (1904), 324–5
—, 'Four and a half months in Crete in search of Pleistocene mammalian remains', *Geological Magazine*, 5: 2 (1905), 193–202
—, 'Further note on the remains of *Elephas cypriotes* from a cave deposit in Cyprus', *Philosophical Transactions of the Royal Society of London*, series B, 197 (March 1905), 347–60
—, 'On the mammals of Crete', *Proceedings of the Zoological Society of London*, 2 (1905), 315–23
—, 'The pygmy hippopotamus of Cyprus', *Geological Magazine*, 5: 3 (1906), 241–5
—, 'On elephant remains from Crete, with description of *Elephas creticus* sp. n.', *Proceedings of the Zoological Society of London* (1907), 238–50
—, 'Excavation of a cairn at Mossknow, on the Kirtle Water, Dumfriesshire,' *Proceedings of the Society of Antiquaries of Scotland*, 8 (February 1909) 165–9
—, 'Preliminary note on a new artiodactyle from Majorca, *Myotragus balearicus* gen. et sp. nov.', *Geological Magazine*, 5: 6 (1909), 385–8
—, 'On a new species of mouse and other rodent remains from Crete', *Geological Magazine*, 5: 9 (1912), 4–6
—, 'The Caves of Crete', in Aubyn Trevor-Battye, *Camping in Crete*, London: Witherby & Co., 1913
—, 'On the Pleistocene ossiferous deposits of the Balearic islands', *Geological Magazine*, 6: 1 (1914), 337–45
—, 'On remains of a gigantic land tortoise (*Testudo gymnesicus*, sp. n.) from the Pleistocene of Menorca', *Geological Magazine*, 6: 1 (1914), 100–7
—, 'On a small collection of vertebrate remains from the Har [sic] Dalam Cavern, Malta, with note on a new species of the genus *Cygnus*', *Proceedings of the Zoological Society of London*, 2 (1916), 421–30
—, 'On a new genus of extinct muscardine rodent from the Balearic islands', *Proceedings of the Zoological Society of London*, 2 (1918), 209–22
—, 'Note on a new vole and other remains from the Ghar Dalam cavern, Malta', *Geological Magazine*, 57 (1920), 208–11
—, 'Notes on the vertebrate remains from the Ghar Dalam cave, Malta, found

by Miss Caton Thompson', in M. A. Murray, *Excavations in Malta*, London: Bernard Quaritch, 1923, 12–13
—, 'On the animal remains obtained from the Mugharet-el-Emireh in 1925', in F. Turville-Petre, *Researches in Prehistoric Galilee*, London: British School of Archaeology in Jerusalem, 1927
—, 'On the animal remains obtained from the Mugharet-el-Zuttiyeh in 1926', in F. Turville-Petre, *Researches in Prehistoric Galilee*, London: British School of Archaeology in Jerusalem, 1927
—, 'The animal remains', in Garrod *et al.*, 'Excavation of a Mousterian rock-shelter at Devil's Tower, Gibraltar', *Journal of the Royal Anthropological Institute*, 58 (1928), 92–110
—, 'Note on the fauna of the Athlit caves', *Journal of the Royal Anthropological Institute*, 62 (1932), 277–9
—, 'A new fossil hedgehog from Palestine', *Annals and Magazine of Natural History*, 10: 10 (1932), 575–85
—, 'A fossil wart-hog from Palestine', *Annals and Magazine of Natural History*, 10: 13 (1934), 120–9
—, 'Two additions to the Pleistocene cave-fauna of Palestine (*Trionyx* and *Crocodilus*)', *Annals and Magazine of Natural History*, 10: 14 (1934), 474–8
—, 'Discovery of a fossil elephant in Palestine', *Nature*, 134: 1 (1934)
—, 'The habits of *Enhydrictis galictoides* with description of some limb-bones of this mustelid from the Pleistocene of Sardinia', *Proceedings of the Zoological Society of London*, 2 (1935), 241–5
—, 'Two new mammals from the Pleistocene of Malta', *Proceedings of the Zoological Society of London*, 2 (1935), 247–64
Gardner, E. W. and D. M. A. Bate, 'The bone-bearing beds of Bethlehem: their fauna and industry', *Nature*, 140 (September 1937), 431–3
Bate, D. M. A., 'New Pleistocene mammals from Palestine', *Annals and Magazine of Natural History*, 10: 20 (1937), 397–400
—, 'Palaeontology: The Fossil Fauna of the Wady el-Mughara Caves', in D. A. E. Garrod and D. M. A. Bate, *The Stone Age of Mount Carmel*, i, part 2, Oxford: Clarendon Press, 1937
—, 'Note on recent finds of *Dama clactoniana* in London and Swanscombe', *Proceedings of the Prehistoric Society of Cambridge*, 2: 3 (1937), 460–3
—, 'Vertebrate Report', in J. d'A. Waechter *et al.*, 'The excavations at the Wadi Dhobai 1937–38 and the Dhobaian industry', *Journal of the Palestine Oriental Society*, 18 (1938)
—, '*Crocidura samaritana* Bate: a correction', *Annals and Magazine of Natural History*, 11: 1 (1938), 78–9
—, 'The fossil antelopes of Palestine in Natufian times', *Geological Magazine*, 77: 6 (1940), 418–43
—, 'The bone-bearing beds of Bethlehem', *Nature*, 147: 2 (June 1941), 783

—, 'Pleistocene *Murinae* from Palestine', *Annals and Magazine of Natural History*, 11: 9 (1942), 465–86
—, 'New Pleistocene *Murinae* from Crete', *Annals and Magazine of Natural History*, 11: 9 (1942), 41–9
—, 'The fossil mammals', appendix to D. A. E. Garrod, 'Excavations at the Cave of Shukbah, Palestine, 1928', *Proceedings of the Prehistoric Society*, 8 (1942), 15–20
—, 'A new Pleistocene deer from Gibraltar', *Annals and Magazine of Natural History*, 11: 10 (1943), 411–24
—, 'Pleistocene *Cricetinae* from Palestine', *Annals and Magazine of Natural History*, 11: 10 (1943), 813–38
—, 'Pleistocene shrews from the larger western Mediterranean islands', *Annals and Magazine of Natural History*, 11: 11 (1944), 738–69
—, 'Note on small mammals from the Lebanon Mountains, Syria', *Annals and Magazine of Natural History*, 11: 12 (1945), 141–58
—, 'The Pleistocene mole of Sardinia', *Annals and Magazine of Natural History*, 11: 12 (1945), 448–61
—, 'An extinct reed-rat (*Thryonomys arkelli*) from the Sudan', *Annals and Magazine of Natural History*, 11: 14 (1947), 65–71
—, 'A new African long-horned buffalo', *Annals and Magazine of Natural History*, 12: 2 (1949), 396–8
—, 'The Fauna', in A. J. Arkell, *Early Khartoum*, Oxford: OUP, 1949, 16–28
—, 'The Fauna of Esh Shaheinab', *The Archaeological News Letter*, 2: 8 (January 1950), 128–9
—, 'A fossil vole from Cyrenaica', *Annals and Magazine of Natural History*, 12: 3 (1950), 981–5
—, articles on 'Archaeopteryx', 'Dodo', 'Moa', 'Solitaire' in *Chambers's Encyclopaedia*, London: George Newnes, 1950
—, 'The Pleistocene fauna of two Blue Nile sites: the mammals from Singa and Abu Hugar', *Fossil mammals of Africa*, BM (NH), 2 (1951), 1–28
—, 'The Pleistocene mammal faunas of Palestine and East Africa', in L. S. B. Leakey and Sonia Cole (eds), *Proceedings of the Pan-African Congress on Prehistory, Nairobi 1947*, Oxford: Basil Blackwell, 1952
—, 'The vertebrate fauna', in A. J. Arkell, *Shaheinab*, Oxford: OUP, 1953
—, 'Vertebrate faunas of Quaternary deposits in Cyrenaica', Appendix A in C. B. M. Burney and R. W. Hey, *Prehistory and Pleistocene Geology in Cyrenaican Libya*, Cambridge: CUP, 1955

B. Secondary Sources (books, articles and unpublished mss)

Adams, A. E., *Mallorcan Geology*, Cardiff: Department of Extra-Mural Studies, University College, 1988

Bellamy, C. V. and A. J. Jukes-Brown, *The Geology of Cyprus*, Plymouth, 1905

Blondel, Jacques and James Aronson, *Biology and Wildlife of the Mediterranean Region*, Oxford: OUP, 1999

Bulleid, Arthur and Harold St George Gray, *The Glastonbury Lake Village*, 2 vols, Glastonbury, i, 1911; ii, 1917

Everett, Major-General Sir Henry, *The History of the Somerset Light Infantry (Prince Albert's) 1685–1914*, London: Methuen, 1934

Garrod, D. A. E., *The Upper Palaeolithic Age in Britain*, Oxford: Clarendon Press, 1926

Garrod, D. A. E., Disney Professor of Archaeology in the University of Cambridge, *Environment, Tools and Man, an Inaugural Lecture*, Cambridge, 1946

Geikie, Sir Archibald, *The first century of Geology in Britain: being two presidential addresses to the Geological Society of London, 26 September 1907, and 21 February 1908*, London, 1908

Hawes, Charles Henry, 'Crete Diary 1904–5', MS Eng. Misc. d. 1449, Department of Special Collections and Western Manuscripts, Bodleian Library

Horowitz, Aharon, *The Quaternary of Israel*, New York: Academic Press, 1979

Ireland, John de Courcy, *Wreck and Rescue on the East Coast of Ireland*, Dublin: Glendale Press, 1983

Molloy, Pat, *A Shilling for Carmarthen: The Town they Nearly Tamed*, Llandysul: Gomer, 1980

'Obituary of the Revd Robert Ashington Bullen', *Geological Magazine*, 5: 9 (1912)

Ovey, Cameron Darrell, *The Swanscombe Skull: A Survey of Research on a Pleistocene Site* (Occasional paper 20, Royal Anthropological Institute of Great Britain and Ireland), London, 1964

Peel, Robert, *Mary Baker Eddy*, 3 vols, New York: Holt, Rinehart & Winston, 1966, 1972, 1977

Rackham, Oliver and Jennifer Moody, *The Making of the Cretan Landscape*, Manchester: Manchester University Press, 1996

Reese, David S. 'Report from Cyprus', *Art and Archaeology Newsletter*, 36/37: 3, New York (1975)

Reese, David S., 'Dwarfed hippos past and present', *Earth Science*, 28: 2 (1975), 63–9

Reese, David S., 'Men, saints or dragons?' *Expedition*, 17: 4 (1975)

Reese, David S., 'The Pleistocene vertebrate sites and fauna of Cyprus', *Ministry of Agriculture, Natural Resources and Environment, Geological Survey Department, Republic of Cyprus, Bulletin*, 9, Nicosia (1995)

Savage, R. J. G., 'Martin Alister Campbell Hinton 1883–1961', *Biographical Memoirs of Fellows of the Royal Society*, 9 (1963), 155–70

Savage, R. J. G., *Mammal Evolution: An Illustrated Guide*, London: BM (NH), 1986

Webb, William, *Coastguard! An Official History of HM Coastguard*, London: HMSO, 1976

Williams, M. A. J., D. L. Dunkerley, P. de Deckker, A. P. Kershaw and T. J. Stokes, *Quaternary Environments*, London: Edward Arnold, 1993

Zittel, Karl A. von, *History of Geology and Palaeontology to the end of the nineteenth century*, translated by Maria M. Ogilvie-Gordon, London, 1901

INDEX

Footnotes are indicated by n after the page number: 262n. DMAB stands for Dorothea M. A. Bate, and Museum for the Natural History Museum.